P9-DTF-279

SOWING MODERNITY

SOWING MODERNITY

America's First Agricultural Revolution

PETER D. McCLELLAND

Cornell University Press

ITHACA AND LONDON

First published 1997 by Cornell University Press.

Printed in the United States of America.

Cornell University Press strives to utilize environmentally responsible suppliers and materials to the fullest extent possible in the publishing of its books. Such materials include vegetable-based, low-VOC inks and acid-free papers that are also either recycled, totally chlorine-free, or partly composed of nonwood fibers.

Library of Congress Cataloging-in-Publication Data

McClelland, Peter D.
Sowing modernity : America's first agricultural revolution / Peter D. McClelland.
p. cm.
Includes bibliographical references (p.) and index.
ISBN 0-8014-3326-6 (cloth : alk. paper)
1. Agriculture—United States—History. 2. Agricultural innovations—United States—History. 3. Agriculture—History. 4. Agricultural innovations—History. I. Title.
S441.M115 1997
630'.973—dc21 97-9030

Cloth printing 10 9 8 7 6 5 4 3 2 1

For Cornell Librarians

WHO HELP US ALL PLACE THIS PIECE NEXT TO THAT

UNTIL PART OF THE PICTURE REVEALS ITSELF

Contents

❦

Preface

This investigation began with an obvious puzzle that, pursued in a predictable manner, led to an unexpected conclusion. The larger problem of which the puzzle is a part is what caused the extraordinary discontinuity in Western economic development in the past few hundred years—arguably *the* central problem of economic history. Part of the answer, of course, is a radical alteration in technological capabilities. This, in turn, suggests that one key question underlying the larger problem of Western modernization is when and why producers in any numbers first began to ask: Is there a better way? Once that question was directed persistently, systematically, and pervasively to existing production technologies, the march to modernization appears to have been virtually assured.

Having selected the puzzle to investigate—when this attitudinal transformation first occurred and why—I had to decide which kind of producers to investigate, and in which country. My primary interest has always been American economic history, and the choice of the United States reflects that professional bias. As for the producers, developments in a nascent manufacturing sector, often featured in histories of America's industrial revolution, for purposes of this study seemed second best—indeed, second best by a wide margin. If the critical question is when and why an attitudinal shift occurred among producers toward technological change, the obvious set of producers to investigate were those who usually dominate in a premodern economy, namely, those who tilled the land.

I was therefore led naturally to an exploration of America's first agricultural revolution—"first" in the sense that changes made in techniques and implements signaled a departure from methods that for generations had been largely unchanged because their use had been largely unquestioned. The facts of this study, while of particular interest to the agricultural historian, have a wider significance insofar as they indicate both the timing and the pacing of a transformation in outlook with monumental implications for American development. Once that quintessential question of modern producers—Is there a better way?—became pervasive among American farmers it marked the arrival of the first of many agricultural revolutions in this country. These successive revolutions were in turn part of a larger upheaval in economic attitudes and practices, one that would fashion a new economic order the likes of which the world had never seen before.

I first set out, therefore, to identify from a welter of details about American agricultural procedures and implements just when and how rapidly the above question began to pervade this country's agrarian sector. What I discovered was totally unexpected. The evidence suggests, and in no uncertain terms, that this attitudinal transformation was accomplished in less than two decades, primarily between 1815 and 1830.

My remaining assignment was to explain why this occurred. Originally, I intended to combine both research agendas in a single book, but it became evident early on that such an enterprise would ultimately require two volumes. The first, which constitutes the present work, focuses on the factual details of the initial agrarian transformation: which new practices and novel implements were developed and when, how they worked, and why they were considered an improvement over traditional tools and procedures. (Not all were considered an improvement.) The question of why they occurred must be the topic of another book, and for reasons that, in retrospect, seem self-evident. If what is to be explored is nothing less than a revolution in the attitudes of American farmers toward production possibilities, the intellectual terrain that must be traversed is both vast and formidable. Not only must close scrutiny be given to economic events—rapid population growth, fluctuations in foreign trade, transportation improvements, and the first stirrings of what would become an industrial revolution, to mention only a few of the more obvious developments of relevance—but also, precisely because what is to be explained is a seismic shift in the way farmers viewed the world, some understanding will be required of how those views were shaped by the political, social, and cultural upheavals taking place as America was transformed from thirteen colonies in revolt to a young republic struggling to survive and flourish in a hostile world.

A single example suggests the complexity of such a task. Between 1783 and 1830, July 4 orations provide a useful window on changing American attitudes. Given annually throughout the country, these addresses were generally crafted by a speaker of some note whose self-appointed task was to articulate the achievements of the past as well as the hopes and fears about the future commonly shared by members of his audience. A survey of these orations reveals a sudden change in content occurring at almost the exact moment when America's first agricultural revolution began in earnest. From 1783 to the end of the War of 1812, these speeches featured two conflicting expectations about the future. One tended to emphasize what the wording of the Great Seal of the United States emphasized: that the founding of the republic signaled "a new order of the ages" (*Novus Ordo Seclorum*), and that this new order was viewed by God both favorably and expectantly (*Annuit Coeptis*). But in many other speeches the tone was anything but optimistic, the nation's prospects portrayed as anything but bright. According to this second set of orators, doom probably lay just around the corner for a people who, they claimed, had lost their revolutionary fervor, were losing their virtue, and, in all likelihood, were losing favor with the Almighty. Their message, in short, was that Americans must mend their ways or face catastrophe. That both bright and dark visions should inform forecasts about the nation's future is not surprising. The nation, after all, confronted multiple threats as well as multiple opportunities. What is surprising is that the dark vision all but disappeared from July 4 orations immediately following the conclusion of the War of 1812, just when the accelerating pace of changes in farm implement design first signaled that an agricultural revolution was under way. Were these two developments related? The tempting inference is that both were symptomatic of larger upheavals in the young republic. But whether that inference is warranted is a subject that must await investigation in another book.

A number of agrarian experts have helped to shape this project in great and small ways. Wesley Gunkel of Cornell's Agricultural and Biological Engineering Department was a patient and indispensable guide to the workings of modern agricultural machinery. Equally patient and similarly expert with the operation of early nineteenth-century tools were Bruce Craven and Jochen Welsch of Old Sturbridge Village, who kindly complemented their explanations with demonstrations in the field. To Wayne Randolph of Colonial Williamsburg I am particularly indebted for sharing both his understanding of early farm implements and his encyclopedic knowledge of American farming literature. Almost from the beginning of this project, he has been a constant guide and critic, sharing illustrations, suggesting sources, and correcting errors in earlier drafts, particularly those parts of the manuscript touching upon complex issues of plow operation.

Two of America's foremost experts on agricultural history read the entire manuscript and offered comments and criticisms that helped to strengthened the final version: Allan G. Bogue of the University of Wisconsin and Morton Rothstein of the University of California at Davis. Through the volunteered efforts of a Cornell colleague and good friend, Ari van Tienhoven, I was able to establish contact with several Dutch scholars knowledgeable about the agricultural transformation of Western Europe. P. G. M. Hoppenbrouwers of the University of Leiden and

G. J. H. Bieleman of Wageningen Agricultural University read a number of chapter drafts, and offered many useful suggestions. My greatest single debt is to J. M. G. van der Poel, also of Wageningen, widely regarded as one of Europe's foremost experts on farm implements. Never before in my professional career have I received such a close reading of an entire manuscript. With scrupulous attention to even the smallest detail, Professor van der Poel evidenced a stunning erudition in qualifying, elaborating, or correcting both text and footnotes. The repeated citation in those footnotes, "J. M. G. van der Poel, private communication," does not begin to capture his contribution to this work.

Like any writer of a lengthy manuscript, I am indebted to legions of librarians for tracking down sources, rounding up books, and—in the case of this project—making available for copying purposes many illustrations contained in ancient tomes and periodicals. For more than two decades the staff of Cornell libraries have been making clear to me that their reputation for outstanding service is well earned. The present work, I think, placed a special strain on both their expertise and their patience, but every query, no matter how esoteric the subject or ancient the source, was fielded with an unfailing competence and good humor. While many helped, two must be singled out for a graciousness, knowledge, and tenacity that puts them among the best of the best: Fred Muratori of Olin Library and Tom Clausen of Mann Library.

The staff of many other libraries, museums, and historical associations also offered repeated and invaluable assistance, suggesting possible sources and tracking down illustrations: Roberta Waddell, Curator of Prints, New York Public Library; Georgia B. Barnhill, Curator of Graphic Arts, American Antiquarian Society; Wendy Shadwell, Curator of Prints, The New-York Historical Society; Peter Cousins, Curator of the Collections Division, Henry Ford Museum; Sally Pierce, Print Curator, Library of the Boston Athenaeum; and from the Smithsonian Institution, Marjorie Berry and Peter Daniel, both of the Division of Agricultural and Natural Resources, and David C. Burgevin of the Office of Printing and Photographic Services.

In the many years this project has taken, I have been helped by research assistants who scoured libraries, pursuing citations and collecting books and periodicals with a diligence and cheerfulness that belied the tediousness of so many of their tasks: To Shan Li, Jennifer Stansbury, Roger Kim, Erick Schonfeld, and Suzanne Wallace. Finally, for consummate care in the face of endless tedium, I am especially indebted to Pat Paucke, who typed the many versions of each chapter, carefully integrating, wherever necessary, text, illustrations, and footnotes.

PETER D. MCCLELLAND

Ithaca, New York

SOWING MODERNITY

[1]

The Problem

Of all the graphs known to social scientists, historians, and demographers, some variant of Figure 1.1 is perhaps the most familiar. With a single line it captures an extraordinary discontinuity in human history

Figure 1.1. World population, 7000 B.C. to 1990 A.D.

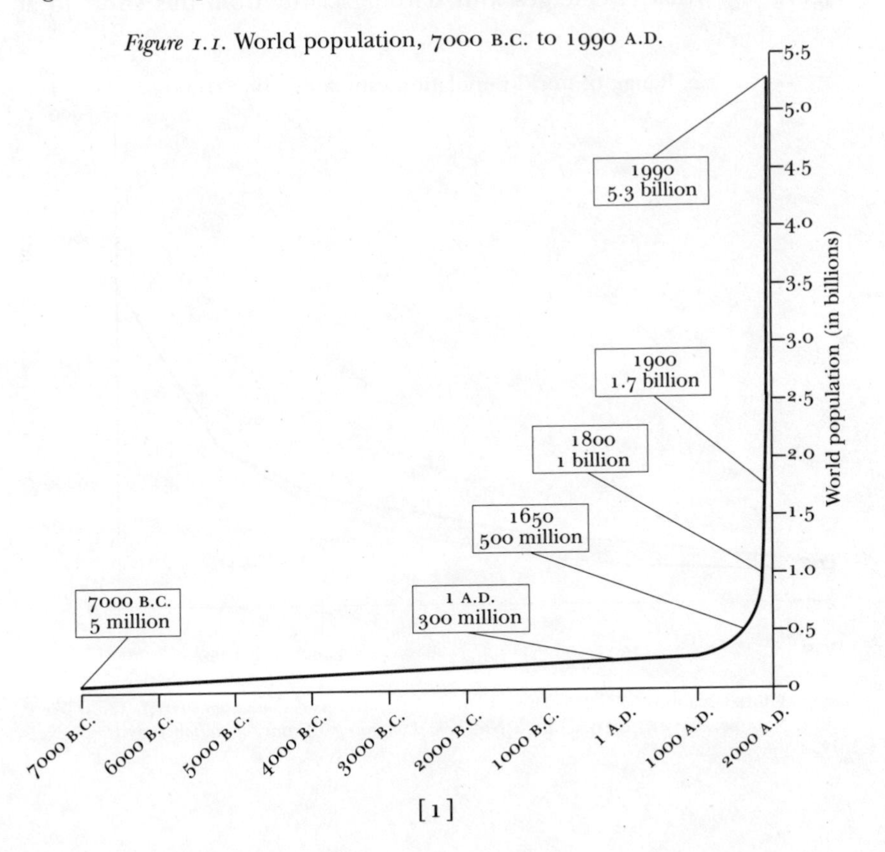

that has attracted the attention of scholars from many disciplines. Something quite without precedent has radically altered the growth of the human species on this planet in the past few centuries. But what?

The cautious expert might argue that given the intrinsic defects of the diagram, that question is premature. How, for example, was total world population for 1650 estimated? The answer, briefly put, is with great difficulty. Rough estimates for regions of the world for which some counts are available must be combined with assumptions about regions for which no counts whatsoever can be found. The resulting total population number must therefore be considered nothing better than a rough approximation. In the single line linking the point estimates of Figure 1.1, the argument runs, there is an unwarranted implication of exactitude. If instead one takes the best of world population estimates, beginning with the year 1650, and plots those numbers as a range, the result is Figure 1.2.

One cannot be certain that the true population count for any given year lies within the range depicted, but at least the odds have been improved of detecting the actual pattern of world population growth. If the range of population estimates of Figure 1.2 are combined with two others (5–10 million for 6000 B.C. and 200–400 million for 1 A.D.),[1] the result is Figure 1.3. What re-emerges with startling clarity from this effort to ac-

Figure 1.2. Range of world population estimates, 1650–1900

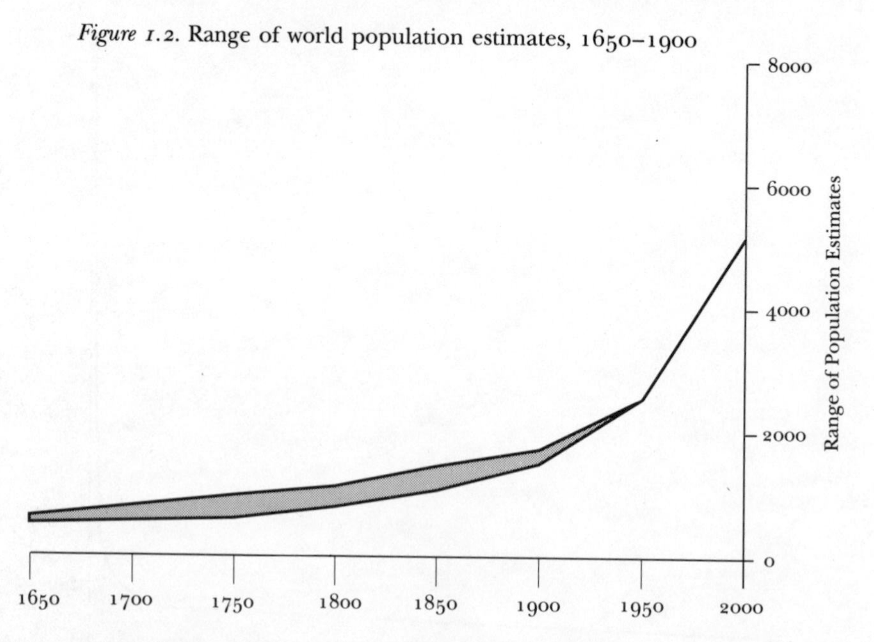

Sources: United Nations, *The Determinants and Consequences of Population Trends* (New York: United Nations, 1973), p. 10; United Nations, *Long-range World Population Projections* (New York, 1992), p. 22.

Figure 1.3. Range of world population estimates, 7000 B.C.–1990 A.D.

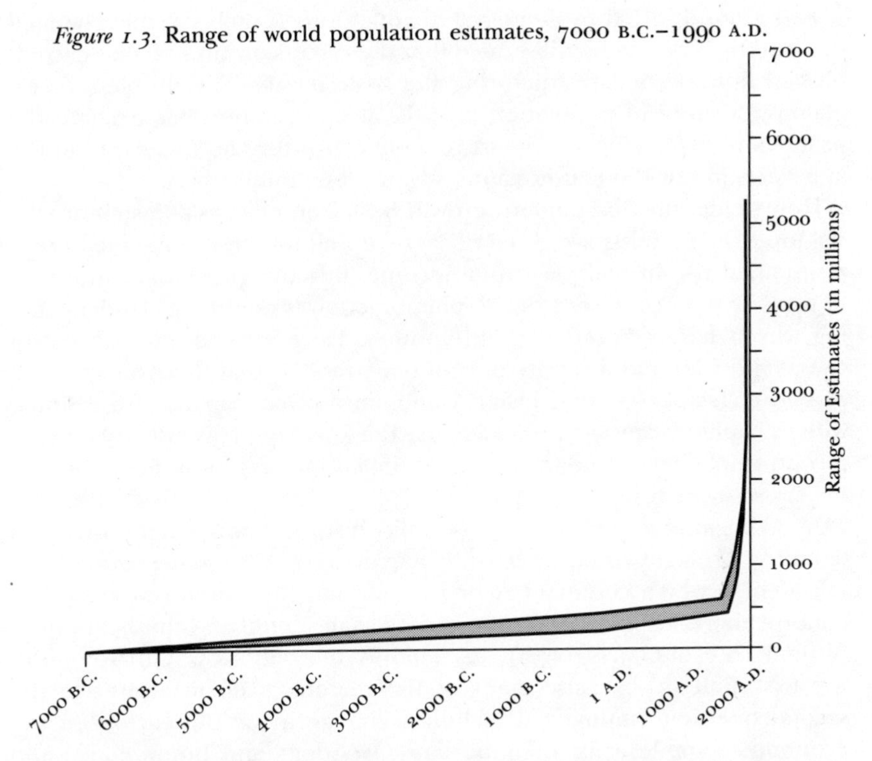

Source: United Nations, *The Determinants and Consequences of Population Trends* (New York: United Nations, 1973), p. 10; United Nations, *Long-range World Population Projections* (New York, 1992), p. 22.

knowledge imperfections in world population estimates is the same discontinuity in the past few hundred years and the same question: Why did it occur?

The search for the answer must begin with an awareness of the nature of the numerical trend to be explained. The problem is a surge in total world population, and the possible causes reduce to two: either death rates dropped or birth rates rose (or both happened simultaneously). Experts seem generally agreed that, at least prior to the present century, the major cause of the observed increase was a decline in death rates attributable to rising living standards (particularly gains in food, clothing, and shelter) plus the increased availability of resources for public health and for "other services relevant for saving and prolonging life."[2] (The odds of surviving, we have learned, are significantly affected by improvements in sewage and water systems as well as by improvements in diet.) In the earliest stages of industrialization, the impact of falling death rates upon population growth may have been reinforced by a rise in birth rates. This latter development is also linked to economic growth.

Those not diverted by demographic questions (which variable changed first and by how much) will notice that the discussion thus far has accomplished little beyond restructuring the basic puzzle. The problem of explaining a surge in population growth, at one remove, has become the problem of explaining a surge in economic growth. And the causes of the latter are, to put the matter cautiously, not well understood.

How to identify that modern growth has taken place is a less controversial topic. Economists are generally agreed that the best single measure is a sustained rise in real per capita income. But an agreed-upon indicator to identify the occurrence of the phenomenon is one thing. Understanding why it has occurred is quite another. Here experts offer a host of causal variables and a variety of economic models that as a collective explanation features partial insights and unresolved debates. Confronted with cacophony among professionals, the nonexpert wonders where to turn to gain a better understanding of the curious discontinuity in human history so evident in Figure 1.3.

As a beginning student of economic history, I was taught that this daunting problem could be usefully narrowed by a three-step procedure: (1) identify which country was first to have an "industrial revolution" (a concept more fashionable in those days than identifying the beginnings of modern economic growth), (2) identify the reasons why that country was first, and (3) because many of the reasons given in answer to the second question emphasized technical change in various sectors of the economy, compile a list of important inventions and innovations,[3] and explain why they occurred. This approach inevitably led to a careful scrutiny of British developments in the eighteenth and early nineteenth centuries, culminating in an innovation-by-innovation review of technological changes that ranged from seed drills to Rotherham plows, from flying shuttles to power looms, from Newcomen's atmospheric engine to Watt's separate condenser.

In recent years the concept of the British industrial revolution has been subjected to withering fire by a new generation of historians. Changes crucial to the transformation of England's economy, some contend, can be traced as far back as the early Middle Ages. Others emphasize the gradualness of growth after 1750 in various estimates of aggregate output. The combined effect is to drain the discontinuity from the transformation of the British economy, replacing the old revolutionary imagery with one that features gradual evolution in both the causes and the consequences of modern economic growth.[4]

The danger is that the smoke of unresolved debates about the British experience will obscure the obvious pattern and the obvious puzzle. If one turns again to the pattern displayed in Figure 1.3, two points seem above dispute. Clearly, a revolution of some sort has taken place in the last few hundred years. And however that revolution is defined, one necessary condition was a radical change in the technology applied to economic activities. To put that second point a second way, if production technology had remained substantially unchanged for the last five hundred years—if goods and services were created today in much the same way as in the Middle Ages—then the sudden and sustained spurt in population growth depicted in Figure 1.3 could not possibly have occurred.

That second insight is the starting point for this inquiry.

In my student days, as previously noted, I was taught to focus upon a long list of inventions, particularly those made in Britain. In later years, the more I wondered about the nature of the transformation lurking in Figure 1.3 (variously referred to as "industrialization," "modernization," or "modern economic growth") the more impressed I became with the importance, not of the inventions themselves, but of a change in the willingness to innovate.[5] Throughout the long premodern history of humanity, inventions were by no means absent. From the birth of Mohammed to the death of Luther, for example, the list of new devices created in the Western world appears, at first glance, formidable in the extreme. The compass and the astrolabe to aid in navigation; the trebuchet and gunpowder to add destructive muscle to military ventures; the horse collar, whippletree, and three-field system to better till the soil; overshot waterwheels and windmills with progressively more complex gearing mechanisms to better harness nonhuman forms of energy; weight-driven mechanical clocks and printing presses with movable type to revolutionize attention to time and the reading of books: these are merely some of the better-known items from an extensive list of inventions, many dazzling in originality and not a few with major implications for the welfare of humanity.[6]

And yet something of crucial importance is missing. What was absent in the Middle Ages and present in the modern age was a tradition of inventing. More particularly, in terms of forcing economic transformation, what did not exist in the premodern world was the persistent, pervasive, and systematic directing of a single question to the processes of production: Is there a better way? Before *that* tradition was established, all inventions can fairly be viewed as flickering lights on a darkened landscape, infrequent in occurrence and with no capacity to effect a general transformation in illumination. But once a tradition of questioning the status quo in production techniques became established, the march to modernization was virtually assured.

The major goal of this book is to identify the origins of this crucial and distinctly modern tradition. The most obvious arenas in which to begin that search are the histories of those countries industrializing first: Britain on one side of the Atlantic and the United States on the other. I chose the latter because it has long been the focus of many of my interests. The central problem, then, is to identify when the question whether there is a better way became a widespread and persistent concern for American producers. A related issue is whether this question's rise to prominence was a sudden or a gradual process.

The first order of business was to decide which producers to investigate. Given that the object of the search was a pervasive change in a tradition, a case could be made for focusing upon that sector of the premodern American economy in which the vast majority of producers were engaged, namely, agriculture.[7] The final problem of a preliminary sort was to decide which kind of activity in that sector to investigate. In retrospect, the choice was obvious. At the outset, it was not.

[2]

The Approach

"The subject of these lectures is the Industrial and Agrarian Revolution at the end of the eighteenth and beginning of the nineteenth century."[1] Thus began Arnold Toynbee's influential analysis of British economic transformation, first published in the 1880s as *Lectures on the Industrial Revolution in England.* As Toynbee's opening remarks make clear, that title omits a crucial part of Britain's mutation to modern growth. Revolutions in industrial production were important, but so too were those in agriculture. Writing eight decades before Toynbee, T. R. Malthus had missed the latter possibilities in his grim forecast that population growth would invariably outstrip any likely increase in the food supply.

But if this assemblage of upheavals that lead to modern growth includes an agrarian revolution, how should that revolution be defined? Which kinds of discontinuities in the farming sector are reasonably regarded as being revolutionary? Not surprisingly, different historians have proffered different answers. In the American case, the characteristics singled out for special emphasis have included the first widespread use of complex machinery or the systematic application of science (particularly chemistry) or the recourse to modern management techniques as American farms, at least in some cases, became big business.[2] Each of these possible definitions of an agricultural revolution necessarily excludes the beginnings of the transformation process. Many changes of the first importance took place down on the farm long before complex machines or modern chemistry or agribusiness were an integral part of American husbandry. All these later revolutionary modifications in agricultural production techniques had to be preceded by a fundamental shift in attitudes about the merits of modification: an attitudinal shift that led first to changes on a small scale, and only later to changes on a grand scale of the sort just noted. The "first agricultural revolution," as that phrase will be used in this book, indicates the rise to prominence of the search among agrarian producers for a better way. Admittedly, not every new device or technique generated by this constant questioning of the status quo was successful. Many failed. But a Darwinian imagery applied to technical progress suggests that, once the change in attitude is firmly entrenched in the farming community, the fittest novelties will survive, with fitness defined primarily as a capacity to increase the profits of those who own the farm. In short, this first agricultural revolution must precede all the rest, and once accomplished, assures that recurring waves of innovations in agrarian techniques will become the norm.

One final matter of method must be resolved. If the attitudinal change to be investigated is a new willingness to search for better ways of farm production, how can "better" methods be identified? At first glance, this appears to be an innocuous issue with an obvious answer. Yet it necessarily raises research strategy problems of the first importance.

Consider the difficulties lurking in those apparent contenders for attention, changes in fertilizer use and crop rotation. Identifying what people tried and which experiments were truly "better" is a complicated task, given the absence of scientific knowledge, the limited evidence on procedures used, and the sensitivity of results to differences in climatic conditions and soil types. Similar difficulties complicate the interpretation of changing livestock practices. Here, too, scanty records and unscientific procedures confound the identification of changes actually made in the breeding and raising of animals. American farmers in the early years of livestock experimentation also evidenced considerable uncertainty as to which results were "best." For some, the objective was to maximize the likelihood that the "improved" animal would win a prize at the local fair, with little or no attention to the costs incurred. Others defined "improved" not in terms of animal appearance (or likely responses of local judges to animal appearance), but rather in terms of net contribution to farm profits.

The remaining obvious contenders for attention are changes in the types of farm implements used. Here available material for research is far more promising. Which implements were considered "new" and how they differed from the old can be studied with the aid of surviving relics in American museums, illustrations dating from the era of initial introduction, and extensive commentary by contemporaries when such implements were introduced.[3] The replacement of the old by the new (always the acid test of superior efficiency) in some cases can be traced with the aid of advertisements, occasionally supplemented by manufacturing and sales data, and, in almost every case involving a new farm implement of any note, by examining the commentary over time of agrarian experts, editors of agricultural periodicals, and the various committees of different agricultural societies specifically charged with investigating the merits of novel implements.

At issue is a point more fundamental than the availability of surviving evidence. The long history of the rise of Western civilization has included many agricultural revolutions, in the sense of changes in farming practices that have had a widespread and beneficial impact upon productivity. But only one set of changes unmistakably heralds a total break with the past—a break so fundamental that it prefigures the declining dominance of agriculture itself as a major occupation and the rise of that poorly understood process, "modernization," with its crucial characteristic absent from all the revolutions that have gone before: a sustained rise in real GNP per capita.

To understand which agricultural revolution is special in this sense, consider all the changes in production techniques that had to take place before the farming practices of, say, the Dark Ages could be transformed into a reasonable approximation of those presently in use. Each set of changes, in crop rotations, for example, may have required a few decades or a few centuries to become established practice. In Western Europe, particularly in those countries on the cutting edge of agricultural change, such as England and the Low Countries, many of these fundamental changes in a basic set of farming practices (such as fertilizer use and crop rotation) took centuries to initiate, where "initiate" indicates not the completion of a process but the end of the beginning: that point in time when the search for ways to improve a particular set of farming practices has become firmly entrenched, and should therefore lead to further modification and improvements at subsequent points in time.[4] The pattern of agricultural development in the New World was markedly different from that of the Old. In the half-century following the American Revolution, modifications first began in earnest across almost the entire range of farming practices: in fertilizer use and crop rotations, in livestock breeding and implement design. The question is which of these signal a definitive change in producers' attitudes toward innovation, and hence America's "first agricultural revolution."

The broad pattern of development seems clear enough. In the decades immediately following 1783, three types of changes in farming practices appear: in applying fertilizer, in designing better crop rotations, and in developing better pastures. Hard on the heels of these changes came a second set in implement design and livestock breeding practices during the period 1815–1830. As a first approximation, then, all of these initiatives seem appropriate to incorporate into a broader concept of a first revolution in America's agricultural sector.

Close inspection of the changes made, however, suggests problems with such an all-inclusive approach. Consider the three types of changes initiated after 1783. All three had in common a number of characteristics, of which five seem particularly noteworthy.

(1) All began, or began in earnest, as noted, at roughly the same time.
(2) All were prompted not so much by a new spirit of inquiry throughout the agricultural sector as by changing relative factor prices reflecting declining land abundance.
(3) The resulting search for preferable practices focused, by and large, upon options already developed in Europe during the previous centuries.[5]
(4) New procedures and practices adopted at this time often reinforced one another. (For example, better pastures increased manure supplies that could then be used as fertilizer, and the fertilizer, gypsum, when combined with clover, often led to dramatic improvements in pastures whose development, in turn, became an integral part of some rotation schemes.)
(5) Last, but of the first importance, all the associated changes in fertilizer use and crop rotations and pasture development *did not, by themselves,* set American agriculture firmly upon the path to modernization. Had only these changes occurred, and nothing more, American agriculture of the early twentieth century in all probability would have closely resembled that of the early nineteenth century.

What was missing—or largely missing—from these developments was (1) a search for "a better way" that signaled a new spirit of inquiry (rather than responses largely generated by changing factor prices), and (2) the inclusion under "better" of new technological possibilities (not the recourse to alternatives already developed and used in parts of Europe). Once these two characteristics, in concert, become firmly established in the agricultural sector, the march to modernization appears to have been virtually assured. Although largely missing from the first set of changes, these were unambiguously present in the second set. The choice of which sets to include in America's "first agricultural revolution," however, is left to the reader. This work concentrates on changes in implement design, because this is the transformation best documented by available evidence (Chapters 3 through 9). The discussion of changes in fertilizer use, crop rotations, pasture designs, and livestock breeding practices are mainly confined to Appendices A and B. Some readers will regard the title of this work as encompassing almost every change discussed. Others, who confine their definition of a first agricultural revolution to changes that unambiguously signal the initiation of modernization, will necessarily have a more narrow focus that gives a special importance (in the American case, at least) to changes in implement design. What should be clear to all is that, however America's first agricultural revolution should be defined, it occurred—and in no uncertain terms—in the half-century following the

Revolution, which by European norms was a transformation of staggering rapidity.

The final task of a preliminary sort is to decide which farming implements to investigate, and in what order. Consider the two collections depicted in Figures 2.1A and 2.1B. To the uninitiated, both will seem remarkably primitive, and the second collection hopelessly cluttered. The challenge is to give a semblance of order to the collection examined and a coherence to their evolution by choosing a subset from a sprawling array of tools and implements that varied by crop raised and region of use. Subsequent chapters focus upon small grains, particularly wheat—arguably the premier crop of premodern Europe and America. The investigation of technical changes in implements used will follow the order in which those tools were commonly used for the raising of small grains from the time of the Pharaohs to the founding of the American republic, beginning with plowing (Chapter 3) and ending with threshing and winnowing (Chapters 8 and 9). The objective of the investigation, to repeat, is to use changes in techniques to identify a shift in attitude toward technical change. The hypothesis to be tested is that the attitudinal change was somewhat abrupt, and therefore the first major technical changes in farm implements should be clustered in a limited time period and affect many of the different procedures employed in the growing of small grains.

This issue of close proximity in time of many different innovations can be reformulated with the aid of imagery once popular in the recounting of Britain's industrial revolution. T. S. Ashton used a schoolboy's answer to an exam question as a half-serious and half-ludicrous characterization of the processes that began the transformation of the British economy: "About 1760 a wave of gadgets swept over England."[6] The imagery was modified and made more apt by economic historian Donald McCloskey: "The gadgets came more like a gentle (though unprecedented) rain, gathering here and there in puddles. By 1860 the ground was wet, but by no means soaked, even at the wetter spots."[7]

What follows is the story of a different revolution in another time and place—a story of a brief period in the history of a young republic in the New World, when the rain of gadgets for the first time began to fall in earnest across the American farming landscape. It is a story of high drama involving lowly implements, a story largely missing from the literature.[8] The stakes involved provide the drama. Had America remained a premodern economy well into the nineteenth century, both the welfare of its citizens and the international bargaining strength of the nation would have been adversely affected in great and small ways. Once the march to modernization was initiated, future economic progress, anchored in abundant natural resources and persistent technological change, seemed assured. The pressing matter was to initiate the march. One indispensable

Figure 2.1. American farm implements

A. Circa 1790

B. Circa 1860

Source: Charles L. Flint, "Progress in Agriculture," in *Eighty Years' Progress of the United States* (New York: L. Stebbins, 1864), pp. 28–29.

requirement was a radical change in attitude that gave the search for new technologies a popularity and respectability never found in premodern economies. Once such a search became pervasive, the inevitable result was a wave of gadgets that swept across the American landscape. Or in McCloskey's imagery of water from the sky, the consequences of such an attitudinal transformation was an initial downpour of gadgets which, by modern standards, was little better than a sprinkle. What mattered was not the density of the downpour but the fact that it signaled an atmospheric shift assuring repeated rains of gadgets in the future.

The foregoing imagery assumes what has yet to be demonstrated: that the first round of gadgets came as something of a cloudburst, and that this close proximity in time of many new innovations heralded a shift in attitude among producers concerning the merits of technological change in farm production. That shift, in turn, was sure to generate what the Great Seal of the United States had promised in a different context: *Novus Ordo Seclorum*—a new order of the ages.

[3]

Plowing

Dissatisfactions with a Complex Implement

In 1784 James Small noted: "The circumstances upon which the goodness of a plough depends, are both many, and difficult to be discovered; and it requires both great ingenuity and great experience to determine that form of a plough which will enable it to perform its work in the best manner, and with the least labour possible." Four years after Small penned these thoughts about the difficulties of designing better plows,[1] Thomas Jefferson was returning to Paris from Amsterdam by way of Germany, and stopped for the night at Nancy, a small town in Alsace-Lorraine. He noted in his journal that the town was "neat" and the surroundings "agreeable," but the wines were bad and the plows deplorable.[2] For Jefferson, both shortcomings deserved to be taken seriously. The low quality of local wines was easily solved by continuing on into France, where he sampled vintages and purchased bottles (the latter figuring prominently in the fifty cases of baggage that accompanied Jefferson on his return to America the following year). The problems of defective plow design were more complex. The common plows of Europe and America at that time were hulking and unwieldy, unappealing to the eye and often inefficient in the soil. Reviewing in his mind that evening "the awkward figure of their moldboards," Jefferson hit upon a plan for a radical improvement. The key was mathematics, or so he thought. More than two centuries before Jefferson's speculations about moldboards, Galileo had articulated an insight that came to dominate the methodology of modern science: that the natural world was subject to laws capable of being expressed in mathematical form discoverable by mankind. Perhaps this general principle could also be used to identify the optimal shape for parts of the plow. Accordingly Jefferson tried to specify the exact geometric form of the moldboard's surface if a plow were to pass through the soil with the least effort and the best results.[3]

One puzzle is why the design of American and European plows of Jefferson's day should differ so markedly from those used in ancient Egypt and Mesopotamia, where so many agrarian practices central to Western civilization had originated. A second puzzle is why American plows of the sort depicted in Figure 3.1 changed so little from the time of the New World's first settlement to the founding of the new republic. A third is why Jefferson was wrong: mathematics would not provide the key to improving plow designs, either in the twenty years following Jefferson's presidency, when a host of new plow designs burst upon the American landscape, and not for the remaining decades of the nineteenth century.

Figure 3.1. Colonial plows, eighteenth century

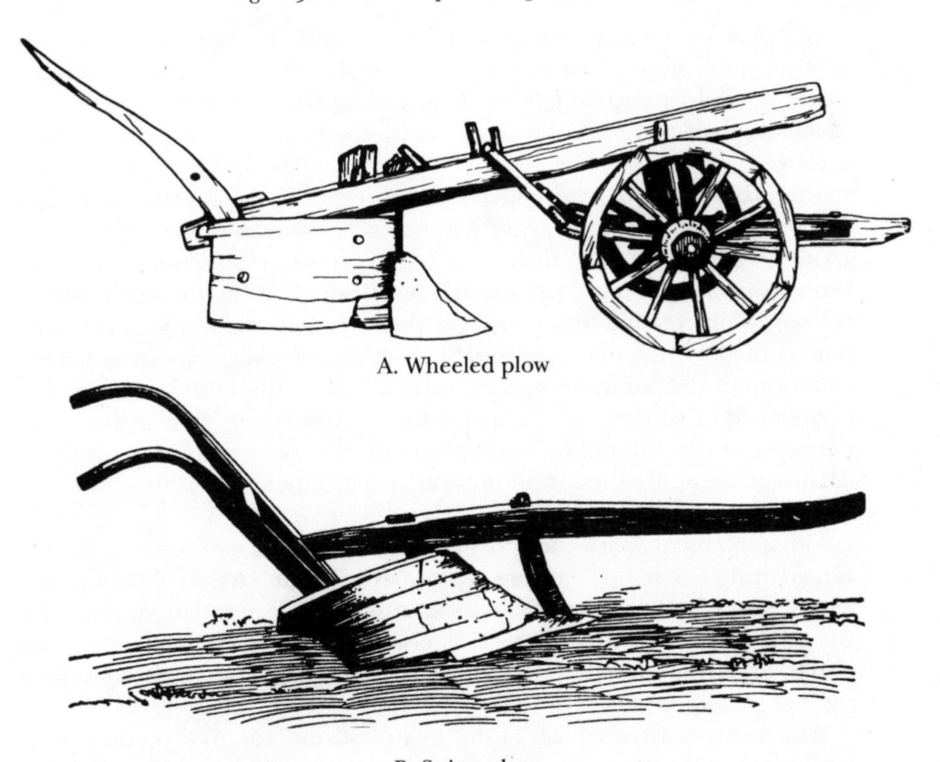

A. Wheeled plow

B. Swing plow

Sources: Peter H. Cousins, *Hog Plow and Sith* (Dearborn, Mich.: Edison Institute, 1973), p. 6; Percy W. Bidwell and John I. Falconer, *History of American Agriculture in the Northern United States, 1620–1860* (Washington, D.C.: The Carnegie Institution, 1925), p. 124.

Of Ancient Ards

The plow, declared the New York Agricultural Society in its review of 1867 plowing trials, "lies . . . at the root of all human civilization."[4] As sweeping as that generalization may seem, it does contain an element of truth. As long as ancient peoples turned the soil with a crude digging stick or hoe, the power directed to that task was necessarily limited to whatever human muscles could generate. Once some ancient version of a plow was devised, the muscles directed to soil preparation could include those of domesticated animals. Tilling capabilities were thus enormously expanded, as were the possibilities of an increase in the food supply. And given that an agricultural surplus is a necessary condition for achieving any form of complex social organization, the plow might fairly be viewed as being, if not "at the root of all civilization," at least a necessary condition for achieving most forms of advanced civilization in the premodern world.[5]

The plow commonly used in the ancient civilizations of the Near East, however, bore little resemblance to the implement familiar to American colonists, and unsurprisingly so. It was designed to solve much simpler problems than those encountered by the early settlers of British North America.

The main objective of any tillage is to prepare the soil for the germination of seeds and the growth of plants. Put another way, the objective of good tillage is good tilth, including a minimum of weeds and soil so broken up that air and water can move freely through it. In arid and semiarid regions, which include those areas of the Near East where the great civilizations of the West were first established, the growing seasons generally feature high temperatures and little rain. Tillage therefore requires only frequent light stirrings of the soil without inverting it. This assures that weeds, once cut, will not be buried but rather remain upon the surface as dead vegetation, limiting wind erosion and facilitating moisture conservation.

The implement well suited to such tasks is the ancient ard. When it is viewed from above, its striking feature is its symmetry, as illustrated in Figure 3.2. Whether broad or narrow, with or without a flattened bottom (or sole), it pushed the soil to either side, and left behind furrows with unplowed strips between them, usually requiring a second plowing at right angles to the first.[6]

Which shapes were typical of the first plows devised and whether these implements, when initially created, were modified designs of hoes or digging sticks, are unsolved mysteries, and likely to remain so.[7] A wooden (or largely wooden) tool built thousands of years ago leaves few traces to aid the archaeologist in its reconstruction. What is clear is that the earliest surviving depictions of plows, which predate the founding of Christianity

Figure 3.2. Ancient ards

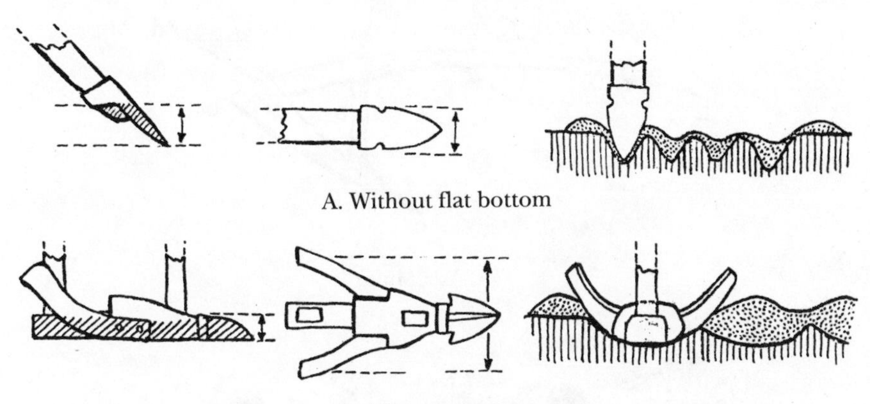

Sources: František Šach, "Proposal for the Classification of Pre-Industrial Tilling Implements," *Tools & Tillage,* 1 (1968), p. 11; H. C. Dosedla, "František Šach's Contribution towards Research on Pre-Industrial Tilling Implements in Austria," *Tools & Tillage,* 5 (1984), p. 45.

by more than three thousand years,[8] all tend to be some variant of the two basic ard designs reproduced in stylized form in Figure 3.2. The oldest illustrations usually depict a plow with no sole or horizontal base of the sort appearing in Figure 3.3. The figure at the top "is copied from an ancient monument in Asia Minor,"[9] and seems to be constructed from the natural crooks of the branches of a tree, with a beam *ab,* a share carved at *c,* and a handle *d,* plus what was probably an added brace at *e.* The bottom diagram, from an Egyptian tomb, shows two implements, somewhat less crude, at least one of which appears to have a sole. (Notice that the plow beam is linked to the oxen's horns. The neck yoke would come later.) The flattened bottom was an obvious advance, simplifying the worker's task of running the plow at a uniform depth in a given direction.[10]

The forward point in contact with the soil was invariably an area of maximum wear, and thus from earliest times builders tried to make the share of the ard from material more durable than wood. The Egyptians, for example, tended to use flint,[11] but the Biblical suggestion about beating swords into plowshares makes clear that iron was a common choice by the time Imperial Rome dominated the civilizations of the Near East.[12]

The difficulty, at least from an agrarian standpoint, was that Imperial Rome also dominated territories in transalpine Europe, and here the ancient ard was a most unsatisfactory implement for tillage. Gone (or largely gone) were the loose and light soils usually encountered in arid or semiarid regions. In their stead were the heavy cohesive or "sticky" soils commonly found in temperate climates with lower temperatures and more

Figure 3.3. Ancient plows

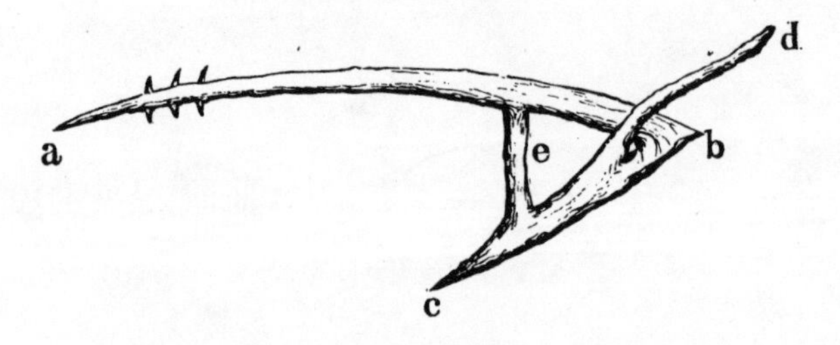

A. Asia Minor

B. Egyptian, circa 1400 B.C.

Sources: New York State Agricultural Society, *Report on the Trials of Plows Held at Utica . . .* (Albany: Van Benthuysen and Sons, 1868), p. 8; Norman D. Davies, *The Tomb of Nakht at Thebes* (New York: Metropolitan Museum of Art, 1917), Plate XVIII.

abundant rainfall. To prepare good tilth in such conditions required a deeper plowing and complete inversion of the soil. Discovering the most efficient plow design to accomplish such tasks was one of the most complicated problems in all of Western agriculture, with the answer not found—indeed, not even approximated—until the beginning of the modern era. How that search to make a better moldboard plow (as distinct from a better ard) proceeded in the West is more easily understood if preceded by some explanation of the requirements for success. What follows is a brief discussion of the main features of the modern moldboard plow. The history of its development from Roman times to the early years of the young republic will be taken up in a later section.

The Moldboard Plow: Function and Form

The subsequent discussion is not intended to be a comprehensive treatise on all parts of the modern plow, nor does it attempt to outline, for the few parts considered, every design possibility and the strengths and weaknesses of each. It tries instead to provide the minimum technical knowledge (and only that) concerning how certain key plow components function, to serve as a background for the exploration of how and why designs of these parts were modified over time, first in Europe, and later in America. (Agrarian experts may wish to skip the text and skim the illustrations, or proceed directly to the next section.)

A useful starting point is to ask what crucial tillage problems the modern moldboard plow was designed to solve. Much of the answer is evident in Figure 3.4. If the soil to be converted to good tilth is heavy and sticky, it must be pulverized and inverted. Moisture conservation is no longer an

Figure 3.4. Furrow turning in heavy soil

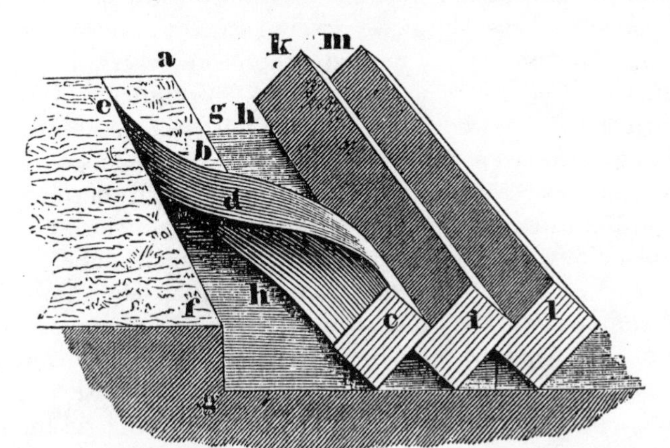

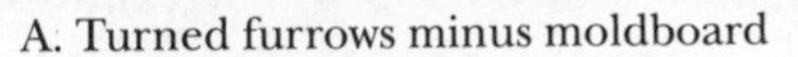
A. Turned furrows minus moldboard

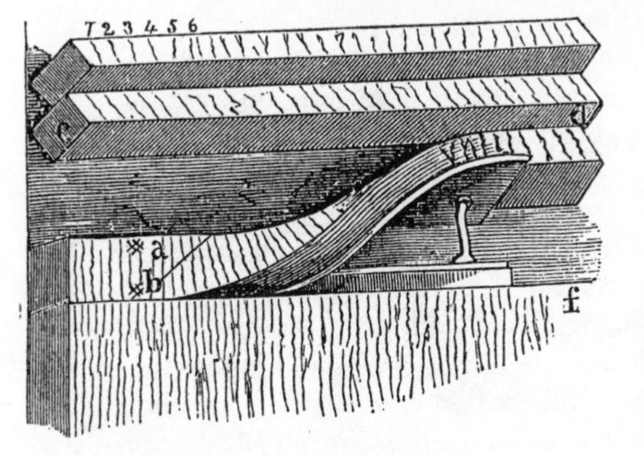

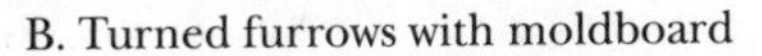
B. Turned furrows with moldboard

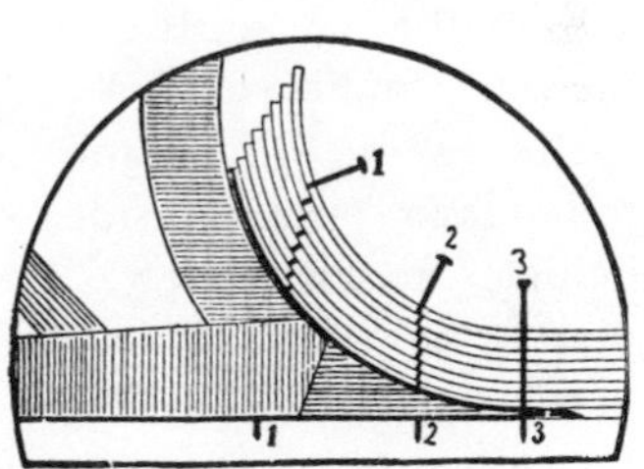

C. Action of moldboard on layers of soil

Sources: New York State Agricultural Society, *Report on the Trial of Plows held at Utica . . .* (Albany: Van Benthuysen and Sons, 1868), pp. 152, 188; F. H. King, “Tillage: Its Philosophy and Practice,” in L. H. Bailey, ed., *Cyclopedia of American Agriculture* (New York: Macmillan, 1907), Vol. 1, p. 385.

important priority, as it is in semiarid regions. (Indeed, after heavy spring rains, a surplus of water is often a problem and good drainage a matter of the first importance.) As indicated in Figure 3.4A, one objective is to flip over the surface sod, turning it under to bury and destroy weeds that will subsequently decompose and add organic matter to the soil. This inversion process requires the application of force in three directions. (1) To carve the furrow from the land, a slicing blade must penetrate along the vertical plane *ef* in Figure 3.4A. (2) Similarly, to carve the soil at a uniform depth requires a slicing action along the horizontal plane at *h*. (3) Finally, to lift and invert the severed furrow, a lateral force must be generated by a twisted surface at *d*. Designing an appropriate shape for this twisted surface is a complicated task. The surface obviously must be made to function as a wedge to thrust the newly severed earth to one side only, as in Figure 3.4B. (The symmetry of the ard is therefore necessarily replaced by the asymmetric structure of a moldboard.) The difficult problem is to identify what shape that wedge should be to push aside and invert the soil that is spilling backward across its surface. One possible answer is to duplicate in the moldboard's twist the curvature evident in the furrow as it is turned over. But as we will shortly see, such a design priority, while limiting the possibilities, does not identify a single optimal shape.

If air and water are to move freely through what heretofore has been heavy and cohesive soil, the other major task beside inversion is pulverization. The process whereby the latter is accomplished is illustrated in Figure 3.4C. Soil is comprised of multiple layers, not unlike the multiple pages of a telephone book. Imagine driving pins through all the pages of an unopened book, and then forcing a wedge into those pages. As the pages lift, they would necessarily slide over one another, causing a shearing of the pins (if they are brittle enough) of the sort depicted in Figure 3.4C. In much the same way, the wedging action of a moldboard plow pulled across a field forces soil particles to slide over one another. The steeper the wedge, the more abrupt the sliding action, and the greater the pulverization achieved.

During the nineteenth century, the implement used to accomplish these tasks on both sides of the Atlantic came to resemble the plow depicted in Figure 3.5. The "furrow side" shows the perspective of the plow if viewed from the "wedging-action" side; the "landside" shows the plow's appearance if viewed from the opposite side. The part that works beneath the soil is indicated by the dotted line in the topmost illustration (and by the part hidden from view in the bottom illustration). Perhaps most revealing to the novice is the view from above, which illustrates the peculiar twist of the metal surface designed to lift and invert the severed furrow.

The main parts of a moldboard plow are identified in Figure 3.6A, viewed from the furrow side, and in Figure 3.6B, viewed from below (or

Figure 3.5. Scotch plow, mid-nineteenth century

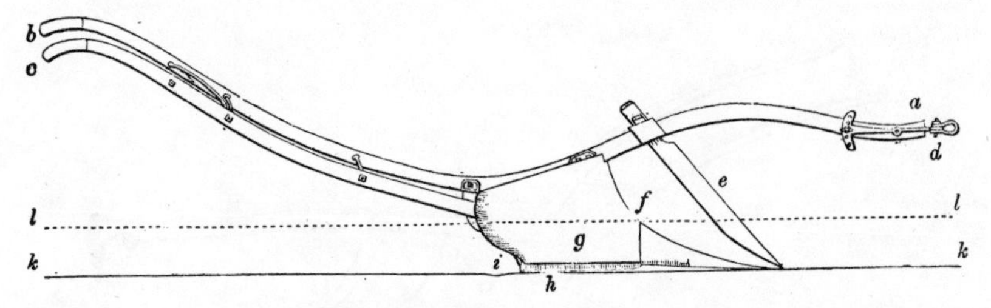

A. Side view: moldboard or furrow-side view

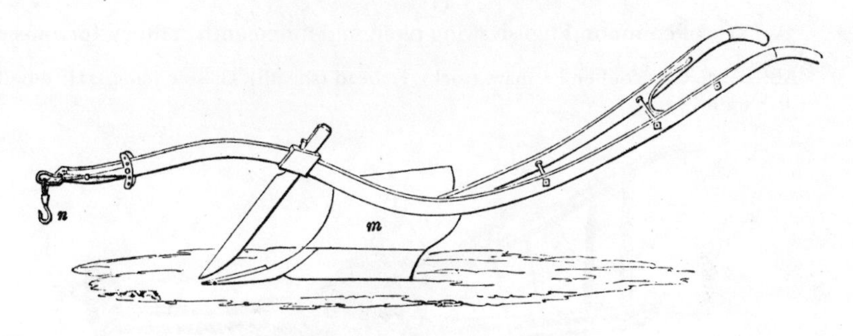

B. Side view: land-side view

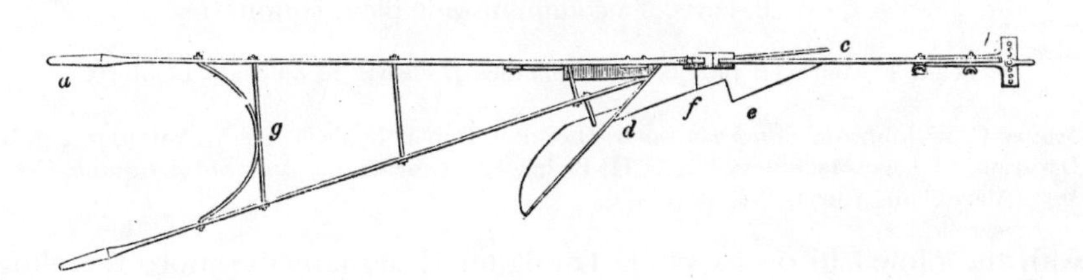

C. Top view

D. At work: furrow-side view

Source: Henry Stephens, *The Book of the Farm* (London: William Blackwood and Sons, 1855), Vol. 1, pp. 150, 151, 158.

Figure 3.6. Swing plows

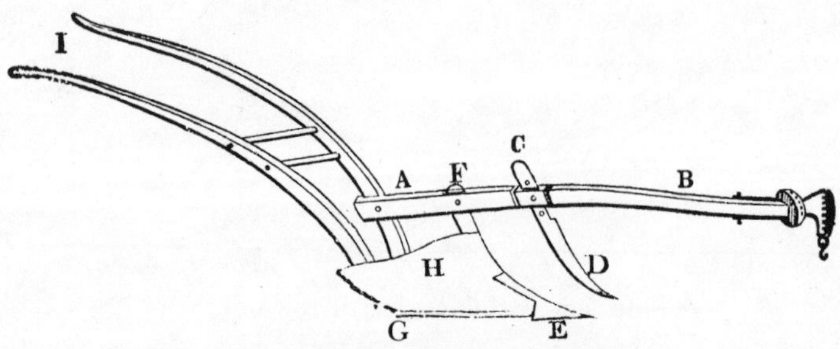

A. Parts of common English swing plow, mid-nineteenth century, furrow-side view

AB: beam. CD: coulter. E: share (sock). F: head (sheath). G: sole (chep). H: moldboard. I: handles (stilts).

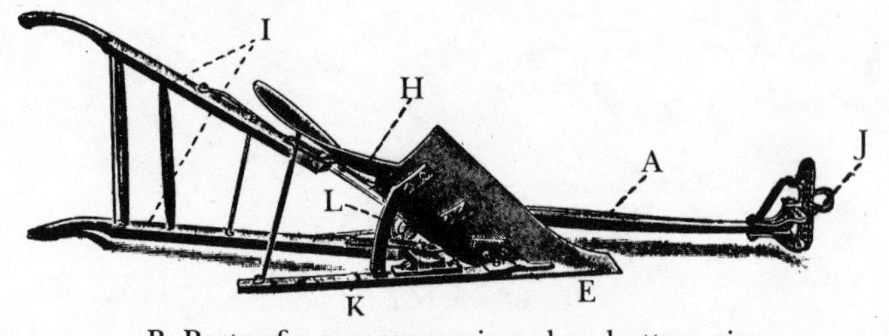

B. Parts of a common swing plow, bottom view

A: beam. E: share. H: moldboard. I: handles. J: cleavis. K: landside. L: brace.

Sources: C. W. Johnson, *British Husbandry* (London: Robert Baldwin, 1848), Vol. 2, p. 2; J. B. Davidson, "Tillage Machinery," in L. H. Bailey, ed., *Cyclopedia of American Agriculture* (New York: Macmillan, 1907), Vol. 1, p. 389.

with the plow laid on its side). The latter is perhaps the more revealing perspective of the peculiar structure and curvature of the moldboard. The working parts of this tillage implement are mounted on a beam (A) that is harnessed in front to a draft animal (with the aid of the cleavis (J)), and guided from the rear by the plowman using "stilts" or handles (I). The main focus for the remainder of this section will be the design and placement of three crucial parts in contact with the soil: the share, the coulter, and the moldboard.[13]

As summarized by a British expert, plowing "is digging on a grand scale."[14] But tilling with a moldboard plow is digging with a difference. Ideally, this implement is designed to pass through soil with a maximum of stability and a minimum of friction. One difficulty is its shape. Precisely because of the asymmetric structure of the moldboard, a number of forces are generated which can throw the implement off line as the plowman attempts to carve a straight furrow of uniform width and depth. One design priority is therefore to structure the plow so that as many of these potentially disruptive forces as possible roughly offset one another.

Consider first that part of the implement which begins the slicing and turning process: the share. By the middle of the nineteenth century in the United States, these often took the general shape outlined in Figure 3.7A and 3.7B.[15] That shape was essentially an elongated wedge sharpened at the point and along a cutting edge on the furrow side (e.g., along *ef* in Figure 3.7B). To perform satisfactorily it had to have three design features, one of which is obvious, and two of which are not.

If the share is to initiate the cutting of the furrow (before passing the

Figure 3.7. English share, late nineteenth century

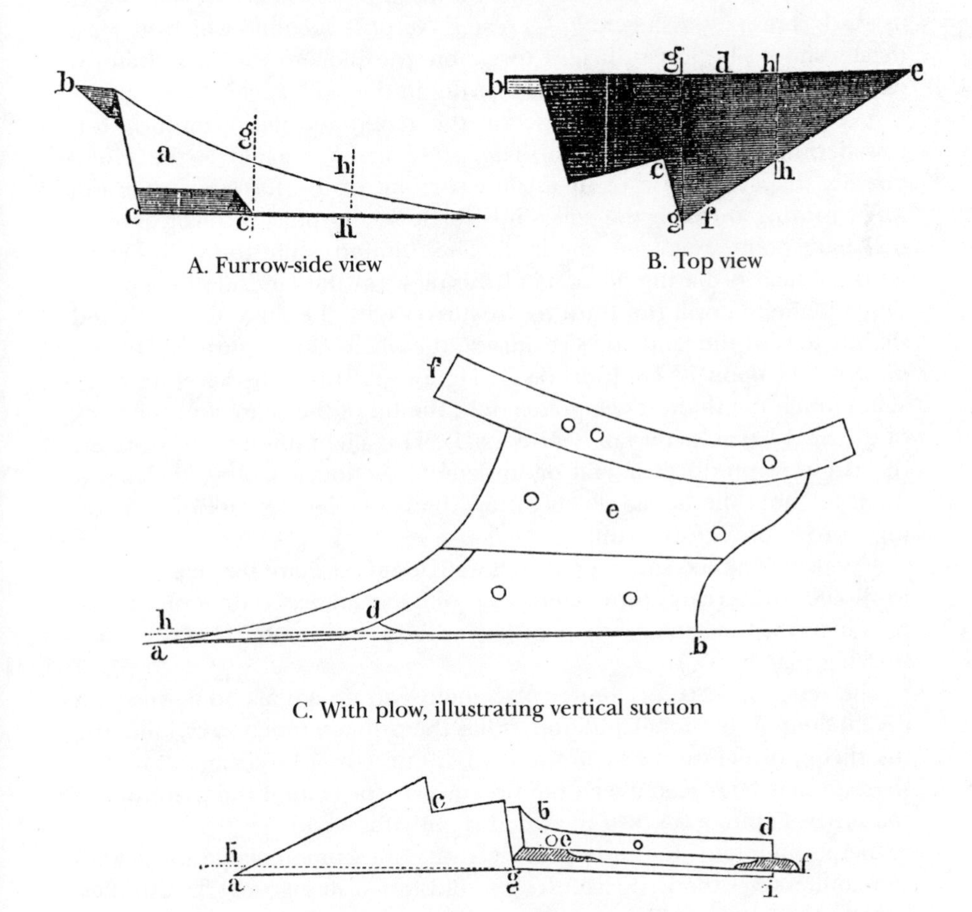

A. Furrow-side view

B. Top view

C. With plow, illustrating vertical suction

D. With plow, illustrating lateral suction

Source: New York State Agricultural Society, *Report on the Trial of Plows Held at Utica . . .* (Albany: Van Benthuysen and Sons, 1868), pp. 41, 45. I am indebted to Wayne Randolph for pointing out that while the source is American, the designs depicted are probably English.

severed earth back to the moldboard's surface), it must have a keen point and a sharp edge for the plow to function smoothly—requirements which, as we shall see in the next section, are not easily maintained when the part in question is subjected to so much pressure and wear. Smooth functioning also requires two other features in share design. A flat-bottomed wedge dragged through the earth has a natural tendency to rise, much as a boat will rise in the water when subjected to acceleration. To hold the plow down so it will carve a furrow of roughly uniform depth, the point of the share must run through the earth at a level slightly below the bottom of the plow. When placed upon a barn floor and viewed from the landside (see Figure 3.7C), the plow will therefore touch that surface only at the tip of the share *(a)* and the back of the heel *(b)*. This slight upward convexity, rising to *(d)*, creates "vertical suction" which at least ideally should be just sufficient to enable the plow to run at a uniform depth, with no tendency to rise up or dig in deeper.[16]

A second design difficulty concerns the generation not of uniform furrow depth, but uniform width. The problem once again is the moldboard's shape. Because of its asymmetry, one of the forces it generates when moving through the soil tends to cause the plow to pivot, pulling the share point away from the land. The solution is illustrated in Figure 3.7D. Instead of having all parts of the plow on the landside lie along a single plane that will run flush to the furrow wall, the share point is tilted slightly toward the land (or just outside the plane *hi* in Figure 3.7D). If a plow is laid upon its landside (as in Figure 3.6B), it will therefore once again touch the floor at two points only, the tip of the share and the back of the heel (or at *a* and *i* in Figure 3.7D). This slight inward curvature on the landside produces lateral or horizontal suction, enabling the implement to "hold the land," which in turn makes easier the task of generating furrows of uniform width.

A poorly designed share, or even a well-designed share that has become worn and dull, creates three kinds of problems: increased drag or friction plus increased instability of operation on both its horizontal and vertical working planes.

The requirements for coulter placement and design are somewhat less demanding. An optional addition, it has the primary function of initiating the slicing of the furrow from the land. Mounted on the beam above the share, it may be placed ahead of, directly over, or behind the point of the share. (In Figure 3.6A, it is mounted slightly forward.)

By the middle of the nineteenth century, a common shape for American coulters was the knife-like design[17] illustrated in Figure 3.8. Also illustrated are typical options for attaching it to the beam, including inserting it through the center or fastening it to the side, the latter sometimes achieved with a circular plate allowing for easy adjustment of the angle between coulter and beam. To cut sod, for example, that angle would be

Figure 3.8. American coulters, mid-nineteenth century

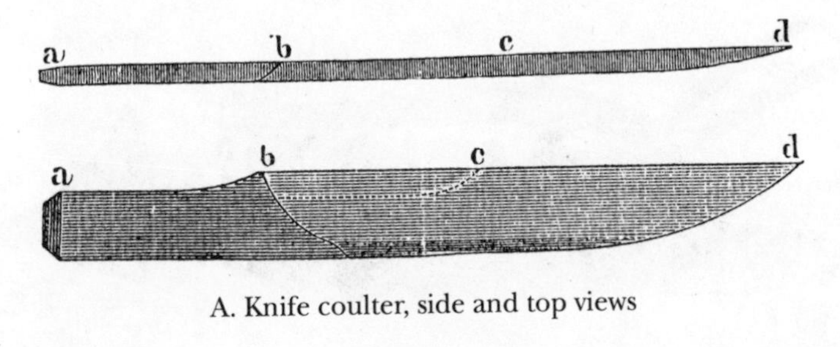

A. Knife coulter, side and top views

B. Knife coulter, mounting options

C. Revolving or wheel coulter

Source: New York State Agricultural Society, *Report on the Trial of Plows Held at Utica . . .* (Albany: Van Benthuysen and Sons, 1868), pp. 43, 180, 181.

made quite acute. For working newly cleared land with many roots, a more obtuse angle was desirable if the coulter was to sever such obstacles instead of uprooting them. Alternatively, in well-tilled land free of such obstacles, the knife design might be replaced with a revolving metal disk (see Figure 3.8), which gave a smoother cut of surface sod and was less inclined to accumulate debris on the coulter shaft just below the beam.[18] To accomplish this slashing function, two of the designs used in nineteenth-century America are illustrated in Figure 3.9. One design essen-

Figure 3.9. Soil-slashing designs

A. American fin share

B. American lock-coulter

Sources: American Agriculturist, p. 5 (April 1846), 108; New York State Agricultural Society, *Report on the Trial of Plows Held at Utica . . .* (Albany: Van Benthuysen and Sons, 1868), pp. 181, 182.

tially scrapped the coulter and extended a cutting edge upward from the share, to produce a "fin share." The other retained the coulter, but for added strength and stability fastened the bottom of the coulter blade to the top of the share, creating a "lock-coulter."

The share and the coulter, while crucial for initiating the shearing of the furrow slice, do not by themselves contribute much to the goals of inversion and pulverization. The latter is effected primarily by the moldboard. The puzzle is which moldboard shape is best for achieving such results. Unfortunately, a bewildering array of variables affects moldboard performance. The most obvious, perhaps, is the type of soil tilled. In the cautious wording of an agricultural engineer, "soil in its natural state is a non-homogeneous material."[19] It can be wet or dry, sand or stubble, sod or clay, and littered with roots and stones or comparatively free of such obstructions. The size of the moldboard—its length and height—affects performance, as does the abruptness of its curvature. A long moldboard with a gradual twist will pulverize less and produce coarser results. A steeper twist has a more disruptive impact upon the soil and pulverizes more thoroughly. In addition, the forward speed of the plow affects the path soil takes across the moldboard's surface, and accordingly also affects performance.[20]

In the exploration of moldboard designs, a useful starting point is to regard the building of a moldboard as analogous to the building of a curving staircase. The objective is to make the soil "climb the stairs," in a manner illustrated in Figure 3.10A. Irregular steps might then be converted into a smooth surface using the mathematics of straight lines (in much the same manner as Jefferson originally proposed). Consider the lines depicted in Figure 3.10B, beginning at the front of the plow and sloping upward to the rear. A uniform curvature can be achieved by constructing the moldboard in such a way that a chalked string laid along any of these parallel lines would touch the surface at every point. (Alternatively, a yardstick laid along *bd* and then, while remaining parallel to that line, moved up or down would remain in contact with the moldboard's surface throughout its entire length.) This approach guarantees a uniform curvature, but is it an optimal curvature? Jefferson's moldboard design, for example, was criticized for being "not as favorable for good and easy plowing" as designs based upon curved rather than straight lines.[21] But which curves guarantee the best results? Should the moldboard's surface, for example, follow the undulating lines of Figure 3.10C (one solution advocated in the nineteenth century), or should it try to approximate the inner surface of a cone-shaped cylinder, as depicted in Figure 3.10D? Or should it try to approximate the vase-like shape of Figure 3.10E (a hyperboloid of one sheet), as illustrated in Figure 3.10F? The answer is none of the above. To be more cautious, the many different shapes that came to dominate moldboards in twentieth-century America cannot all be represented by any one of these designs. Nor is this surprising. Optimal shape varies with type of soil, and soil types, as previously noted, vary enormously. The associated complexities have defeated all efforts to use mathematical principles to identify optimal moldboard shapes. Thus, the radical improvements in design achieved in nineteenth-century America were the product of trial and error.

Not that mathematics was neglected. Some would-be inventors tried to follow Jefferson's lead, beginning with the mathematics and ending with a moldboard shape, including (with patent dates) Gideon Davis (1818), James Jacobs (1834), Samuel Witherow and David Pierce (1839), Samuel A. Knox (1852), and F. F. Holbrook (1867).[22] If their erudition was impressive, their plows were not. A pioneer in American agricultural engineering suggests why.

> To be sure, geometrically exact moldboards furnished the basis in many instances for more perfect developments, but the results obtained by empirical plow designers who worked in the field were so far superior to the results obtained by the men who worked in the laboratory that the theorists were

Figure 3.10. Possible moldboard shapes

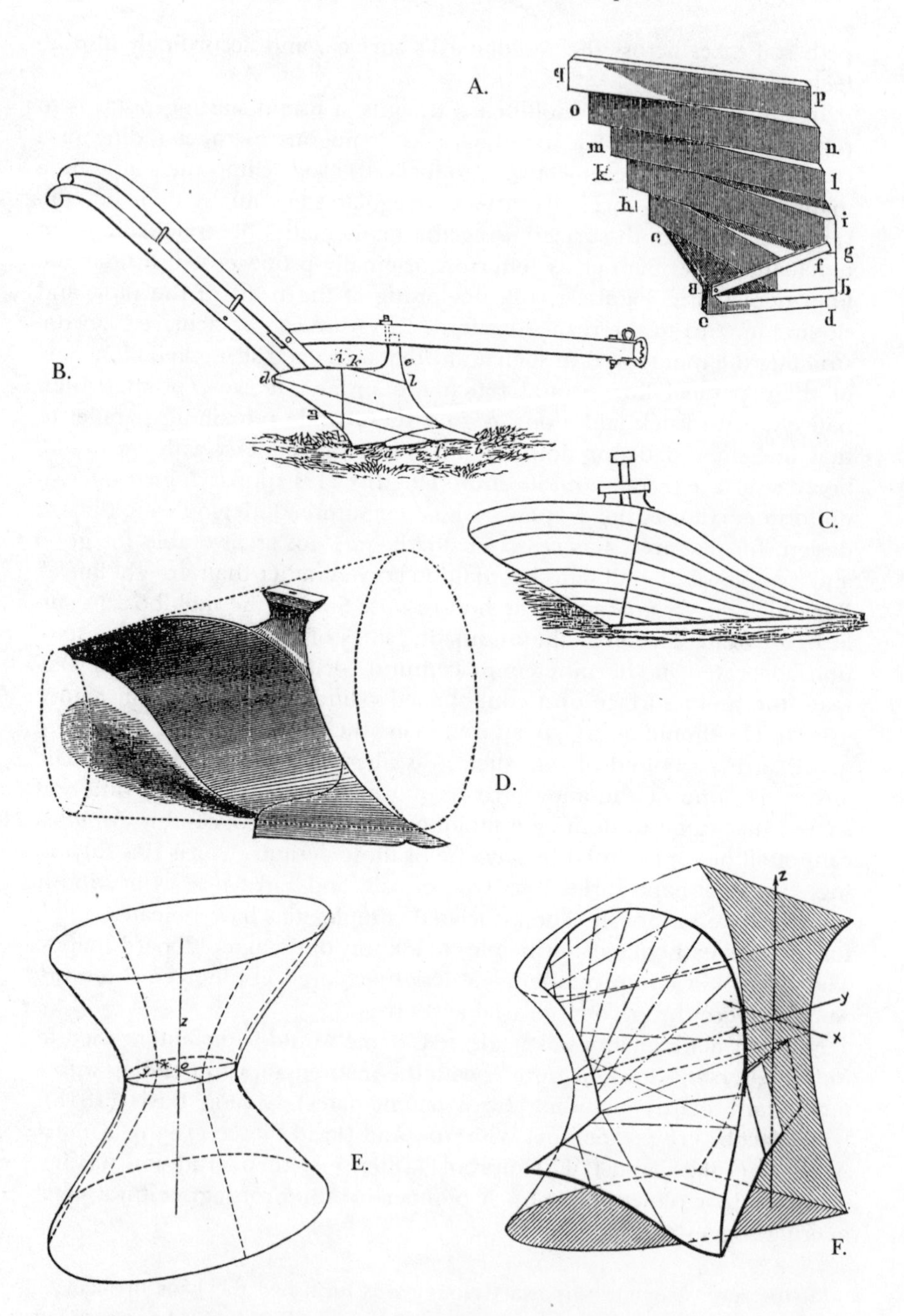

Sources: New York State Agricultural Society, *Report on the Trial of Plows Held at Utica . . .* (Albany: Van Benthuysen and Sons, 1868), pp. 57, 85, 98, 128; E. A. White, "A Study of the Plow Bottom . . . ," *Journal of Agricultural Research,* 12 (January 28, 1918), pp. 155, 160.

soon completely outstripped and even held up to ridicule by the men who developed their machines in the hard school of experience.[23]

Even from a modern perspective, "tillage is more an art than a science."[24] Our task is to study how that art evolved; in particular, how plow designs were modified, not in response to scientific insights, but in response to new tillage challenges for which old plow designs were unmistakably unsatisfactory. Given the complexities of a well-functioning modern moldboard plow (only a few of which have been discussed above), this evolution in design from primitive times to the present, in its early stages can be expected to be halting and imperfect, spanning many centuries during which negligible advances will be the rule and significant progress the exception. The history of this evolution begins, in the West, with the challenges would-be tillers of the soil faced in transalpine Europe during Roman times. Here the ancient ard was clearly an inadequate implement. The question was how quickly an alternative would be developed.

Evolution of the Moldboard Plow: From Roman Times to the Eighteenth Century

Because our ultimate concern is the evolution of American plow designs from first settlement to the 1830s, and developments in Europe are of interest only insofar as they contribute to improvements on the other side of the Atlantic, this section focuses on two questions. One concerns the process whereby the ard design of the ancient world was modified to produce a plow capable of tilling the heavy soils of northern Europe, because most American soils would require a similar plow. The other concerns the pace of European technological progress. Was the development of the moldboard plow from Roman times to the Napoleonic era a story of gradual design improvements over the intervening centuries that ultimately produced the sophisticated tilling implement of Figure 3.5? Or was negligible change the rule over most of these centuries, with the tradition of searching for, and finding, better plow designs for heavy soils a late development, leading to a rush of innovations centered in the eighteenth century? The answer will condition our expectations about the likely pace of agricultural change in colonial America. If gradual evolution was the rule in Europe, then immigrants to the New World would bring from the Old a tradition of design improvement. If instead Europe experienced little change for centuries followed by an explosion of novelties in the eighteenth century, early American settlers would be unlikely to have a tradition of constantly striving to improve their agricultural implements. Moreover, being on the periphery of the British Empire, they might well remain immune for decades to an initial burst of agrarian experimentation in Europe during the eighteenth century.

To identify the pacing of European change, our focus will continue to be those parts of the plow that cut the furrow and invert and pulverize the soil: the share, the coulter, and the moldboard. The European changes of particular interest are the major design modifications in the country that most influenced American agricultural practices, namely, Great Britain.

Experts generally agree that the Romans can be credited with first incorporating the coulter and the wheel into plow designs, as well as with several significant improvements in share design.[25] The Roman contribution to moldboard development is a more controversial topic. They did develop a "ridging board": a flat projection to one side only, to move the earth in that direction. While this board could help to create a furrow, it did little to invert and pulverize the soil. If a moldboard is defined as having the "complicated double curvature" depicted in Figure 3.10, then one might conclude with K. D. White that the "true" moldboard plow "did not appear in Europe before the latter part of the eighteenth century."[26] But the issue here is not when the modern design first appeared, but which steps were crucial in its early evolution.

The ridging board of Roman times would seem an obvious beginning. A double curvature is difficult to envision and hard to design, whereas a flat board pushing earth to one side is not. As a design modification to meet the challenge of the heavier soils of Northwest Europe, the ridging board had obvious defects which were overcome to some extent by other changes in plow design. Asymmetric coulters were developed both to cut the soil and to move it well to one side.[27] Shares were made larger and one innovative shape had a vertical cutter to create a furrow and two extended horizontal edges to cut roots and weeds.[28] These larger shares were often curved slightly downward to provide vertical suction. But imperfect curvature could cause a cumbersome and primitive implement to dig into the earth, while wider shares driven through heavier soils sharply increased the drag upon the implement. A partial remedy for both problems was to add a wheeled forecarriage somewhat similar to those depicted in Figure 3.11. The resulting Roman "heavy" plow, by modern standards, was the very model of awkwardness and inefficiency, but it did make the tilling of heavy soils possible.[29]

What remains obscure is the influence of Roman plow designs upon the plows that first appeared in European illustrations about 1000 A.D. The problem is an absence of evidence. From the Roman era, few plows parts have survived (and only those made of metal), depictions of plows on monuments are generally poor, and the handful of literary descriptions by Roman contemporaries do not include any supporting illustrations.[30] From the fall of Rome to the rise of Charlemagne, almost nothing is known about plow development. No documents have survived from the

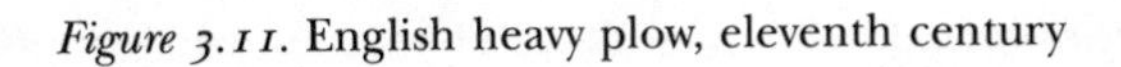
Figure 3.11. English heavy plow, eleventh century

Source: James Carson Webster, *The Labors of the Month* . . . (Evanston, Ill.: Northwestern University Press, 1938), Plate XIX.

early Middle Ages explaining how this implement was constructed or should be constructed. When illustrations of heavy European plows first become available around 1000 A.D., the drawings are poor and design details are therefore difficult to interpret. Consider, for example, the eleventh-century illustration in Figure 3.11. Most of the earliest depictions of plows, like this one, show a heavy implement drawn by a number of oxen, with wheels and a coulter. But whether these features reflected Roman or other influences (such as the Saxons in the case of English plows) cannot be known with any certainty.[31] Whether the ridging board of Roman days had evolved into a better approximation of a moldboard also remains obscure, in part because these earliest illustrations were so badly drawn. In Figure 3.11, for example, a small open moldboard seems to be appended to the furrow side of the plow, but whether it functioned well or badly cannot be gauged by close inspection of the picture.[32]

With the passage of time, illustrations of the plow improve, and several features of the medieval plow become more clear. It usually has a coulter of some type, but frequently does not have a wheeled forecarriage.[33] The shading that soon appeared on depictions of the moldboard suggest that, at least by the fourteenth century (if not long before), the flatness of the Roman ridging board was giving way to curvature in one dimension, from top to bottom. (See Figures 3.12 through 3.14.)[34] Alternatively, most or all of this shading may have been meant to suggest not a curvature but the addition of iron strips to the wooden surface.[35]

Over the next few centuries patterns of change are difficult to detect because designs were generated by legions of carpenters and blacksmiths in numerous villages, each following his own whim. The different tilling requirements of different soils added further diversity to what was intrinsically a heterogeneous array of implements. England would ultimately be

Figure 3.12. English moldboard plows, fourteeth century

Sources: Thomas Wright, ed., *Piers Ploughman* (London: John Russell Smith, 1856), frontispiece; Eric George Millar, *The Luttrell Psalter* (London: Bernard Quaritch, 1932), f. 170, Plate 92.

among the leaders in the search for better and standardized plow designs, but that search and that standardization are difficult to detect prior to the eighteenth century. By the mid-seventeenth century, for example, an officer in Cromwell's army who tried to generalize about English husbandry practices conceded that the number of plow designs in use was "too endless to discourse."[36] Half a century later, the author of *The Whole Art of Husbandry* conceded that, with respect to English plows (as elsewhere in Europe), "There is a great difference in most places in their make and shape . . . every place . . . being wedded to their particular fashion, without any regard to the goodness, conveniency or usefulness of the sort they use."[37]

The literature attempting to describe this diverse reality is, by and large, unhelpful to those searching for signs of central tendencies and evolutionary trends. The first English treatise on practical husbandry dates

Figure 3.13. Steps to construct an "ordinarie plough," early seventeenth century

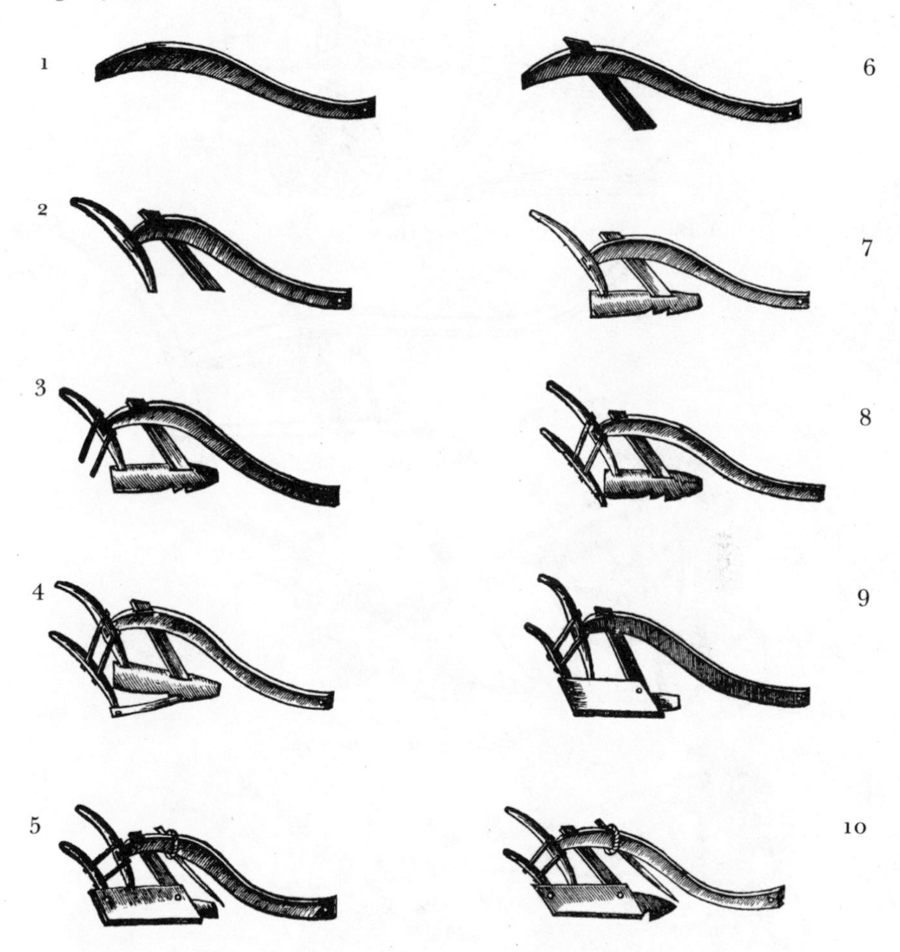

Source: Gervase Markham, *The English Husbandman* (London: John Browne, 1613), chapter III.

from 1523,[38] but neither this work nor those that followed in the next two centuries give much information on how plows were actually made. One notable exception is the sequence of pictures reproduced in Figure 3.13. Writing at the beginning of the seventeenth century, Gervase Markham described construction of a plow as beginning with a beam (1), to which a "sheath" or "standard" was attached (2) and then a left-hand handle or "stilt" (3). To the bottom of the sheath a "share-beam" or "plough head" was fixed (4). After spindles (5) were joined to the first handle, a second (right-hand) handle[39] was added (6), the latter also joined to the share-

Figure 3.14. Notable English plow designs, 1650–1750

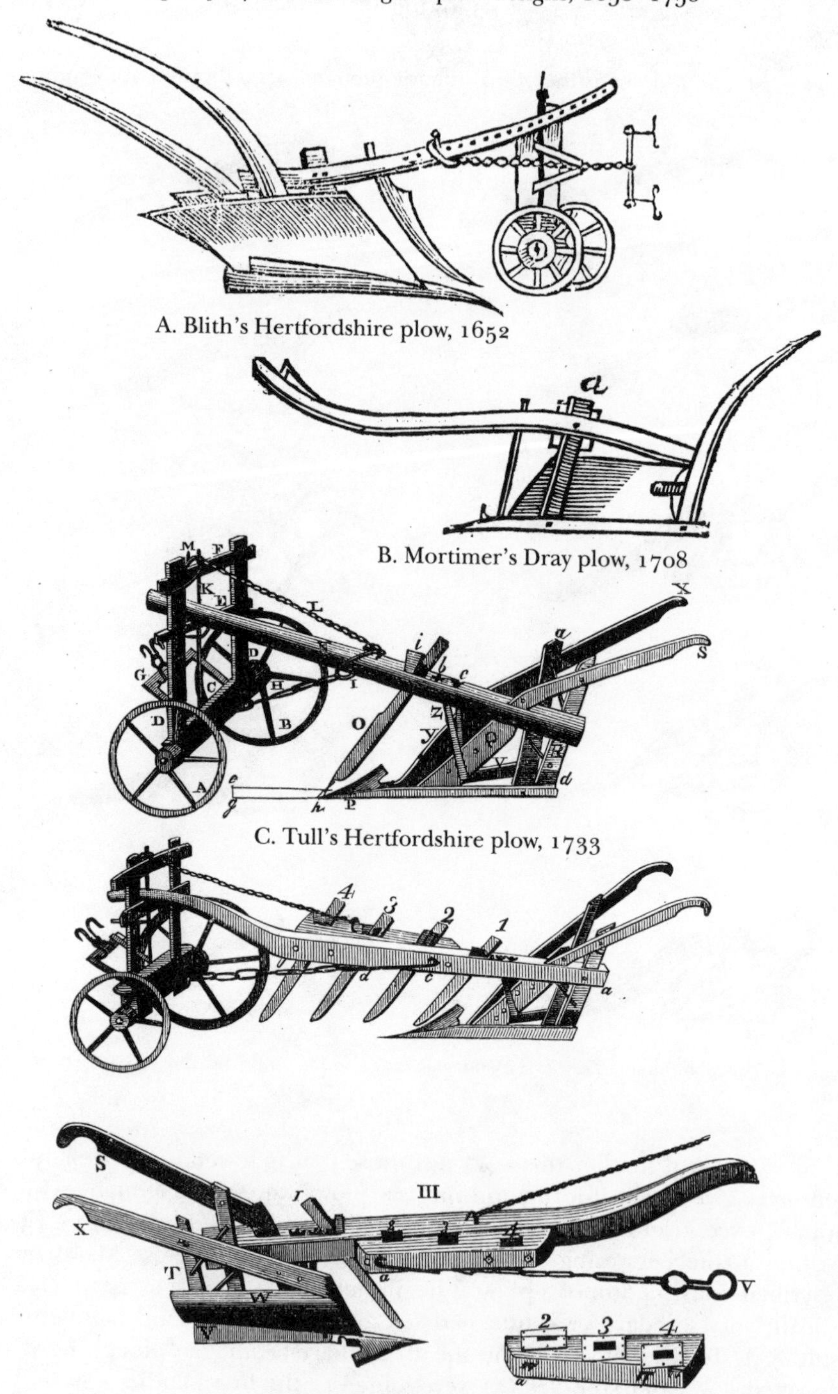

A. Blith's Hertfordshire plow, 1652

B. Mortimer's Dray plow, 1708

C. Tull's Hertfordshire plow, 1733

D. Tull's four-coulter plow, 1733

Sources: Walter Blith, *English Improver Improved* (London: John Wright, 1652), plate facing p. 203; John Mortimer, *The Whole Art of Husbandry* (London: J. Robinson, 1708), p. 41; Jethro Tull, *The Horse-Hoeing Husbandry* (Dublin: A. Rhames, 1733), Plate I.

beam with a brace (7). A moldboard (8) was then fastened to this brace and the right-hand handle. Finally, a coulter (9) was attached to the beam, and a share (10) attached to the front of the share-beam.[40]

Missing in Markham's accompanying comments, and absent from any of the English husbandry literature published between 1523 and 1730, is a detailed discussion of how these different parts *should* be made to perform effectively. A survey of English plows in 1652, for example, found much to criticize: handles tended to be too short or too long, plows were too heavy, and moldboards were too straight (the latter Walter Blith deemed "a very great fault").[41] But the author's recommendations for change were couched in the most general terms; for example, "plough irons should be well steeled, sharp and well pointed."[42] The same was true of John Mortimer's recommendations, made in 1708: the plow for "stiff black clays" should be "long, large and broad"; for "white, blue, or grey clay [it] need not be so large as the former"; and for "red, white, sands or gravel . . . [it] may be lighter and nimbler than the former."[43] Commentary this vague without supporting detailed illustrations offered little help to those trying to build plows, and gives little guidance to historians searching for evidence of improvements in design.

A scanning of the illustrations (as distinct from the written commentaries) occasionally published in the two centuries following 1523 does suggest that modest improvements in some parts of the plow were being made by at least some of the English plow designers. The 1652 illustrations of Walter Blith, for example, include a circular coulter not unlike the one depicted in Figure 3.8, and a broader, flatter share that, in appearance, somewhat resembles the share in Figure 3.7B.[44] Between 1400 and 1730, in general appearance plows seem to become less primitive and cumbersome. (Compare, for example, the plows of Figure 3.12 with those of Figure 3.14.)[45] And in the final century of this extended period certain general types begin to emerge, such as the Hertfordshire, which appears in husbandry books published in 1652, 1708, and 1733.[46]

To modern eyes, however, what is noteworthy is what is missing. While the plows of the early eighteenth century seem less primitive than those of the fourteenth century, they still differ strikingly from the modern plow.[47] By 1730 the Hertfordshire, for example, may have been, in J. B. Passmore's words, "the highest achievement in [English] plough design which was possible along the old orthodox lines,"[48] but British expert Arthur Young, writing in 1804, saw it as "a heavy, ill-formed, and ill-going plough," distinguished for its "miserable construction." As for its performance in the field, "worse work can scarcely be imagined."[49] What British plow designers had failed to accomplish over the centuries is well summarized by Percy Blandford: "Cumbersome and crude ploughs, but showing some understanding of the principles involved, were in use before the

Norman Conquest in 1066, and continued with only slight improvements up to the eighteenth century."[50]

In support of that judgment, consider what is missing in Figure 3.14D. This plow design was created by one of the foremost agrarian innovators of the eighteenth century, Jethro Tull. But aside from the addition of three more coulters, it bears a marked resemblance to the Hertfordshire plow of 1733 (Figure 3.14C), which does not differ significantly from the Hertfordshire plow of 1652 (Figure 3.14A).[51] Moreover, the moldboard of Tull's experimental plow duplicates the one-dimensional curvature (from top to bottom) evident in fourteenth-century illustrations, with no hint of the double curvature that is central to the superior performance of nineteenth-century designs.[52] Perhaps most distinguished by its absence from all the English husbandry literature published in the two centuries before 1733 is any attempt to link mechanical principles to the priority of plow improvement, identifying in the process specific, detailed instructions for modifications in design.[53] Put another way, a would-be inventor of a better plow, after scanning available husbandry literature for helpful hints, would be about as much in the dark and on his own in 1725 as in 1625 or 1525.

All this changed, and quite dramatically, beginning in 1730. A radically new plow design appeared with which Jethro Tull was evidently still unacquainted in 1733. But note the implications, at least to this point in time, of English theorizing for American agrarian practices. Insofar as colonists were recent immigrants from the British Isles or acquainted with the English husbandry literature of their day, they would find little in their knowledge of the Old World to encourage experimentation with plow designs in the New.

The story of subsequent changes in English implements, while fraught with drama and complexity, can be quickly summarized. In the judgment of *The Encyclopaedia Britannica*, "The quarter of a century immediately following 1760 is memorable for the introduction of various important improvements"[54] in British agriculture. That judgment misses a crucial improvement by three decades. In 1730 Joseph Foljambe and Disney Stanyforth patented a plow that became known by the name of the Yorkshire town where it was first manufactured: the "Rotherham." Who designed the plow and whether the structure of contemporary Dutch implements influenced the designer are still topics of dispute.[55] Scholars commonly acknowledge, however, that for English agriculture the Rotherham represented a technological breakthrough. (According to one analyst not given to understating the case, it was "the greatest improvement in plow design since the late Iron Age."[56])

What made the Rotherham so revolutionary is almost impossible to grasp from written descriptions, but easily understood by comparing illustrations. The starting point is Figure 3.15, which reproduces one of the

Figure 3.15. Rotherham plow

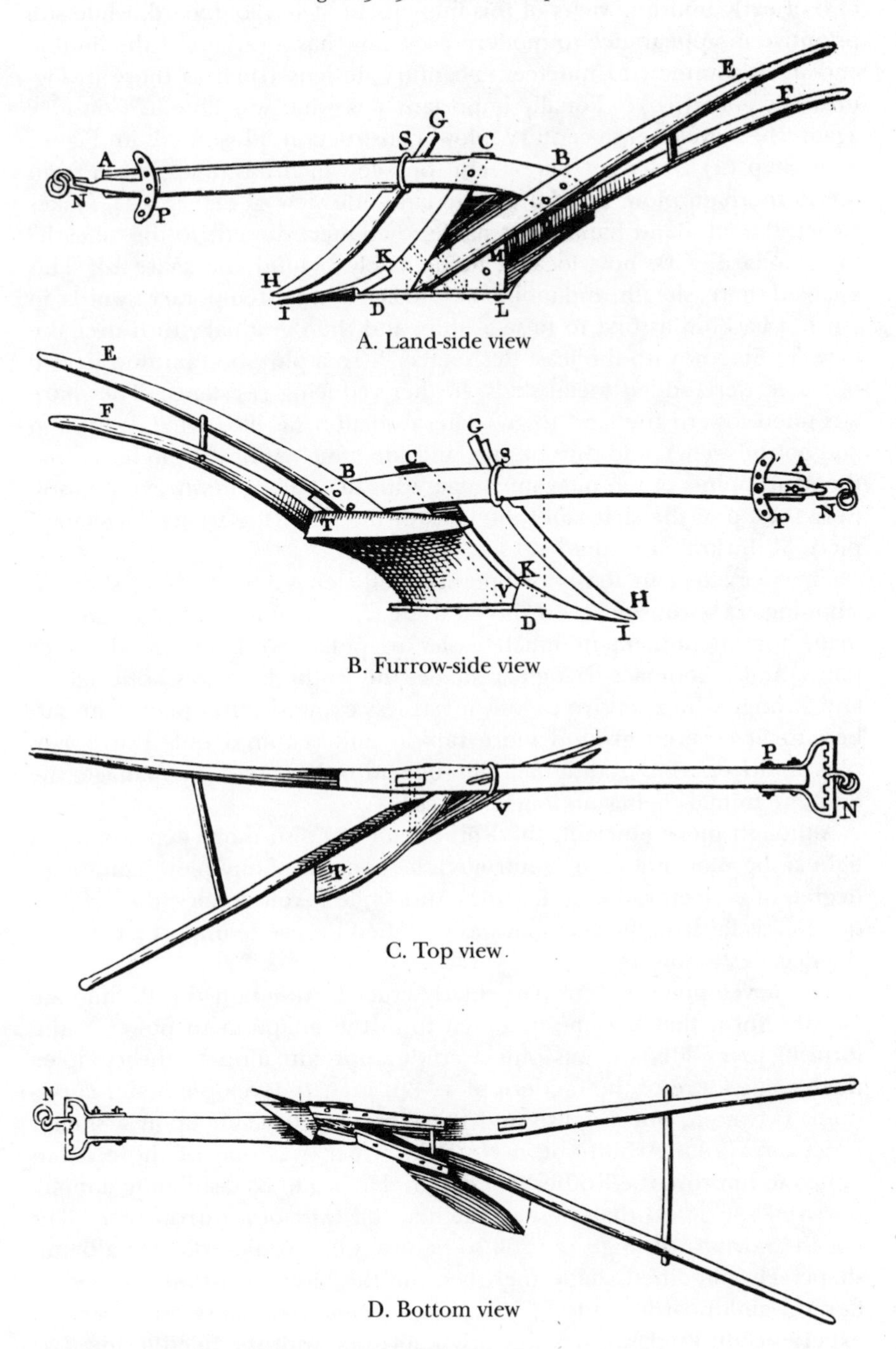

Source: Robert Maxwell, ed., *Select Transactions of the Society of Improvers . . .* (Edinburgh: Sands et al., 1743), plate following p. 458.

best of early multiple views of this implement. The moldboard, while still primitive in appearance to modern eyes, now has a variant of the double curvature common to nineteenth-century designs (such as those in Figures 3.5 and 3.10).[57] Equally important was what was absent. Consider again the seventeenth-century plow construction illustrated in Figure 3.13. Step (4) added a "share-beam" or sole which has disappeared from the Rotherham plow. Instead (as the land-side view of Figure 3.15 makes clear), the left-hand handle was made to connect directly to the "sheath" or "standard" CD, now located immediately behind the share KI. This enabled share, sheath, and moldboard to act, in a contemporary's words, in "such a Fashion as first to raise a little, and then gradually turn over the new cut Furrow with the least Resistance."[58] To a plow bottom now devoid of a sole were added metal skids, further reducing resistance. The share was tilted toward the land to give lateral suction (as illustrated in the top and bottom view), and thus on the landside, much as in the modern plow, the main points of the plow in contact with the wall of newly exposed soil were the tip of the share and the back of the heel (the triangular-shaped piece M illustrated in the land-side view).

The net effect in terms of operating efficiency was nothing short of stunning. Less resistance on the landside, furrow side, and plow bottom made for an implement much easier to pull.[59] With less wood, fewer joints, and a compact triangular shape, the Rotherham was both lighter and stronger. Inexpensive to buy, it was less expensive to operate, in part because it covered ground more rapidly and required only two horses plus a driver, the traditional plow-boy with his goad to manage the draught animals being no longer necessary.[60]

Although more efficient, the Rotherham was also more demanding. A light swing plow not easily controlled, it required of any plowman a high degree of concentration and skill.[61] And while revolutionary, the original design was far from perfect, but was modified in several important ways in the decades following 1730.

Two developments were particularly crucial. Although the Rotherham "set the form that became accepted until the adaption to power,"[62] the form of its moldboard was only a crude approximation to the complex double curvature of the modern plow. But how to develop a better curvature? Two men, one English and the other Scottish, hit upon a similar procedure. John Arbuthnot, a Norfolk farmer, was one of those determined to improve the Rotherham design. He began by fashioning a moldboard of soft wood that closely matched the twist of a furrow slice. This was then driven through the soil so natural wear would establish a better shape. The modified shape then became the basis for drawings used to fashion moldboards made of iron.[63] James Small, a Scotsman with work experience in England, was similarly impressed with the need to improve moldboard designs, and began producing superior models in his native

Scotland in the 1760s. As Small tells the story, he began with what he termed a "simple" principle: "As the best way of shifting the earth to one side, or of raising it, is to shift it, or raise it, by equal degrees for every inch that the plough advances, so the best way of turning it over, is to give it equal degrees of twist for every inch that the plough advances."[64] Using this principle as a guide, Small built a moldboard from soft wood and then, like Arbuthnot, drove it through the soil to identify, by the process of wear, desirable modifications. When satisfied, Small cast the resulting shape in iron.[65] Accordingly, his plows, while similar in general appearance to the Rotherham, featured a moldboard more closely approximating the shape that came to dominate nineteenth-century British agriculture. (Compare, for example, Figure 3.16 with Figures 3.5 and 3.15.)[66]

Of the two men, Small had by far the greater impact upon British husbandry literature.[67] The plows of both, however, suffered from a major flaw that another Englishman soon remedied.

The problem was the share. A point of maximum wear, it continually lost its cutting edge during plowing. To resharpen it required special skills, which at best meant lost time in the field using a file, and at worst necessitated repeated and costly time-consuming trips to the local smithy. Robert Ransome, the son of an English schoolmaster, stumbled upon the solution by accident. Having established a small foundry at Norwich, he took out a patent for tempering cast-iron shares in 1785. Almost two decades later, as a result of spilling molten iron on the foundry floor, he discovered that its contact with a cold surface made for a harder metal. Using this principle, he created a share with a bottom and landside of hard iron and a top of softer iron.[68] The top therefore wore more quickly than the bottom, thereby preserving a sharp bevelled edge[69] until the share itself had to be replaced. Ransome also pioneered the use of interchangeable parts (for which he received a patent in 1808),[70] but his plows, in general appearance, were quite similar to those of James Small. (Compare, for example, the two models illustrated in Figure 3.16. Although the Ransome model illustrated has wheels, he also produced swing plows.)

The plow innovations discussed thus far are but a few of the better-known design modifications instituted between 1730 and 1810.[71] Agrarian innovation was becoming the fashion in Britain, if not the rule, and novel implements first appearing during this period included trenching plows and mole-plows, plows with exceptionally long or exceptionally short moldboards,[72] double-furrow,[73] three-furrow, and even six-furrow plows. To test the efficacy of novel designs, a new form of competition, the plowing match, developed, complete with a new measuring device, the dynamometer, to identify the force required to pull different plows.[74]

Insofar as plowing practices in British North America followed the lead of plowing developments in the British Isles, one might expect the evo-

Figure 3.16. Improved English plows

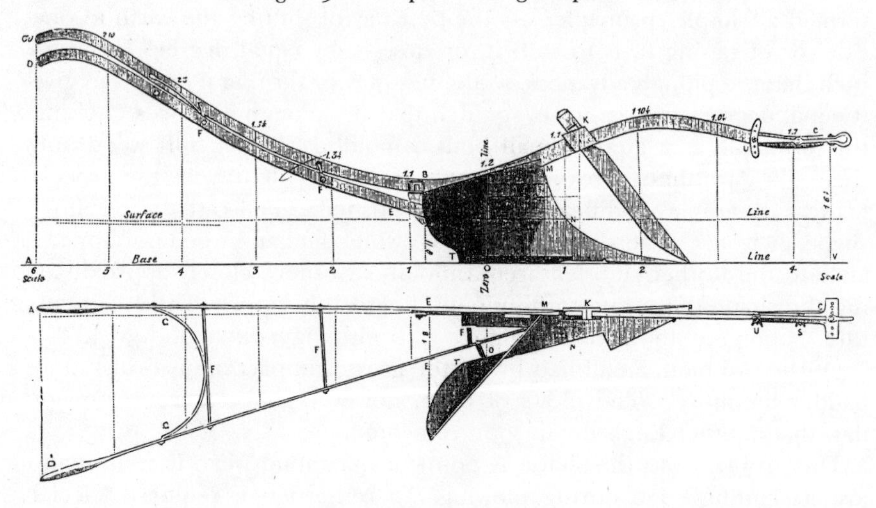

A. James Small's plow

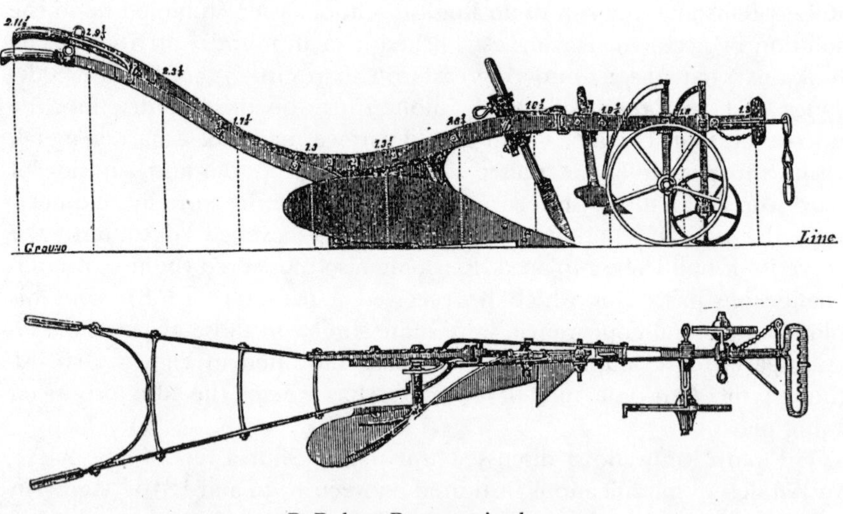

B. Robert Ransome's plow

Source: New York State Agricultural Society, *Report on the Trial of Plows Held at Utica* . . . (Albany: Van Benthuysen and Sons, 1868), Plates II, V.

lution of colonial implements prior to the American Revolution to be a two-part story.[75] Seventeenth-century plow designs should remain comparatively static, while those of the eighteenth century should improve radically (if with a lag) in response to the British technological breakthrough dating from the 1730s that ushered in a new age of superior plows and a new willingness to experiment with designs. Surviving evidence of American farming practices, sparse and imperfect as it is, suggests a totally different story.

Evolution of American Plows: Colonial Times to Independence

Farming practices in colonial America were not, of course, the exclusive product of British influences. Settlers came from many countries—Germany, France, Holland, Switzerland, and Ireland, to mention only a few—and from their native lands they brought agrarian knowledge and implements. (The influence of the Dutch, as we shall see throughout this work, was particularly important.) But when European immigrants encountered New World conditions, they were often compelled to modify their farming practices.

The central economic fact was abundant land. Cheapness of this factor of production, as every beginning economics student learns, should encourage those production techniques that are, in some sense, "land-intensive." Colonists therefore farmed the New World in ways unsurprising to the economist and yet offensive to European visitors accustomed to European ways. James Madison put the point as well as any and better than most. "Whilst there was an abundance of fresh and fertile soil, it was the interest of the cultivator to spread his labour over as great a surface as he could. Land being cheap and labour dear and the land co-operating powerfully with the labour, it was profitable to draw as much as possible from the land."[76]

An uncommon sight in the settled areas of Europe but the norm in most farming regions of British North America was a newly cleared field strewn with stumps, roots, and stones. This uncleared residue was the result of a tree removal system that sacrificed thoroughness to save labor while gaining quick access to potential crop land. The usual practice was to burn the undergrowth and kill the trees by girdling.[77] Crops could then be sown, first among the dead trees, and later among the rotting stumps. By the time the stumps had finally rotted away, the farmer often had moved his crops to other fields that were recently cleared, and for that reason promised higher fertility. The resulting tillage problems are immediately apparent in Figure 3.17. Faced with potential crop land strewn with so many obstacles, those wishing, in Madison's phrase, "to draw as much as possible from the land" would often use hoes or spades instead of plows to break up fields prior to sowing. The owner of a Virginia plantation in the eighteenth century, for example, did not use a plow when he first began to farm, and his father had never used a plow at all.[78] While

Figure 3.17. Virginia tobacco field, seventeenth century

Source: E. R. Billings, *Tobacco: Its History, Varieties, Culture* . . . (Hartford, Conn.: American Publishing, 1875), p. 51.

common in the seventeenth century, such choice of tillage implements persisted well into the eighteenth, as the above case illustrates.[79]

When fields were finally clear, or at least relatively clear of obstacles (primarily due to rotting, often supplemented by the burning of stumps), those with sufficient fertility to warrant further tilling were often put to the plow, but here too the plowing techniques chosen commonly reflected the cultivator's desire, as Madison would have it, "to spread his labor over as great a surface as he could." A popular method to increase this spreading was shallow plowing, often less than three inches deep, and seldom more than four inches.[80] The end result, which invariably struck European visitors as work "carelessly performed," was judged by most Americans to be "a satisfactory job of plowing," even if the fields were "covered with clods" and "not all the ground touched by the plow."[81]

Cheap land and dear labor were not the only reasons for the colonists' preference for shallow plowing. Another of the first importance was the primitive nature of the plowing implements used. To force a cumbersome device to carve a deeper furrow required additional draught animals and possibly an additional worker riding the plow beam to hold it down, and most colonists were reluctant to incur the associated expenses.[82]

Conversely, the relatively low price of land encouraged the development of plow designs considered satisfactory for shallow plowing. Little is known about this process of development, or which designs were "typical," particularly in the early period. The same set of problems confounding generalizations about plow types in the Old World complicates analysis of these implements in the New. Colonial settlements sprawled along the eastern seaboard, from the gravelly loams of the Maine coast to the sandy loams of upland Georgia, and the plows in use, if not imported, were usually the product of local carpenters and village blacksmiths. No two looked exactly alike, but the question of importance at this juncture is whether central tendencies in design existed among what was necessarily a heterogeneous array.

The answer cannot be found among written descriptions of the implements. The classic account, often reproduced in agricultural studies of eighteenth-century America, actually dates from 1846.

> A winding tree was cut down, and a mould-board hewed from it with the grain of the timber running as nearly along its shape as it could well be obtained. On to this mould-board, to prevent its wearing out too rapidly, were nailed the blades of an old hoe, thin straps of iron, or worn-out horseshoes. The landside was of wood, its base and side shod with thin plates of iron. The share was of iron with a hardened steel point. The coulter was tolerably well made of iron, steel-edged, and locked into the share nearly as it does in the improved lock-coulter plow of the present day. The beam was usually a straight stick. The handles, like the mould-board, split from the crooked trunk of a tree, or as often cut from its branches: the crooked roots of the white ash were the most favorite timber for plow handles in the Northern States. The beam was set at any pitch that fancy might dictate, with the handles fastened on almost at right angles with it, thus leaving the plowman little control over his implement, which did its work in a very slow and most imperfect manner.[83]

Such descriptions support the unsurprising proposition that in premodern economies, particularly frontier economies, production techniques tend to be primitive. But they leave largely unanswered whether colonial plows had common design features, and if they did, whether those features were similar to, or different from, the plow designs in European countries from which the settlers came. Few plow descriptions are available in colonial literature, and these tend to be too general to be of much use. Depictions of American plows dating from the eighteenth century are relatively rare, and more importantly, are usually so badly drawn they reveal little beyond the outline of the implement.[84] By a process of elimination, then, those seeking to learn about eighteenth-century American plow designs must rely primarily upon the few from this period still preserved in museums.[85]

Although limited in number, these relics are consistent with the hypothesis that many (and possibly most) colonial plows were designed for one of two quite different tasks, and accordingly tended to feature one of two sets of design characteristics. A "heavy" or "strong" plow was needed for fields strewn with roots, stumps and stones. (Recall again the landscape sketched in Figure 3.17.) In his 1790 study of New England agriculture, Samuel Deane summarized the problem and outlined the popular solution.

> When land is to be ploughed that is full of stumps of trees and other obstacles, as land that is newly cleared of wood, or that is rocky, the strong plough should be used; and the strength of the team must be proportioned to the strength of the plough.
>
> It is sometimes advisable, to cut off close to the bodies of stumps, before ploughing, the horizontal roots which lie near the surface; especially if there be no stones, nor gravel in the way, to hurt the edge of an axe. When this is done the strong plough will be apt to take out . . . most of the roots so parted.[86]

A "light" or "horse-plough" was often used either to cross-plow fields initially broken up by the strong plow, or to till fields already used for a number of years and accordingly relatively free of major obstacles.[87]

One version of the heavy or strong plow is illustrated in the top diagram of Figure 3.18,[88] and another in the bottom diagram of Figure 3.1. Its triangular share, relatively flat, was designed to cut on a horizontal plane. A coulter provided a second slashing surface. Typically anchored to the beam above and the share (or landside) below, this was an appendage literally braced for a collision. A plow without such a coulter, when run under a heavy root, "had to be relieved" by cutting the root with an axe.[89] (The absence of fields strewn with roots, stumps, and stones in Britain may help to explain the absence of the lock-coulter in many British plow designs.)[90]

The light or horse-plow was not built for slashing or collisions. It was designed for use in soils relatively free of large obstacles. One version is illustrated in the bottom diagram of Figure 3.18, and another in the top diagram of Figure 3.1.[91] (The latter is somewhat atypical, in that it includes wheels. Given the rough fields and rough-and-ready plowing tech-

Figure 3.18. Late eighteenth-century American plows

A. New England bar share plow

B. Pyramidal share hog plow

Source: From the collections of Henry Ford Museum & Greenfield Village, Negatives B61020 and B64760.

niques typical of the New World, colonists had a marked preference for the swing plow.)[92] Compared with the heavy plow, the most notable features of the light plows depicted are the absence of a lock-coulter and the presence of a "pyramidal" share that ran through the soil much as a blunt-nosed ship runs through water.[93] The use of this odd shape of share in British North America probably reflected the influence of Low Country designs on American implements (and cannot be traced to British influences, where such pyramidal shares were virtually unknown).[94] The bluntness of this type of share, although not well suited to inverting the soil, worked well for shallow plowing, or so later generations claimed who sought to explain the American preference for little depth in furrows turned.[95]

The defects of these eighteenth-century implements went far beyond design features encouraging shallow plowing, but such imperfections were to be expected. The product of primitive production techniques in a premodern world, colonial plows, by modern standards, could hardly be anything but defective. (Even Americans who used them conceded, on occasion, they "plowed most wretchedly.")[96]

One problem, both pressing and pervasive, was the scarcity of skilled workers to build and repair farm implements, a scarcity that persisted as the colonies became more settled. In 1619, for example, a resident of Virginia complained about the absence of "carpenters to build and make Carts and Ploughs." One hundred and fifty years later the governor of the same colony was urging plantation owners to train slaves for such work because he did "not know that there is a white-smith or maker of cutlery" in all of Virginia.[97] Repair by the inept of what was intrinsically a defective implement could make a bad situation worse. One dissatisfied customer spoke for legions when he noted: "The share came back in a different shape. It no longer runs like the same plough."[98]

At this juncture one should bear in mind the problems of plow improvement yet to come, in which trial-and-error was to yield results superior to the rigorous pursuit of design modifications based upon mathematical formulae. Precisely because designing the optimal plow shape for a particular setting was so complex, the many individual designs of colonial days occasionally produced a wooden product that performed extraordinary well in the field.[99] But the majority of eighteenth-century American plows, by modern standards, performed all of the basic functions badly because they lacked the design features to perform them well. Poorly shaped shares and moldboards with little or no curvature[100] were not likely to invert and pulverize the soil with any thoroughness. This is evident in repeated references to the need for cross-plowing fields already plowed once, and in the occasional outburst characterizing a tilled field as a "miserable mass of clods, unbroken in whole or in part."[101] Poor design produced poor suction—in some designs so poor that a man or boy had to ride the beam of a plow "easily thrown out of the ground." (The "hog" plow, by one account, acquired its nickname because of its "propensity to root into and out of the ground.")[102] Perhaps the single defect most conducive to shallow plowing was the heavy friction generated by all American designs. Partly this was the result of rough landsides and of shares quickly blunted by use.[103] Partly it reflected moldboards that, if wooden, scoured poorly, and, if "plated with iron," had a wooden surface covered not with a single smooth sheet, but rather with a rough assortment of metallic odds and ends: blades from old hoes and saws and worn out horseshoes, all pounded into thin strips by the local blacksmith and "then nailed or bolted onto the face of the plow."[104]

Despite its lengthy list of glaring defects, one must say of the colonial plow what Galileo said in a different context: And still it moves. While new designs were being pioneered in Great Britain and the Low Countries, cumbersome and inefficient implements remained the norm throughout the Western world in the eighteenth century, as demonstrated by Jefferson's dissatisfactions noted at the outset of this chapter. Moreover, while American plows were primitive by modern standards and colonial plowing techniques were slipshod by European standards of that day, the plows and the techniques did a tolerable job of tilling soil in a premodern world of abundant land. What is striking is how little the inventive stirrings in the Old World had touched agrarian attitudes in the New by 1776. A survey of American husbandry in 1775, for example, concluded that typically the beam of colonial plows was too low, the share too narrow, and the wheels too low. But the writer offered no concrete and detailed suggestions for design improvements.[105] Instances admittedly can be found of efforts to improve. Jared Eliot in the 1750s advocated deeper plowing.[106] George Washington tried to import a Rotherham plow from England in the 1760s, and so did Benjamin Franklin's son William. (Washington's Rotherham "was ruined by a bungling country smith" in the process of repair, and was soon abandoned for want of spare parts.)[107] But colonial farmers on the eve of the Revolution differed little from the generations that preceded them: their vision of what constituted appropriate agrarian techniques was focused squarely on the past, their preference for the status quo was overwhelming, and that status quo in plow designs could be fairly characterized as cumbersome, unchanging, and premodern.

Evolution of American Plows, 1783–1814

Compared with the agrarian inertia of colonial America, the period between the close of Revolution and the outbreak of the War of 1812 appears, at first glance, to be animated by a lively spirit of inquiry about plowing technology and techniques. The practice of shallow plowing began to be challenged by a small but growing minority who advocated deeper tillage in the manner "well known to the experienced agriculturist of Europe."[108] The importation of British plows, although still comparatively rare, was an option more likely to be considered by a handful of agrarian elite wishing to experiment with new devices on their own farms and plantations. George Washington, for example, ordered two from Arthur Young in 1786. By 1810 Jefferson was anxious to test the performance of his own plows by importing a dynamometer, but encountered difficulties because of "the present state of non-intercourse" with Britain.[109]

Finally and most crucially, as Jefferson's search for a testing device indicates, between 1783 and 1812 on this side of the Atlantic efforts were being made to improve the design of locally produced plows quite without precedent in American history.

In the late 1780s, Thomas Jefferson designed a moldboard "of least resistance" (or so he hoped), and by 1794 was using plows incorporating that design in his own fields.[110] (See Figure 3.19.). Although Jefferson

Figure 3.19. Thomas Jefferson's moldboard, 1803

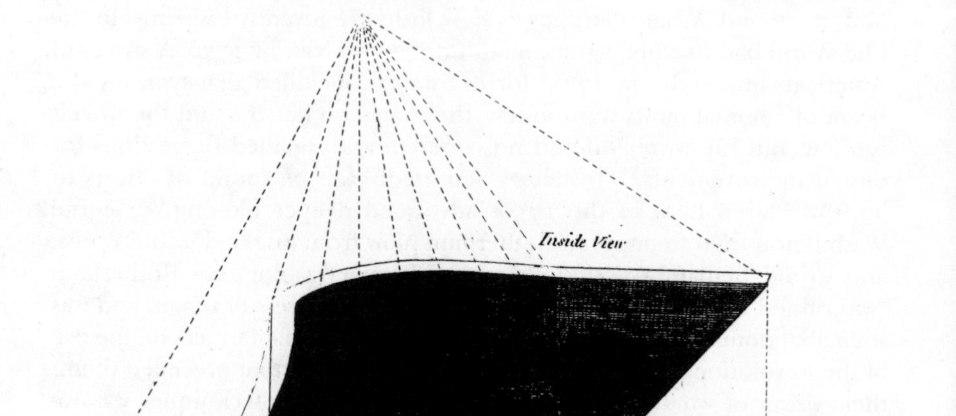

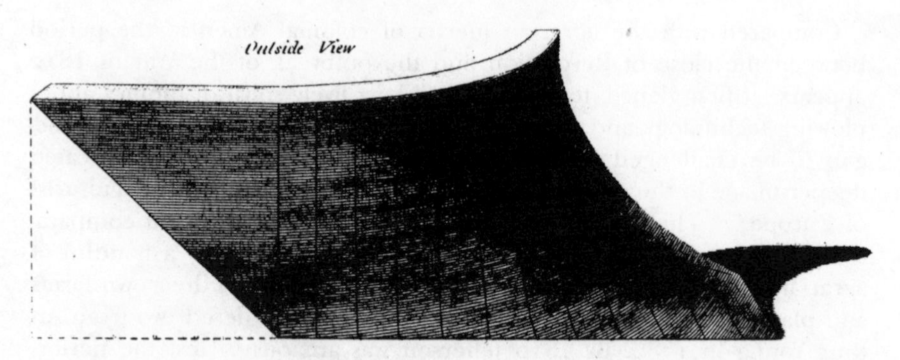

Source: A. F. M. Willich, *The Domestic Encyclopaedia* (Philadelphia: William Young Birch and Abraham Small, 1803), p. 288.

initially used wood to make these models, he was among the first Americans to suggest the merits of making moldboards from cast iron.[111] Others would explore a similar substitution of materials in plow construction. In the 1790s, John Smith of New York experimented with using cast-iron shares, and in 1800 Robert Smith of Pennsylvania received a patent for a cast-iron moldboard.[112] This replacement of wood with iron was carried, if not to its logical conclusion, at least to one possible conclusion by Charles Newbold of New Jersey. For much of the 1790s he experimented with cast iron, and in 1797 took out a patent for a plow in which moldboard, share, and landside were all cast in a single piece.[113] (The beam and handles remained of wood.) Having all three parts constructed as a single piece of iron had its drawbacks, as we shall shortly see. Another and more workable option, patented by David Peacock in 1807, was to cast each of the three pieces separately, but in such a way that moldboard, share and landside could be attached easily to one another. (Peacock's ideas were evidently an extension of Newbold's design, as indicated by the financial compensation paid by the former to the latter.)[114]

Not only new materials possibilities but new design possibilities were explored by would-be inventors of a better plow. Between 1802 and 1805, for example, Jefferson's son-in-law,[115] Thomas Randolph, designed a "hillside" plow to facilitate contour plowing. In listing the advantages, Jefferson focused more upon maximizing water retention than minimizing soil erosion: "Every furrow thus acts as a reservoir to receive and retain the waters, all of which go to the benefit of the growing plant, instead of running off into the streams."[116] As illustrated in Figure 3.20A, Randolph's plow incorporated a simple mechanism whereby a reversible moldboard would throw the furrow "always down hill." Two wings were welded to the same bar at right angles to each other, one "laid to the ground" and the other, "standing vertically," acted as a moldboard.[117] Reversible plows for contour plowing had long been used in Europe.[118] One example in use during Randolph's day was the Flanders plow of Figure 3.20B. (Its moldboard could be flipped to the opposite side by the removal of a pin.)[119] Jefferson knew about such designs from his European travels, but whether he helped in the development of Randolph's version is not known.[120]

A first impression of plowing developments in the young republic therefore suggests an unsurprising pattern: that following the Revolution, a tradition of experimentation and improvement, already well under way in parts of Europe, began to be taken up on this side of the Atlantic.[121] A closer scrutiny of the evidence, however, suggests that what was taken up was, by and large, undistinguished in the extreme, particularly when compared with American plow developments following the War of 1812.

First, and of the first importance, the impact of the new designs upon established practice, by and large, was negligible. Jefferson's moldboard of least resistance, from all accounts, was never used by anyone but Jefferson,[122]

Figure 3.20. Reversible plows, early nineteenth century

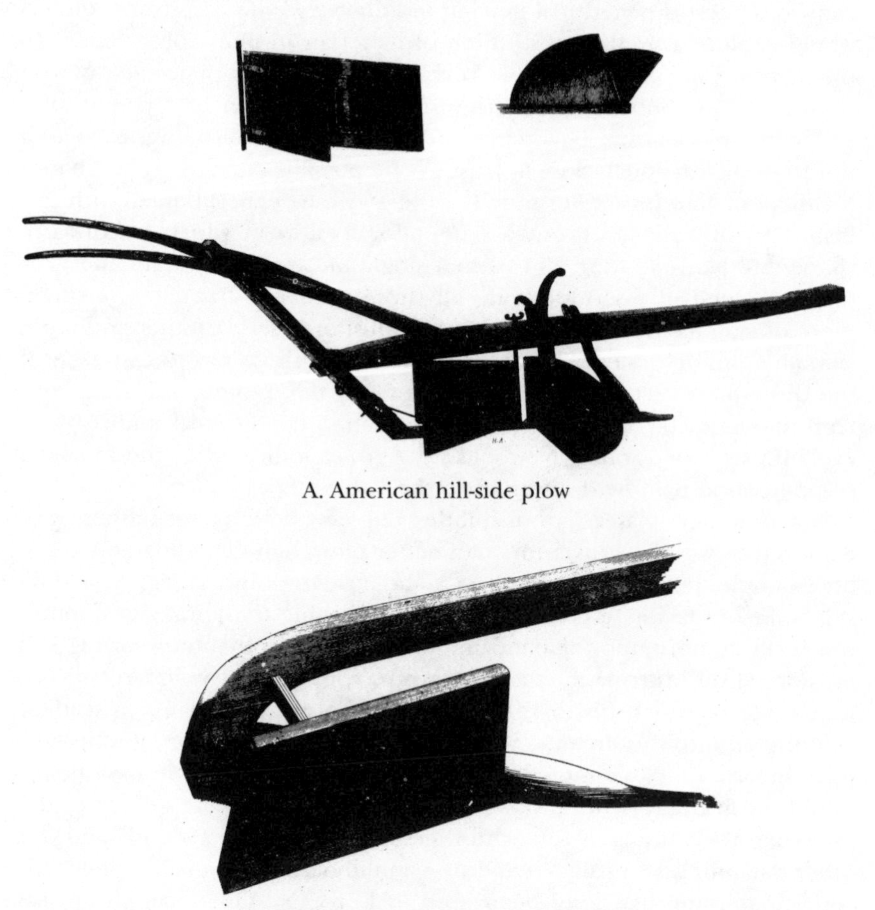

A. American hill-side plow

B. Flanders moveable moldboard plow

Sources: Philadelphia Society for Promoting Agriculture, *Memoirs*, 4 (1818), page facing p. 18; New York Society for the Promotion of Useful Arts, *Transactions* (1807), p. 169.

and with good reason. Even by contemporary standards, it did a poor job of inverting the soil, tending instead "to crowd off rather than raise and turn over the slice."[123] (Compare the moldboard twist in Figures 3.5 and 3.10 with the absence of twist in Figure 3.19.) Even Jefferson's son-in-law, when designing his hillside plow, opted for a simple flat surface for his moldboard rather than the curvature developed by Jefferson. Newbold's cast-iron plow was similarly ignored, but for different reasons. Some farmers were convinced that the cast iron poisoned the soil and encouraged the growth of weeds, evidently untroubled by the fact the second assertion seemed to contradict the first. The major barrier to adoption, however, was not suspected soil chemistry but self-evident economics. A one-piece implement, once broken, had to be scrapped. The share of Newbold's plow was particularly vulnerable, as his own field trials demonstrated. (The plowman "broke the point, and it was never used afterwards.")[124] An expensive implement used only briefly was destined to be ignored. Admittedly not all efforts ended in failure. Others used Randolph's hillside plow, and it remained in use until displaced by more sophisticated versions of reversible plows.[125] Peacock's three-piece cast-iron model anticipated the principle of interchangeable parts that Jethro Wood and others used to revolutionize plowing practices after 1815. But these isolated examples of limited success should not obscure the general lack of experimentation and innovation during the decades immediately following the Revolution.

The most telling evidence indicating the limited importance of the period 1783–1814 in the evolution of American plow technology is what was missing.

The curiosity of the inquisitive few contrasted sharply with the monumental indifference to innovation that pervaded the farming community, including most members of that community who could fairly claim to belong to the agrarian elite. Even the advocates of experimentation were forced to concede that the best and brightest of their associates, by and large, had no interest in their cause. In the 1790s, for example, Washington conceded that he had labored in vain to encourage the establishment of boards of agriculture in this country.[126] After designing a novel moldboard in the same decade, Jefferson was reluctant to forward the details of his innovation to the American Philosophical Society because he "supposed it not exactly in the line of their publications."[127] James Mease in 1804 assured him that his original suspicions were correct. Upon exhibiting a model of Jefferson's moldboard to this group, the latter found "none of the attending members . . . think a moment on the importance of the subject," and accordingly the new moldboard design "did not excite that interest which it deserved, and which I hope to be expressed when a public experiment shall have been made, before men who are practical judges of the merit of a good plough."[128]

Even among those few committed to improving farming practices in this country, most showed little interest in improving plow designs. Thus, for example, nothing whatever was said about the existence of new plow designs or the desirability of developing them in Samuel Dean's 1790 review of experimental possibilities in New England, in Thomas Jefferson's 1798 letter to British expert Sir John Sinclair reviewing local farming practices, or in "Hints for Improvement" written by Thomas Moore of Maryland in 1802.[129] (Moore was preoccupied with deeper plowing, better crop rotation, and the use of manure; Jefferson, with "the new chemistry

. . . the subject of manures, the discussion of the size of farms, [and] the treatise on the potatoe.")[130] John Beale Bordley, a recognized American expert on agrarian practices in other countries, in his 1803 survey entitled *Gleanings from the most Celebrated Books on Husbandry, Gardening, and Rural Affairs,*[131] did review a number of British plow designs, but did not include (here or in anything else he wrote at this time) a plea to import foreign models or to redesign those available in America. Also in 1803 the first American edition of Willich's *Domestic Encyclopedia* was published, providing a more detailed review of British plow designs. In his editorial asides in the plowing section, however, the American James Mease had only two American innovations he considered worth citing—Robert Smith's cast-iron moldboard and Jefferson's moldboard of least resistance—the first of which he mentioned but did not recommend, although the second he considered (incorrectly) worthy of adoption.[132] Even those who advocated deeper plowing usually did not argue in favor of new plow designs to accomplish this objective, preferring instead more extended use of existing models, occasionally modified in trivial ways.[133]

Little was being recommended in large measure because little was being invented. The minutes of the Philadelphia Society for Promoting Agriculture, in its early years following the Revolution, rarely included references to new plows received and were noticeably devoid of discussions of new plow designs.[134] Nor was the society particularly concerned with remedying this lack of interest. In its early lists of premiums offered to stimulate agrarian experimentation it did not offer a single prize for a better plow design.[135] The society's survey of Delaware farming practices in the late 1780s not surprisingly gives no hint of revolutionary plow designs in use or even under consideration. The same is true of Samuel Deane's two surveys of New England practices in 1790 and 1797.[136]

This is not the image of an agrarian society in the throws of an agricultural revolution. At best, the evidence suggests that events between 1783 and 1812 signaled the first stirrings of a revolution to come. And come it did, with extraordinary abruptness and vitality.

Evolution of American Plows, 1815–1830

On May 12, 1819, addressing the Agricultural Society of Albermarle, Virginia, for the first time as its president, James Madison reviewed the recent history of American agriculture and speculated about the future. As talks of this sort went, it was exceptionally long, requiring eight pages in three successive editions of the *American Farmer* to report.[137] Despite the breadth of his topic and the length of his talk, Madison was hard pressed for examples of implement improvement. The present generation, he conceded, "for the first time seems to be awakened to the necessity of reform."[138] But that awakening, in Madison's opinion, had yet to generate much progress. Accordingly he used most of his talk to enumerate "some of the most prevalent errors in our husbandry."[139] Plowing was a case in point. He had much to say about the present "bad mode," including shallow plowing and too little use of contour plowing. But the only implement improvement he could cite was Randolph's hillside plow.[140]

And yet the fact that the initial volume of an agricultural newspaper that would survive for many years reported Madison's address suggested that something was afoot. One harbinger of change appeared in another agricultural newspaper in the same month that the *American Farmer* reported Madison's address. Richard Harrison of Front-street, New York, ran an advertisement announcing that at "the only establishment of the kind in the United States," he now had available for sale, in addition to American farm implements, "the best Models of the English and Scotch machines and improvements."[141] His list of plows included eighteen models, most of them imported. In the years immediately ahead, the variety of American plows multiplied with startling rapidity and their prices plummeted.

At the core of a design revolution under way even as Madison spoke was the substitution of iron for wood in most plow parts. With that substitution came a host of design modifications which, once discovered, could be retained and replicated by a casting process that enabled each plow designer to build upon improvements of the past.

As the story is commonly told, in the first stages of this revolution a crucial role was played by Jethro Wood of Scipio, New York, who filed two patents for cast-iron plows in 1814 and 1819.[142] As outlined in the previous section, the use of cast-iron parts in plows had been anticipated by several early experimenters in plow design, including Newbold and Peacock.[143] Wood's major claim to novelty, according to some historians, was the introduction of interchangeable parts—a crucial innovation if replacement for normal wear was not to be prohibitively expensive (as Newbold's experience demonstrated with a vengeance). Others, however, had anticipated this idea, perhaps most notably R. B. Chenaworth of Baltimore, who received a patent in 1813 for a cast-iron plow with a separate share, moldboard and landside.[144] The use of interchangeable parts was not, in Wood's opinion, one of his claims to novelty, as indicated by his failure to note that feature as grounds for receiving a patent. He did emphasize as innovative his system of linking the separate parts without the use of screws or bolts.[145] Other unique features included a moldboard design with a "spiral wind"[146] that more closely approximated the double twist shape used in modern plows. (Compare, for example, Figures 3.5 and 3.21B.) Wood based his shape upon a series of straight lines, a few of which appear in Figure 3.21B, and more of which can be seen in Figure 3.10B. Jefferson was sufficiently impressed with the results to write a letter

Figure 3.21. Early American iron plows

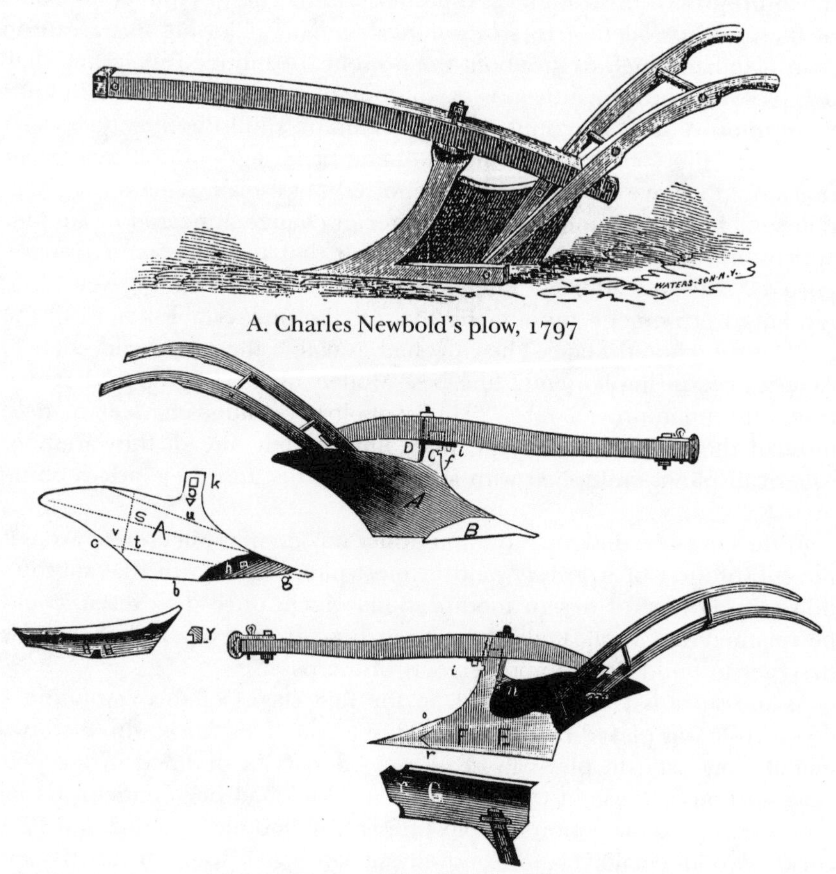

A. Charles Newbold's plow, 1797

B. Jethro Wood's plow, circa 1819

Sources: New York State Agricultural Society, *Report on the Trial of Plows Held at Utica . . .* (Albany: Van Benthuysen and Sons, 1868), p. 67; Frank Gilbert, *Jethro Wood, Inventor of the Modern Plow* (Chicago: Rhodes and McClure, 1882), frontispiece.

of congratulations to Wood although the latter's moldboard design differed radically from his own.[147]

Whether Jethro Wood manufactured his own plows in any volume is a subject of dispute.[148] What seems beyond dispute is that this new design, which could claim "simplicity, strength, and durability" as well as the convenience of interchangeable parts,[149] was sold in unprecedented numbers by manufacturers who bought production rights from the inventor. In New York City alone, 3,600 were reportedly sold in 1819, and an even larger number in 1820.[150]

Others were not slow to follow. In the rush of implement experiments, innovations, and refinements that characterized the 1820s, three different trends in plow development can be detected.

The first involved minor improvements in basic designs in use for decades. The shovel plow provides a case in point, but since this implement functioned as both a plow and a cultivator, consideration of this case will be postponed until cultivator development is discussed in Chapter 6. In the South the Carey plow gradually replaced the shovel plow, as a plow, but the newer design was little more than a modest improvement on the old colonial pyramidal share hog plow.[151] Compare, for example, the two Carey plows of Figure 3.22—quite similar although produced a quarter century apart[152]—with the late eighteenth-century hog plow illustrated in Figure 3.18.

Carey and shovel plows were known and used by 1800, and later modifications in their designs did little more than delay their pending obsolescence. In 1821 the speaker to a Maryland agricultural society noted that while the Carey plow, which was "getting fast into general use," was "much superior to the country ploughs formerly in use," it was "by no means as good as several others readily within our reach."[153] What he had in mind were implements made of iron, many of which in general appearance resembled the cast-iron plows of Jethro Wood. From inventors and manufacturers came a flood of innovations in the 1820s, modifying and improving (or so they claimed) every part of the plow, from the beam to the share.[154] Most of these, however, retained the general shape first popularized by Wood, although they frequently differed in detail. Compare, for example, Figure 3.21B with the plows of Figure 3.23 which date from the early and later years of the 1820s.

If the second trend of the 1820s consisted of innovations similar in appearance to Wood's plow, the third involved innovations that radically departed from that design. Three examples of the latter are reproduced in Figure 3.24. The "five coulter plow," as its name implies, merely added four additional coulters in a staggered row. The same idea (with four coulters rather than five) had been anticipated almost a century earlier by Jethro Tull,[155] and the American version seems to have had no more success than its British antecedent. The plow cleaner was a new design to solve an old problem: the clogging of the coulter with refuse scooped up in the course of plowing.[156] By pulling a handle *(b)* backward, the operator caused an upright *(e)* to swing forward past the coulter *(d)*. When the handle was released, a spring *(g)* would restore it to its original position.[157] The "coulter plow," although similar to implements long known and used in Europe,[158] was typical of many American design experiments of this era. Precisely because it involved a minimum of inventive creativity and mechanical complexity, it was easy to concoct, inexpensive to produce, and cheap to maintain. The tilling problem it was designed to solve was how to stir (but not invert) the soil to a depth beyond the reach of the ordi-

Figure 3.22. Carey plows

A. Circa 1800, furrow-side view

B. Circa 1800, land-side view

C. Circa 1825, furrow-side view

D. Circa 1825, land-side view

Source: Smithsonian Institution Photo Nos. 45064, 45064E, 42163B, and 42164A.

Figure 3.23. American cast-iron plows, 1820s

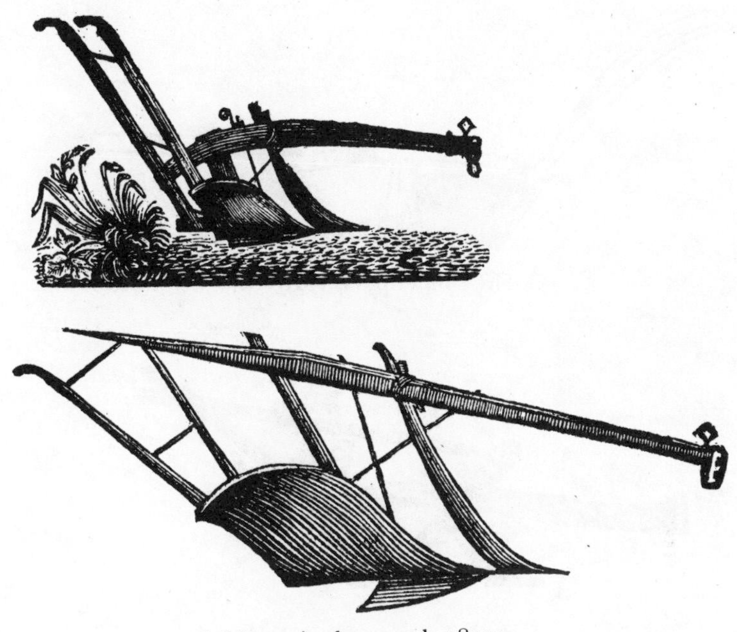

A. Murray's plows, early 1820s

B. McCormick's plow, late 1820s

Sources: American Farmer, 2 (August 11, 1820), p. 160, and 12 (October 8, 1830), p. 240.

nary plow. A later generation of American inventors devised special plows expressly for this purpose.[159] The solution offered here merely combines a plow beam with a coulter-like bar, now with a pointed foot protruding to the front, to churn through the soil at the depth desired.[160]

These three by no means exhaust the list of plowing innovations of the

Figure 3.24. American plowing design innovations of the 1820s

A. Five coulter plow, 1820

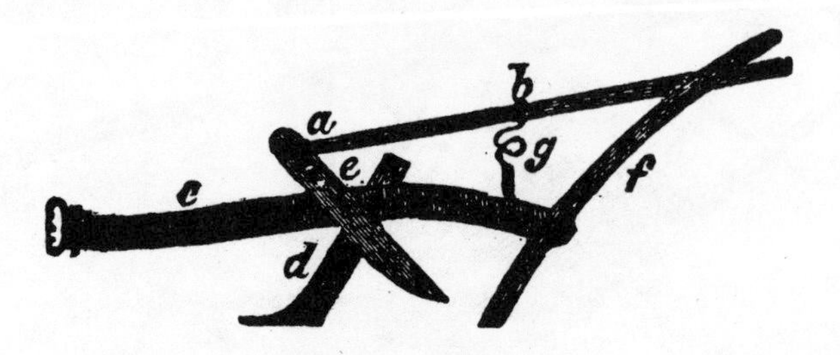

B. Plow cleaner, 1822

C. Coulter plow, 1821

Sources: American Farmer, 2 (November 24, 1820), p. 275; 3 (September 21, 1821), p. 205; and 4 (August 16, 1822), p. 168.

1820s which, in appearance, represented radical departures from Wood's cast-iron plow with interchangeable parts. Others included paring plows, three-furrow plows, and five-furrow plows.[161] Still others were designed to complement the act of plowing: improved mule and horse collars, new hitching devices, levels to assist with contour plowing, and gauges to determine the depth of furrows with greater accuracy.[162]

As innovations multiplied, the prices of some models plummeted. Between 1819 and the mid-1820s prices did not change significantly for many of the more expensive plows, particularly two-, three-, and four-horse models.[163] But in the same six-year span, the prices of low-cost plows dropped almost by half, from about $10 in 1819 to $4.50–$6.00 by 1825.[164] By the latter date, the economic advantage of the new technology over the old was, in many instances, overwhelming. Older models of wooden plows were both difficult and expensive to maintain. Repeatedly in need of resharpening, they usually were taken to the local blacksmith, a trip of several miles requiring (there and back) as much as half a day. The plow returned sometimes noticeably differed from the plow delivered because of the uneven quality of the blacksmith's work. A poorly sharpened share could easily make plow performance worse[165] (a nonexistent problem for cast-iron models with replaceable shares all made from the same mold). Per year, wooden plows cost $3 to $5 more than the iron ones to maintain.[166] Thus, the money saved from lower maintenance costs would cover the cost of buying one of the less expensive new iron plows in less than two years, even if that purchase meant abandoning a wooden plow already owned.

The economic advantages of the newer iron plows over older wooden ones did not end here. With better design features that sharply reduced friction, cast-iron plows could till more acres with the same draught animals, or the same number of acres with fewer animals. The amount of this saving was estimated to range as high as one-third to one-half; that is, 50 percent more acres could be plowed with the same team, or the same number of acres plowed with two horses instead of three.[167]

As models improved and prices fell, potential buyers could assess new options using better sources of information. Agricultural periodicals expanded in number and circulation. More important as a source of information about new implements for most farmers was the rise to prominence of plowing matches, particularly those held in conjunction with agricultural fairs and exhibitions. These gave potential customers with pragmatic instincts and limited resources a chance personally to inspect new cast-iron plows and witness their performance matched against other models—an assessment opportunity vastly superior to viewing pictures or reading testimonials in the farming literature. One of the earliest (and possibly the first) field trials for farm implements was sponsored by the Columbian Society of Georgetown, D.C., in May of 1812. By the end of that year, however, "war with England overshadowed everything else," and the society in question was dissolved.[168] In the immediate postwar era, fairs and exhibitions and plowing matches appear to have revived quickly and spread rapidly, with contests among competing plow models now made more systematic by the use of the dynamometer.[169]

Several effects, all unsurprising, followed from this diffusion of knowledge about new cast-iron models that, by the mid-1820s, cost less to buy and operate than old wooden plows cost to maintain, even if already owned. In region after region, growing familiarity with the cast-iron option led to rapid adoption.[170] British imports, often featured in the advertisements and commentaries of agrarian periodicals in the years immediately prior to 1820, by the middle of the 1820s were seldom the subject of analyses, endorsements, or popular discussions. Their ability to compete in the American farm implement market by that time had been seriously undermined by the improving performance and falling prices of newer American models.[171] Those newer models also affected American plowing practices, insofar as implements featuring less friction encouraged deeper plowing, although the merits of deep plowing remained a topic of discussion and dispute throughout the 1820s.[172]

The Long View: Colonial Times to 1830

In the course of the fitful and protracted evolution of the moldboard plow in the Western world, American developments feature the predictable and the unexpected. To the New World from the Old settlers brought the farm implements of their native region, and primitive implements they were. The plows of 1600 or 1700, while perhaps marginally better than those of the early Middle Ages, were still by modern standards abysmally inefficient and cumbersome. During the eighteenth century, however, agrarian innovation was afoot in Europe, and plow designs began to change as a few inventors in a few regions of a few countries—most notably Great Britain and Belgium—began to modify this implement in ways that, for the first time, foreshadowed the main features of the modern plow. When ripples from this larger movement reached North America, the dominant reaction was indifference. A few, like Washington, tried to import improved models from Great Britain. A handful of others, like Jefferson and his son-in-law, tried to modify existing American plow designs. But these were exceptions to the general rule that the status quo was good enough: that plows of the present (on occasion with trivial modifications) should resemble plows of the past. Between 1783 and 1812 admittedly one can find the first stirrings of what would subsequently become a flood of proposed changes in design. But the three decades that intervened between the two armed conflicts with Britain are noteworthy mainly for what was not accomplished. By the outbreak of the War of 1812, the plows used in America, and the manner of their use, were, by and large, not significantly different from those in evidence on the eve of the Revolution.

All this changed, and with a rush, in the decade and a half beginning with 1815. Between 1797 and 1814 only thirteen patents for new plow designs were filed with the Patent Office in Washington, and more than half of these were registered in the three-year period 1812–1814. By

1830 the number of patents for plows had increased almost tenfold to 124.[173] In 1819 James Madison, as previously noted, singled out for special praise only one new plow design, and that one (Randolph's hillside plow), with its flat moldboard and wooden parts, seemed more closely linked to technologies of the past than to any incipient revolution in material and design. A scant four years later, when John James addressed another southern agricultural society, a shift in attitude was evident in his expectations and his advice. While conceding (as Madison did) that many plows still in use were "very defective," James hastened to add that he was "happy to observe we are beginning to find out their defects," and gave his approval to both the "new modeling" and "purchasing the patent kind."[174] By the middle of the 1820s most experts favored such purchases, and not a few made a point of publicizing where the newer models could be found. William Drown's exhortations are typical: "For the benefit of those who are anxious to avail themselves of the great improvements which have already been made in agricultural implements, it may be proper to state, that Repositories for Agricultural Implements and Machinery, have been established in the cities of New-York and Boston, and we sincerely hope, that the exertions of the proprietors, may, by a judicious public, be liberally rewarded."[175] As the 1820s drew to a close, most analysts with any claim to expertise had some sense that at least the first stages of a revolution in design had been accomplished. Thus by 1835, the editor of *The New England Farmer* seemed to be laboring the obvious when he noted that the plow "has undergone of late years a wonderful change in all its most essential parts, and has been greatly improved. The cast-iron plough is now most generally used among the best farmers, and considered decidedly the best."[176]

If much had been accomplished, much remained to be done. Yet to be developed were many specialized plows for particular tasks: stubble plows and subsoil plows, mole plows and ditching plows, gang plows and steel plows for breaking prairie sod of the sort made famous in the ante-bellum era by John Deere.[177] Also absent as of 1830 was a fundamental insight about how plowing should be done, which had implications for how plows should be designed. From Jefferson's earliest experiments to the cast-iron plows of the late 1820s, the commonly acknowledged objective in plow design was to minimize friction while inverting the soil.[178] In 1839 two would-be inventors pointed out what was missing in this reasoning, and in no uncertain terms. "The main object [of plowing] is *to pulverize the soil*, and the only way in which it can be effected is by bending a furrow slice on a curved surface so formed that it shall also be twisted somewhat in the manner of a screw."[179]

The cast-iron plows of the 1820s, to be sure, did have some twist in their moldboards. What was still needed was perhaps best illustrated by the Eagle plow first invented in the early 1840s, with improved variants of its basic design thereafter becoming widely popular. That design included "more turn" in the moldboard at the rear "and a greater intensity of twist beyond the perpendicular"[180]—all features well calculated to improve the plow's ability to pulverize the soil.

By the middle of the nineteenth century, the main features of the modern moldboard plow had been developed in America.[181] That a revolution in design had been accomplished by the coming of the Civil War is a commonplace of agrarian histories. What is not commonly understood is that the first phase of this revolution came with a rush, most of it during the fifteen years following the War of 1812. This first stage was the crucial stage: that period when allegiance to the status quo in plow designs was shattered and the search for improvements became the norm among an ever-growing number of would-be inventors and farmers. Once that approach became well established, new techniques and new designs swept across the landscape in unprecedented numbers.

Also overlooked is the prevalence of common features among the various design revolutions that took place between 1815 and 1830. These particularly include the following.

(1) Antecedents to the first great surge in design innovation can generally be found during the late colonial period, as well as between the Revolution and the War of 1812, with activity in the second period usually outstripping that of the first by a significant margin. But the experiments tried and inventions made during both periods were fitful and infrequent attempts by the isolated few, usually with no impact beyond the farm of the inventor, and almost invariably with no lasting impact upon the landscape or the literature.

(2) Beginning in the second decade of the nineteenth century, sometimes in the years immediately following 1810, and in all cases unmistakably present after 1815, the number of innovations multiplied rapidly, as did their manufacture, sale, and use.

(3) While these innovations occasionally included esoteric designs extravagantly priced, the vast majority tended to appeal to the pragmatic concerns of hard-headed farmers with limited incomes. Relatively simple in design and easy to comprehend, these new implements were usually inexpensive to buy and cheap to maintain. Above all, they offered the promise of superior efficiency that would translate into higher earnings, compared with the implements they sought to replace.

In sum, this agricultural revolution burst upon the American farming community with unexpected suddenness and, as it gathered momentum, had ever-broadening implications for accepted farming practices. From first to last it was little affected or directed by the American analogue of the British squire, with idle time and abundant funds to investigate elaborate contrivances with questionable economic pay-offs. This revolution was dominated by the market, and its participants, by and large, were preoccupied with profits: inventors and manufacturers striving to make money, and potential customers anxious to save money in the operation of their farms and plantations. It was, in short, a revolution that was quintessentially capitalistic in motivation and American in character. This will be the theme—or, to put it more cautiously, the hypothesis—that will dominate all subsequent chapters.

[4]

Sowing

The Problem

The Bible is not commonly cited as a source of agricultural insights, but many of its authors indicate a firm grasp of the farming techniques of their day. That understanding is evident in the parable of the sower, which provides a succinct summary of the wastage problems inherent in broadcast sowing.

> Behold, there went out a sower to sow:
> And it came to pass, as he sowed, some fell by the way side, and the fowls of the air came and devoured it up.
> And some fell on stony ground, where it had not much earth; and immediately it sprang up, because it had no depth of earth:
> But when the sun was up, it was scorched; and because it had no root it withered away.
> And some fell among thorns, and the thorns grew up, and choked it, and it yielded no fruit.
> And other fell on good ground, and did yield fruit that sprang up and increased; and brought forth, some thirty, and some sixty, and some an hundred.[1]

Roughly eighteen centuries later, an American assessment of the same process offers a more technical compilation of problems in less graceful language. "Much seed is wasted in the common [broadcast] way of sowing; for some of the seeds will be so deeply covered, that they will not vegetate; some left on the surface, which is a prey for birds, and perhaps leads them to scratch up the rest: Some will lie so near the surface as to be destroyed by variation of weather, being alternately wetted and scorched. And of those seeds that grow, some rise earlier, and some later, so that the crop does not ripen equally. The seeds will fall from the hand of the sower, too thick in some spots, and too thin in others, by means of the unevenness of the surface; and the harrowing will increase the inequality; so that many will be so crowded as to be unfruitful, while the rest have more room than is necessary."[2] In sum, throwing seeds from a sack as a method of planting has always been inherently inefficient. The puzzle is why this technique should have dominated western agriculture from the rise of the Pharaohs to the defeat of Napoleon. (See Figures 4.1A to 4.1D.)

The puzzle deepens with the realization that the above descriptions understate the wastage problems. Smaller seeds, such as carrots or clover, required even greater skill to maintain a semblance of even distribution with each toss. As a British writer of the nineteenth century explained: "This egregious fault [wastage] is peculiarly observable in mixed rye-grass and clovers, which require very small pinches of seed to each throw, and, when not equally distributed by a judicious twirl of the hand in throwing them out, a third, or even more, of the land is often left unprovided with plants, and the bare intervals are apt to grow up full of weeds."[3]

The lurking inefficiencies of broadcast sowing do not end here. However perfect the seed distribution, the end result obviously could never be a field of uniform rows. And only uniform spacing between rows of the standing crop allowed any possibility of using a cultivating implement to run between the rows and rip up the weeds, thereby dramatically reducing the drain on farm resources from the labor requirements of hoeing crops by hand from first sprouting until final harvest. Gains from mechanizing the sowing of most seeds therefore promised to be enormous. The persistence of the broadcast technique can be explained primarily by the difficulty of devising a superior alternative. The major problems were mechanical. A sowing machine, or "drill," as they would commonly come to be called, would have to

(1) guarantee a uniform flow of seed from a seed hopper to the ground,
(2) place the resulting flow at a uniform depth in a previously prepared furrow, and
(3) ensure that furrows thereby planted were equally spaced, thus enabling a weeding implement (preferably horse-drawn) to pass between the rows.

Americans would have more success solving the seed flow problem than the other two, but not without a number of false starts. Early British innovations were not much help. The first widely publicized design for a seed drill attempted to solve all three problems simultaneously. The would-be inventor and sometime farmer was Jethro Tull. His experiments date from the turn of the eighteenth century, although his book describing his drill, complete with drawings, did not appear until 1733. The machine was

Figure 4.1. Broadcast sowing

A. Egyptian broadcast sowing, circa 1400 B.C.

B. European broadcast sowing, circa 1340 A.D.

C. American broadcast sowing, circa 1820

D. English broadcast sowing, circa 1850

Sources: Norman DeGaris Davies, *The Tomb of Nakht at Thebes* (New York: Metropolitan Museum of Art, 1917), Plate XXI; Eric George Millar, *The Luttrell Psalter* (London: Bernard Quaritch, 1932), Plate f.170b; Alexander Anderson Scrapbooks, Vol. 1, p. 54, Print Collection, Miriam and Ira D. Wallach Division of Arts, Prints and Photographs, The New York Public Library, Astor, Lenox and Tilden Foundations; Henry Stephens, *The Book of the Farm* (London: William Blackwood and Sons, 1855), Vol. 1, p. 553.

justly famous as an example of inventive genius directed to a practical problem. But as a farm implement destined to precipitate a revolution in American sowing techniques, the insignificance of its immediate impact cannot be exaggerated. But this is to get ahead of our story.

Defective Solutions: From Ancient Babylon to Eighteenth-Century Britain

The time span noted is vast, but the story is quickly told because not much happened. The first known efforts in the West to mechanize sowing, as agricultural historians never tire of emphasizing, involved a curious implement of ancient Babylon, depicted in Figure 4.2. Pioneering it may have been, but influential it was not. The essence of simplicity, the structure combined a cutting device in front to open up a furrow with a tube (fed by hand) to convey seeds into that furrow. Beginning about 2000 B.C., variants of this implement were used for about two millenniums in the lower valley of the Tigris–Euphrates river system.[4] What puzzles most historians is why its use failed to spread beyond the geographic area indicated. But consider the workers depicted in Figure 4.2. With one man to control the oxen, a second to drop the seed, and a third to steady the implement from the rear, the wonder is that such a labor-intensive device giving such crude results was used at all. The most probable explanation concerns the exceptional fertility of the Tigris–Euphrates valley—a region with such productive soil that it prompted "astonishment and envy" among the Greeks.[5] Rich soil plus extensive irrigation ensured, on the average, high yields from heavy sowing, reportedly as much as four or five bushels of seed per acre. These exceptional returns appear to be the main

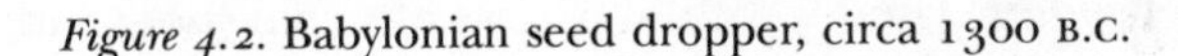
Figure 4.2. Babylonian seed dropper, circa 1300 B.C.

Source: Russell H. Anderson, "Grain Drills through Thirty-Nine Centuries," *Agricultural History,* 10 (October 1936), p. 160.

reason why the Babylonian seed drill made economic sense, at least in Babylon. That it did make economic sense seems indisputable, given the evidence of its continued use for almost twenty centuries. Where such high yield possibilities were absent—that is, throughout most of the remainder of the ancient world west of Babylon—this seed drill was seldom if ever tried, and certainly never became a fixture of agrarian techniques. To put it more cautiously, surviving evidence is consistent with the hypothesis that the limited diffusion of the Babylonian seed drill beyond the Tigris–Euphrates valley is to be explained primarily by the absence of similar soil fertility in other regions.[6]

Whatever its merits, this mechanical sowing implement seems to have disappeared about the time that Christianity began, although the reasons for its abandonment remain obscure. For the next 1500 years, no efforts of consequence to devise alternatives to broadcast sowing appear to have been made. In the later Middle Ages, although where and when is unclear, attempts began anew to solve the problems sketched above. By the latter part of the sixteenth century, some of these attempts were achieving at least a modest reputation in their own locality, particularly implements designed in Italy.[7] Mechanical experimentation was one thing. Economic success was quite another. Given the complexity of the technical problems to be solved, the failure of these pioneering efforts is not surprising. An obvious starting point for early experimentation was some variant of the device depicted in Figure 4.3. The old Babylonian idea of a device to carve a furrow was revived in the form of the Spanish plow. (See Figure 4.3B.) To this common implement was added a box at the rear for sowing seed. The key problem, not satisfactorily resolved for centuries, was how to deliver seeds from the box into the furrow in a uniform and continuous flow. Locatelli's solution consisted of placing at the front of the seed box a solid wooden cylinder with four rows of metal spoons which, as they revolved, scooped up seeds from the hopper and dropped them out the front of the box. Power for moving the seed-dispensing mechanism came from gears driven by one of the machine's land wheels, the latter to become the power source for sowing mechanisms in almost all subsequent seed drills of any importance. Although praiseworthy for mechanical ingenuity, all these early sowing machines were so defective in performance that they never made an impact of any consequence upon established sowing practices, and accordingly they quickly disappeared from the agricultural literature.[8]

As the history of sowing devices is usually told, the year 1733 is a watershed, marking the beginning of the end for the old technique and the end of the beginning for experiments designed to replace sowing by hand with mechanical implements. In that year Jethro Tull published *The Horse-Hoeing Husbandry,* which included a detailed description plus accompanying plates of his revolutionary "drill".[9] As a sign of an incipient revolution

Figure 4.3. Plow and grain drill, seventeenth century

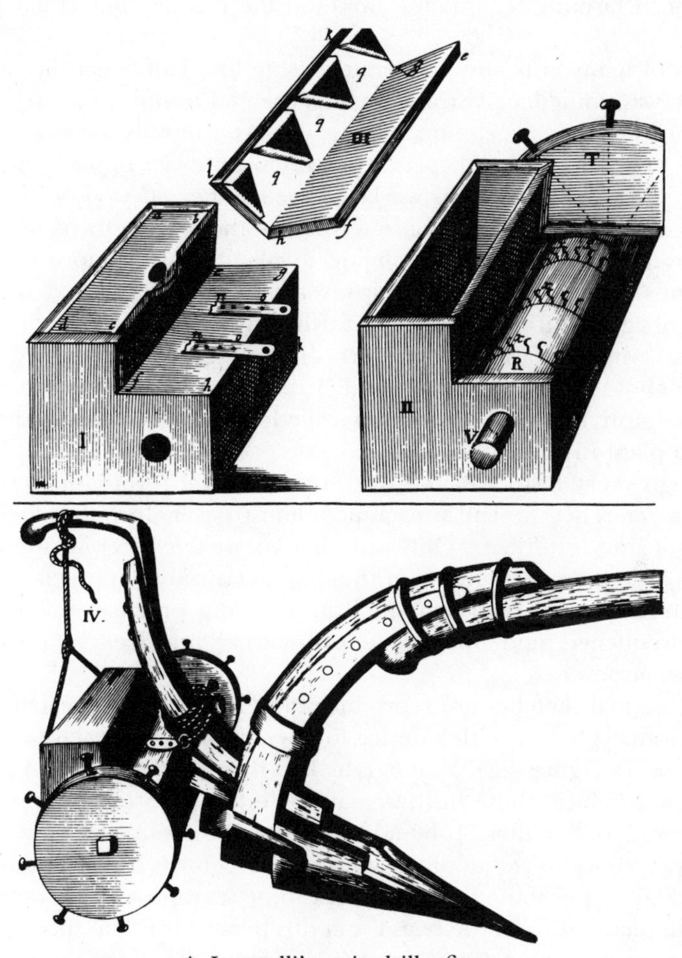

A. Locatelli's grain drill, 1670

B. Spanish plow, circa 1700

Sources: Royal Society of London, *Philosophical Transactions,* Number 60, June 20, 1670, Plate I; R. Bradley, *A Survey of the Ancient Husbandry and Gardening . . .* (London: R. Motte, 1725), Plate II.

in American farming techniques, however, there is less here than meets the eye.

A man of many gifts and many interests, Jethro Tull began farming in 1699. He was immediately struck by what seemed to him unconscionable wastage in the sowing techniques of his day. As Tull tells the story, about 1701 he experimented with placing seed "to an exact depth" and then covering them "exactly."[10] His workers greeted these new procedures by walking off the job. Tull's response was to fire the lot and focus his inventive energies on designing "an engine to plant sainfoin more faithfully than hands would do."[11] Some thirty years later he published both his theories of agriculture and the plans for his "drill" in a book that remains a landmark in the farming literature of the eighteenth century.[12] The name he chose for his machine ("drill") reflected the common terminology of English farmers. They often called rows "drills," and "drilling" meant to plant in rows.

Although every history of the British agricultural revolution is sure to include a reference to Tull's machine, almost never does that literature make clear how it worked.[13] Only with that knowledge can one appreciate its flaws. And an understanding of the flaws is crucial to appreciating why the machine's initial impact upon British sowing practices was of such little consequence, and upon American sowing practices, of no consequence whatsoever.

Tull's original sketches are reproduced in Figure 4.4. He began where the Babylonians began, with a device to open a furrow to receive the seed (see "share" in Figure 4.4), although he had three of these instead of one. Immediately behind these furrow-creating devices (subsequently referred to as "shares" or "coulters"), he added a funnel to convey the seed from the hopper above to the newly created furrow below. (In the bottom diagram of Figure 4.4, these three funnels can be seen protruding below the seed drill, pictured from the rear. From this perspective, the shares immediately in front of each funnel remain invisible, but the diagram at the top left of Figure 4.4 indicates their shape.) The seeds went into one of three hoppers—a short one in front and two taller ones toward the rear. The funnels then conveyed the seeds to newly prepared furrows, where the stirring action of the two wooden teeth, or tines, located at the rear of the drill subsequently covered them, or such was the hope. (As described in more detail in Chapter 5, in Tull's day this seed-covering task was normally done with crude harrows, many of which had wooden tines of the sort Tull used here.)[14]

To this point, what remains totally obscure is how the seeds got from the hopper to the funnel. For all seed drills, this is the mechanical moment of truth. Tull described his solution using the diagrams of Figure 4.5. The crucial part of his feed mechanism is a flap labeled *C* in Figure 4.5A, a flap depicted from the rear and from the side in Figure 4.5B. (To

Figure 4.4. Jethro Tull's wheat drill, 1733

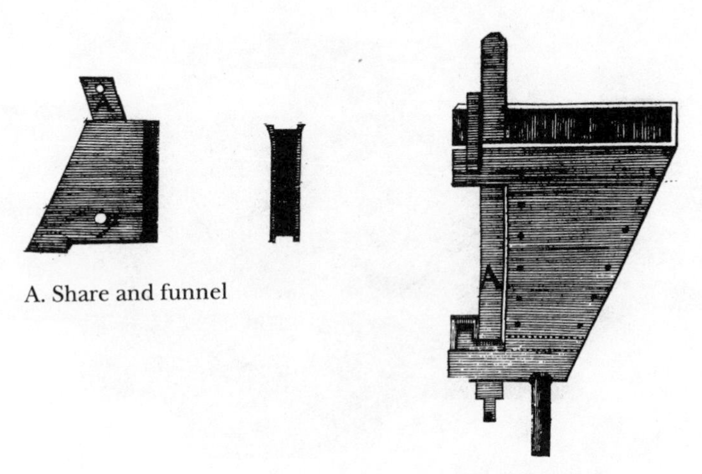

A. Share and funnel

B. Seed hopper

C. Rear view

Source: Jethro Tull, *The Horse-Hoeing Husbandry* (Dublin: A. Rhames, 1733), Plate 4.

Figure 4.5. Jethro Tull's seed dispensing mechanism, 1733

A. Hopper, top view

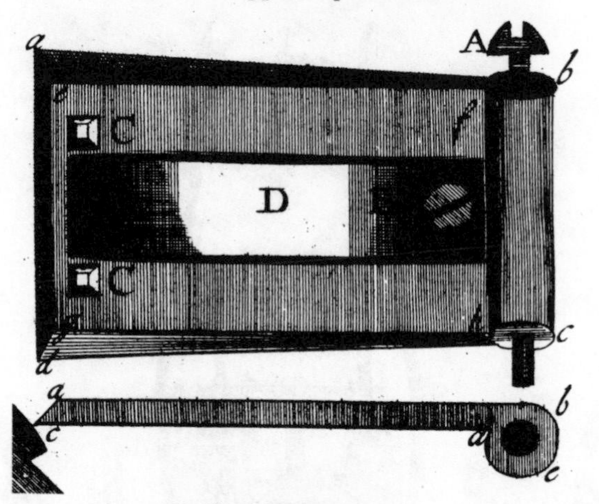

B. Flap, bottom view.

C. Wooden cylinder, side view

Source: Jethro Tull, *The Horse-Hoeing Husbandry* (Dublin: A. Rhames, 1733), Plate 3.

get the correct relationship between these two different perspectives, notice that the screw head labeled *A* in Figure 4.5B is also labeled *A* in Figure 4.5A.) When subjected to pressure from inside the seed box, the bottom of this flap opened outward by rotating at point *A*. When the pressure was released, the flap snapped shut because of a spring mechanism mounted on the back of the flap (*D* in Figure 4.5B).[15] Pressure on the flap to open was created by the forward rotation of a solid wooden cylinder with notches on its circumference, as indicated in Figure 4.5C. As the seed drill was rolled forward, the large wheels in contact with the ground caused the notched cylinder to rotate; the notched cylinder gathered seeds from the hopper and created pressure on the flap; the flap opened and the seeds were deposited into the funnel.[16] Further rotation eased the pressure, the flap snapped shut, and the entire process began again. As Tull himself was quick to emphasize, this feed mechanism was merely a variant of "the groove, tongue, and spring in the sounding board of the organ," prompting the inventor to concede: "I owe my drill to my organ."[17]

Whether this seed drill worked well or badly depended crucially upon the spring central to the feed mechanism. As Tull pointed out, too little tension resulted in too many seeds dropped, while too much tension could strangle the flow.[18] Primarily because the correct spring tension was difficult to construct and almost impossible to maintain, Tull's seed drill was a farming innovation destined for the scrapheap. The only question was how quickly it would disapper.

By the early 1760s, the French writer Duhamel du Monceau considered the implement so defective that he omitted its description from his *Éléments d'Agriculture* "for fear some unfortunate might try to make one."[19] By 1765 it was not clear that any workman could make one.[20] Subsequent efforts to devise a better seed drill emphasized feed mechanisms quite different from the spring-and-flap of Tull's original design. Two options seemed particularly promising, and that handful of Americans concerned with farming innovations prior to 1800 explored both of them. Why they failed is not difficult to understand, once the requirements for success are clarified. They failed primarily because of their inadequate feed mechanisms.

These requirements are easily grasped when one examines the feed mechanisms of those American seed drills among the first to achieve unambiguous success. Consider again the tasks that Tull set out to solve: to design a sowing machine that would simultaneously cut multiple furrows and drop seeds into those furrows in a uniform and controllable flow. Two mechanical problems proved to be particularly troublesome. One was how to ensure uniformity in results while sowing many furrows simultaneously. The more numerous the shares or coulters, the wider the grain drill had to be, and thus the greater the likelihood that one or more of the shares would lose contact with the ground while passing over an uneven field. Tull saw the difficulty and recommended that no grain drill exceed four feet in width. "For should the Drill be broader, some of the Shares might pass over hollow places of the Ground without reaching them, and then the Seed falling on the Ground would be uncovered in such low Places".[21] A wider grain drill raised a second problem also related to imperfect conditions in the field—imperfections no doubt more prevalent in the New World than the Old. Any obstacle in the field that was immovable or almost so, such as a root or boulder, when struck by a single share could throw the entire sowing implement off line. One American solution was to build each share to recoil separately, as depicted in Figure 4.6.

The most important and most difficult mechanical challenge lay elsewhere. As emphasized above, from the outset of seed drill experimentation in the Western World during the premodern period, would-be innovators sought to replace the human hand with a mechanical device to dispense the seeds. The implement in question required a means to feed seeds piled in a hopper (preferably through tubes) in an even flow from the hopper to the ground. Americans did not find a satisfactory solution until more than a century after Tull's path-breaking efforts. One of the

Figure 4.6. Palmer's wheat drill, 1849

Source: John J. Thomas, *Farm Implements and the Principles of Their Construction and Use* (New York: Harper and Brothers, 1855), p. 150.

earliest successful American designs is depicted in Figure 4.7. Acclaimed as "one of the best and most widely known" grain drills in 1869, it featured eight tubes the ends of which were shod with steel. These tubes were "made to pass any desired depth into the mellowed soil," where deposited seed were "immediately covered by the falling earth, as the drill

Figure 4.7. Grain drills, 1860s

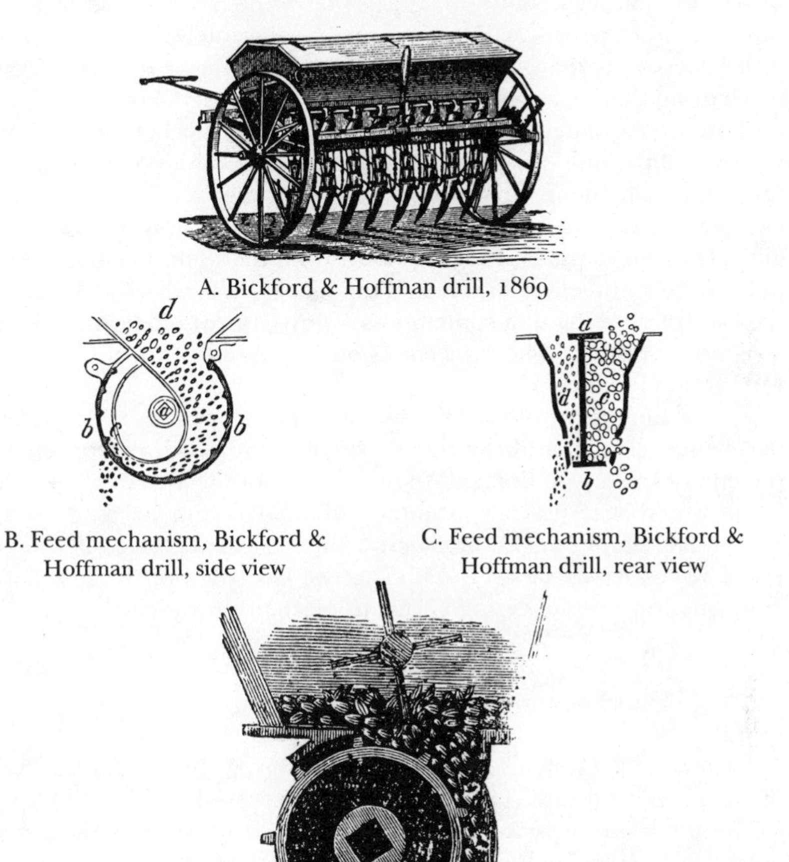

A. Bickford & Hoffman drill, 1869

B. Feed mechanism, Bickford & Hoffman drill, side view

C. Feed mechanism, Bickford & Hoffman drill, rear view

D. H. L. & C. P. Brown drill feed mechanism, side view, 1866

Sources: John J. Thomas, *Farm Implements and Farm Machinery* . . . (New York: Orange Judd, 1869), p. 153; S. Edwards Todd, "Improved Farm Implements," in U.S. Department of Agriculture, *Report of the Commissioner of Agriculture for the Year 1866* (Washington, D.C.: Government Printing Office, 1867), p. 262.

passes."[22] The feed mechanism is depicted from the side (Figure 4.7B) and from the rear (Figure 4.7C). (To establish the relationship between these two diagrams, notice that the circular feed mechanism *b* in Figure 4.7B is the partition *b* in Figure 4.7C, and the axle turning this feed mechanism is labeled *a* in both diagrams.) Two features of this mechanism are less than clear in these diagrams dating from the 1860s. One is that while both diagrams show the seed hoppers as filled (with large and small seeds), the farmer would normally use only one at a time, sealing off the other. Another poorly marked feature, one absolutely central to this grain drill's success, is the projections on the inner rim of the revolving feed mechanism that delivered the seeds from the hopper above to the hole at *c*. Figure 4.7D better illustrates this "force-feed" mechanism in a similar machine. In it one can see how the shape and positioning of baffles or projections on the inner rim of a drum that revolved inside a stationary cast-iron casing ensured uniformity of flow and reduced the risk of clogging. The baffles pulled the seed down from the hopper and around to a hole above a funnel which directed the seeds to a newly created furrow. In the careful phrasing of a contemporary government publication designed to appraise results: "The projections on the periphery turn out the grain with desirable uniformity."[23]

That achievement—and this device—would not crown American experimental efforts until the middle of the nineteenth century. Between the end of the Revolution and the beginning of the War of 1812, a handful of would-be American inventors followed British leads in seed drill design, but all for naught. After 1815 Americans adopted a different strategy. The result was almost instant success, albeit not of the sort Tull had envisioned in his pioneering efforts to mechanize sowing.

Faulty Solutions: American Experiments to 1815

American experiments with grain drills during this period were mainly of three types, all unsurprising and none successful.

The tradition of asking "Is there a better way?" was, by modern standards, astonishingly absent from the thoughts of those who worked the soil from the Atlantic seacoast to the fult line of the Appalachians, from Nova Scotia in the north to the hinterlands of Georgia in the south. Jared Eliot, a New England clergyman, physician, entrepreneur, and inventor, thought otherwise. The first colonist to publish extensively on agricultural topics, he drew in part on Tull's writings in his investigations of farming practices. Following Tull's lead, Eliot wanted to devise a colonial drill which, with slight variations on one basic design, would be suitable for sowing seeds or spreading manure. Not satisfied with Tull's machine (which he condemned as "intricate" and with "more wheels, and other parts, than there is really any need of"),[24] Eliot produced a sowing implement that did little more than simplify Tull's wheat drill.[25] (Compare Figures 4.8 and 4.4.) Like Tull's machine, Eliot's version was little used and soon ignored. From London in 1754, Eliot's friend Peter Collinson summarized the prospects on both sides of the Atlantic: "I am afraid [your new machine] is to do too many things which makes it more complex and will be liable to be often out of order so will never become of general use like Tull's Husbandry, which I never yet saw practiced anywhere."[26] He was right. Part of the problem was the shortage of skilled implement builders. William Logan described the Eliot grain drill shipped to his farm near Philadelphia as "one of the worst pieces of workmanship I ever saw put together." The absence of several key parts was troubling, but more troubling still was the intrinsic complexity of the implement. Logan correctly anticipated its American fate when he judged that Eliot's drill would "prove of little service."[27]

Figure 4.8. Jared Eliot's model of a drill plough, circa 1750

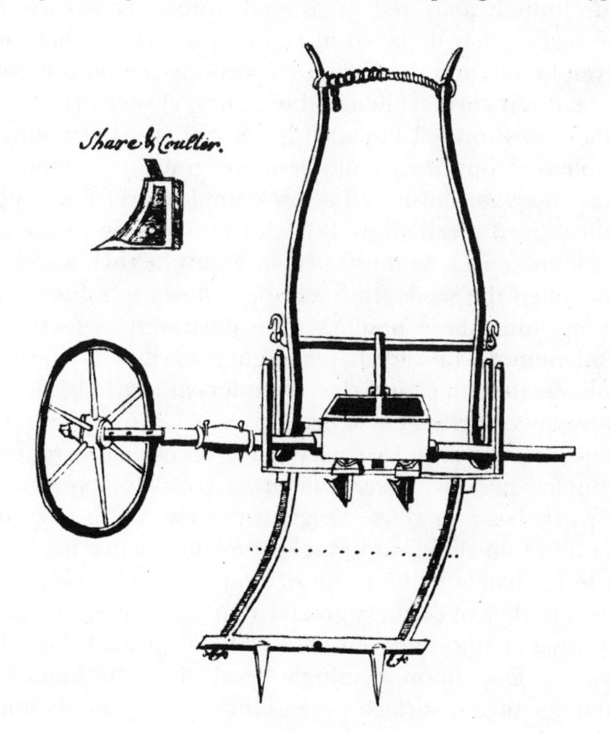

Source: Jared Eliot, *Essays upon Field Husbandry in New England and Other Papers, 1748–1762* (New York: Columbia University Press, 1934), p. 117.

That forecast dates from 1755. During the sixty years that followed, few Americans tried to invent a better seed drill, and when they did, the resulting implement seems to have had little or no impact upon prevailing practices beyond the farm of the inventor.

These efforts at innovation took two main forms. One is evident in the "cluster drill" of John Beale Bordley. Like Jared Eliot, Bordley was one of the rare advocates of agrarian experimentation in a world of farmers adamantly committed to the status quo. Among Bordley's inventions was an implement for sowing wheat in clusters. Ignoring Tull's warning about making seed drills too wide, he designed a six-foot sowing machine with nine feeding tubes. (See Figure 4.9A.) The feeding mechanism upon which success or failure hinged is aptly illustrated by Bordley in Figure 4.9B. A solid wooden cylinder with notches on its circumference rotated within a second cylinder (the latter initially made of wood lined with leather, but the wood-plus-leather was soon supplanted, first by tin and then by sheet iron).[28] As the cylinder spun, these notches filled with seed to move them from the hopper above to a funnel below. (The direction of rotation is indicated in Figure 4.9B by the notches containing seeds.) Nine separate funnels delivered these seeds into furrows carved by nine shares of the sort crudely depicted in Figure 4.9A. Given that each notch delivered several seeds at a time, what was dropped upon the ground was not a single seed but clusters; hence the name, "cluster drill."

In 1786 the newly formed Philadelphia Society for Promoting Agriculture first publicized Bordley's multiple-share grain drill. One year later *The Columbian Magazine* illustrated a less complex drill featuring a variation of Bordley's feed mechanism, but with no mention of the inventor's name. (See Figure 4.10.) As illustrated in Figure 4.10D, a solid cylinder with notches pulled the seeds from a hopper down to a funnel. Whereas in Bordley's machine these notches were portrayed as teeth, now they appear as indentations or carved rectangular cavities in the wood. But their purpose was the same, and this "cylinder-with-cavities" design would dominate subsequent delivery systems.

One example can be seen in Figure 4.11, which dates from this same post-Revolutionary period. Whereas Figure 4.10D shows two rows of cavities, Figure 4.11C has four rows, designed to move seeds from four separate hoppers into four different funnels, the latter delivering the seeds to the ground immediately behind one of four coulters carving furrows in the soil.[29] Any number of coulters greater than one raised the prospect of uneven performance upon uneven fields. The associated difficulties soon became apparent. Even upon seemingly "level" fields "in fine tilth . . . the trivial inequalities of the surface were sufficient to prevent some of the coulters from sinking sufficiently deep for the mould [soil] to fall into the furrows and cover the grain."[30] But once again the most important problem was the reliability of the feed mechanism. The notches tended to

Figure 4.9. J. B. Bordley's cluster drill, 1786

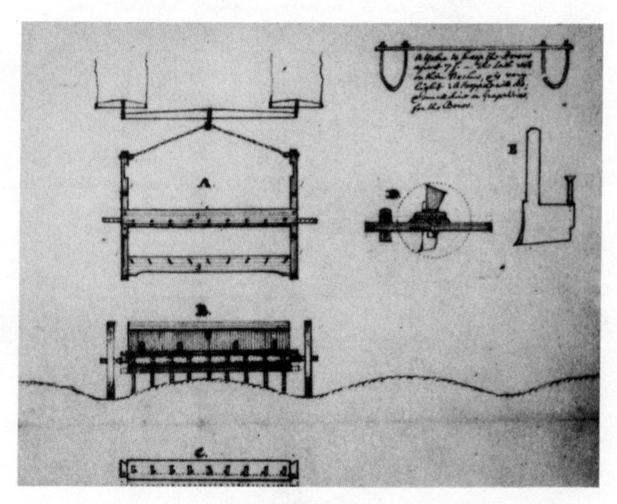

A. Top and rear view

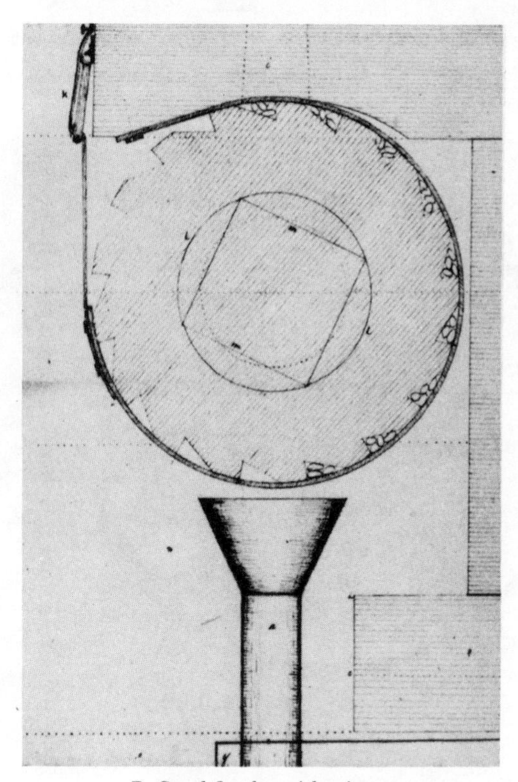

B. Seed feeder, side view

Source: Philadelphia Society for Promoting Agriculture, *Memoirs*, 6 (1938), p. 113.

Figure 4.10. Box for drilling seed, 1787

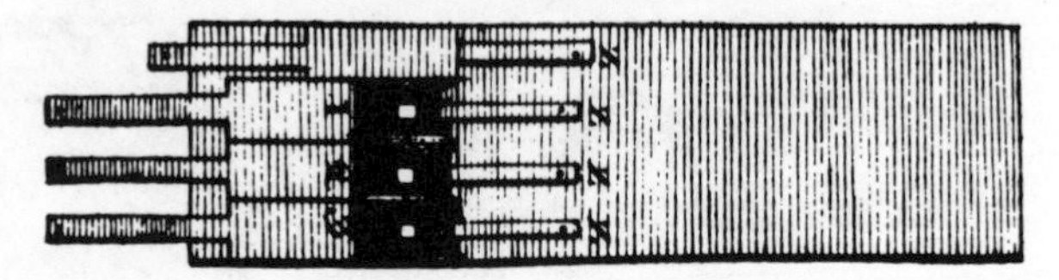

A. Bottom of box from inside, showing sliders

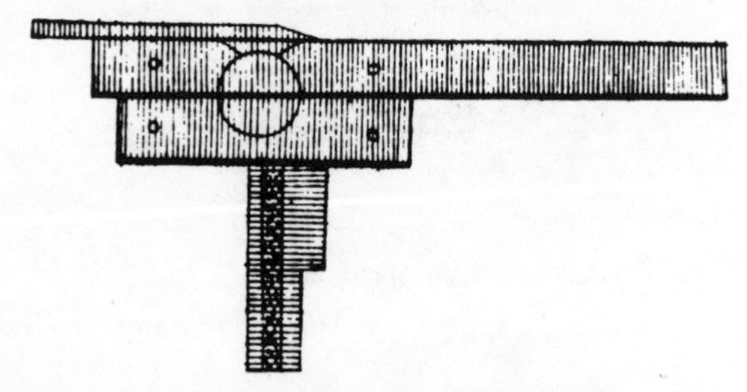

B. Section of bottom of seed box (axis and spout)

C. Side view

D. Axis with holes to receive seed from seed box and deliver same to spout

Source: The Columbian Magazine, 1 (December 1787), p. 827.

Figure 4.11. Lorain's wheat drill, circa 1792

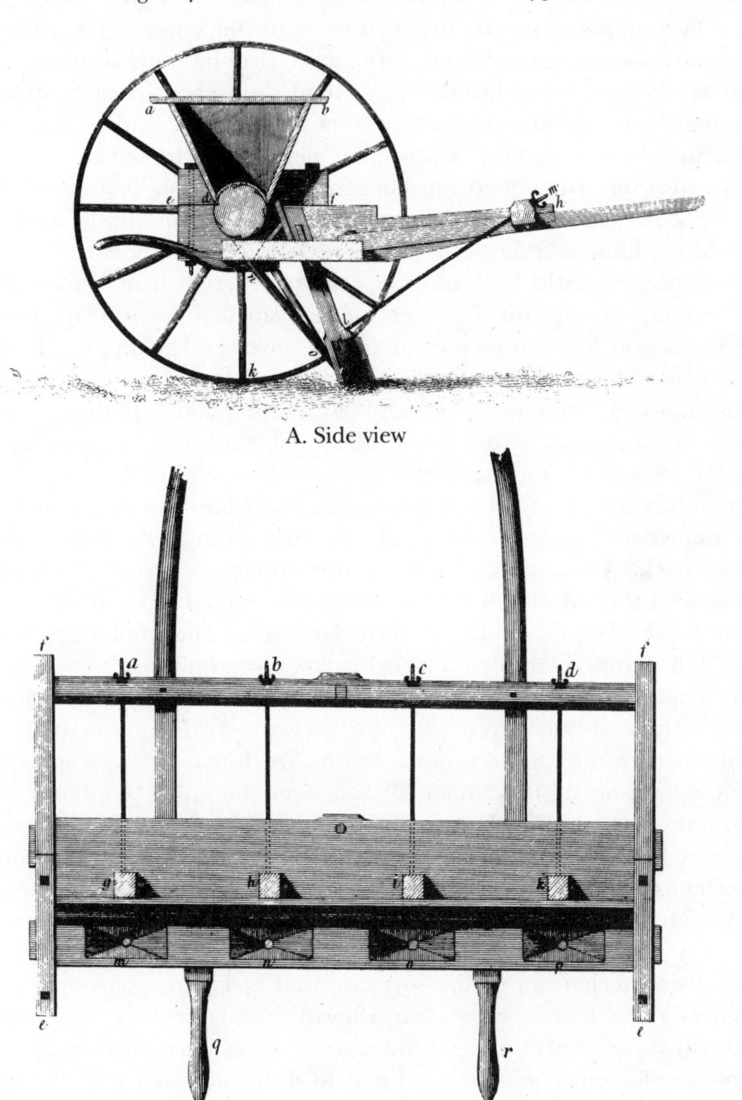

A. Side view

B. Top view, showing four hoppers

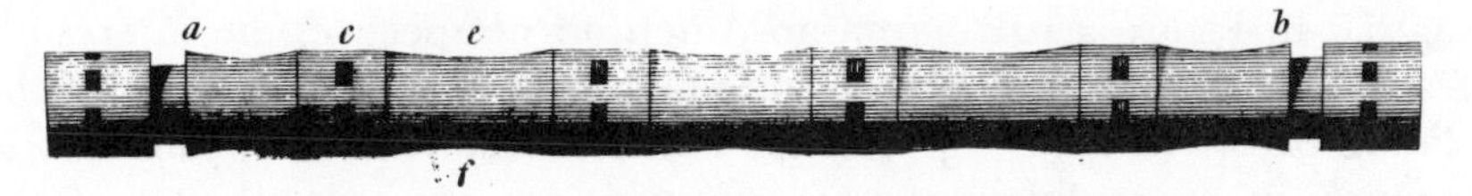

C. Axle with row of mortices to receive seeds

Source: Philadelphia Society for Promoting Agriculture, *Memoirs*, 3 (1814), p. 35.

clog.[31] Once clogging occured, the result was an uneven distribution of seed and empty spaces in the furrows until the sower cleared the implement and began again. Whether the grain drill had one or many coulters, this tendency to clog ensured a general disinterest in such implements among American farmers in the years between the end of their Revolution and the resumption of hostilities with Britain in 1812.

To sum up, Tull's feed mechanism would not do, at least not for the conservative and pragmatic American farming community, and neither would the kind of "cylinder-with-cavities" device that appeared in the post-Revolutionary world. A third design, quite different from these other two, was no more successful. The feed mechanism, less scientific in conception and probably less satisfactory in results, involved shaking a pile of seeds above a small number of holes. This principle evidently underlay George Washington's "seed drill," which was nothing more than a barrel with holes to be rolled across newly tilled soil.[32] John Beale Bordley used a similar idea in his design of a hand-carried sowing implement. He began with a box for sowing clover seed seven feet wide (Bordley was never one to stint when it came to width), divided into seven compartments with two holes each. A "seed man" carried this apparatus across the field to be sown with the aid of two leather straps that served to facilitate the "agitation" of the box en route.[33] A third kind of sowing implement which, in size and shape, resembled a wheelbarrow also employed a shaking principle. This implement's seed hopper had holes drilled in its bottom, above which lay a sliding cover, with the back-and-forth motion of the cover generated "two or three times a second" by linking it "to a spring that is moved by one of the wheels."[34] American literature contains no good illustrations of this mechanism, but pictures of a similar British grain drill suggest how it functioned and why it never functioned well. The basic mechanisms can be seen in Darwin's drill plow. (See Figure 4.12A, B, and C.) Widely cited in British agricultural writing of the early nineteenth century, one version of this sowing implement had the "cylinder-with-cavities" feed mechanism of the sort depicted in Figures 4.10 and 4.11. (As Figures 4.12B and C make clear, Darwin's device had six rows of notches that conveyed seeds into six funnels of a rather peculiar shape.[35]) But the same implement could be modified to distribute seed into the same six funnels by the back-and-forth motion of a sliding bar that fed seeds to these funnels. The six hoppers are depicted in Figure 4.12D and the sliding bar in Figure 4.12E. From an American perspective, the interesting feature of this design is the means whereby the hook at the end of the sliding bar (*h* in Figure 4.12E) generates the back-and-forth motion by running in an undulating groove carved into the axle of one of the wheels in contact with the ground. (See *Y* in Figure 4.12D.) The actual feed mechanism appears to differ somewhat from the American sowing device described above. In that case, the only holes appear to have been

Figure 4.12. Darwin's drill-plough, 1803

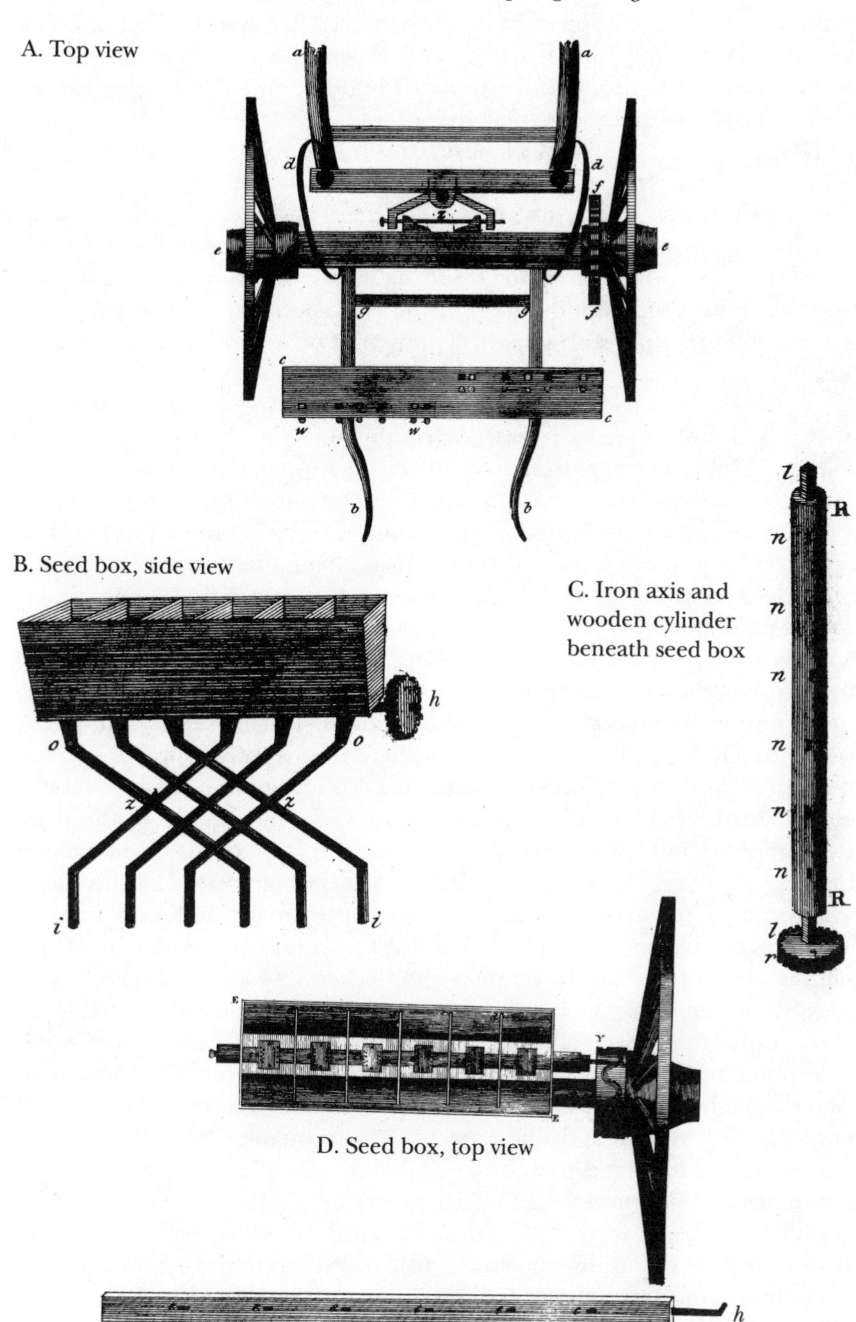

Source: A. F. M. Willich, *The Domestic Encyclopaedia* (Philadelphia: William Young Birch and Abraham Small, 1803), pp. 373, 381.

in the bottom of the hopper. In the British machine, a second set of holes was drilled into the sliding bar to collect seeds and deliver them to a position above the holes in the hopper's bottom, en route passing under a set of brushes (the rectangular shapes, one per seed hopper, in Figure 4.12D). As a reliable feed mechanism, this hook-on-groove device was less than a triumph. In the cautious wording of a contemporary observer: "The reciprocating motion of this slider must be quick, as it necessarily acts once every time the circumference of the carriage-wheel passes nine inches forward, which may not be so easy to execute as the cog-wheel, with the uninterrupted movement of the axis and cylinder in the preceding machine [Darwin's original drill-plough depicted in Figures 4.12A, B, and C]."[36]

All these efforts to design mechanical sowing implements left virtually unchanged the American reliance upon the ancient practice of broadcast sowing. Admittedly experiments with sowing implements were not confined to those discussed above. Contemporaries occasionally reported on a device that reputedly had achieved "some success," although usually the reporters had not seen the implement or evaluated its performance.[37] Admittedly a handful of the agrarian elite tried to encourage others to experiment, and sought to publish any promising results. The Philadelphia Society for the Promotion of Agriculture, for example, offered prizes "for the best comparative experiments on the culture of wheat, by sowing it in the common broad-cast way, and by drilling it, and by setting the grain, with a machine, equi-distant."[38] The society received few applications for this prize. From first to last—from Jared Eliot's pioneering essays on agriculture in the late colonial period to the premiums offered in the post-Revolutionary era by various societies founded by the few to encourage improved farming by the many—the American agricultural community evidenced an unshakable indifference to the prizes and publications. Their continued preference for broadcast sowing over mechanical alternatives was a product of more than mindless conservatism. Machines that were cheap and easy to build were not markedly superior to sowing by hand, while the more elaborate grain drills were expensive to buy, difficult to maintain, and unreliable in performance. Even the most informed American observers condemned British machines as "complicated and costly."[39] After viewing a drilling machine from England, the Philadelphia Society for Promoting Agriculture concluded that it "could not be used to advantage in this country," and accordingly "agreed not to purchase it."[40] A publicist of American drills, for all his enthusiasm for mechanical contrivances, had a scathing view of British practice where that practice, in his opinion, was "so complicated, machinery so expensive, and success so various, that it has never come in general practice [in Britain], nor attempted in that method in America."[41]

British observers echoed American doubts about British machines, but usually in less strident and more tentative language. Arthur Young, for example, had tried four grain drills as of 1779, but was not sure whether to recommend them.[42] About the turn of the century, a series of county surveys in England found that mechanical drilling was "far from common" in most parts of the country. This gradually changed. In Britain, new drills multiplied, old drills were modified, and progress in mechanical sowing after 1800 appears to be a story of the gradual diffusion of improved machines, perhaps the most prominent being the spoon-fed drill of James Cook.[43]

In America, inertia remained the rule and experimentation the exception, at least until the end of the War of 1812. After 1800, little was either attempted or publicized. In the quarter century between 1790 and 1815 the American experimental sowing machines previously discussed all received from the American farming community the most telling condemnation: they were ignored by most, and quickly disappeared from the agrarian literature as subjects worthy of discussion. A farming expert of a later era put his finger on the central difficulty: "seed-drills generally were furnished with a revolving cylinder, in the surface of which small cavities were made, for carrying off and dropping measured portions of the grain; these often broke or crushed the seed, and were liable to derangement."[44] The derangement that most inhibited adoption was the tendency for the indentations to clog with seeds. The typical American farmer of the new republic, pragmatic and impatient with mechanical flaws, wedded to the old and suspicious of the new, was not likely to be much impressed by farming implements such as these.[45]

All this would change after 1815, and with startling rapidity.

A Revolution Begun: Sowing Inventions in America after 1815

If the solution to a problem proves difficult to find, one alternative is to redefine the problem. And that, to some degree, was what Americans did who wanted to dispense with broadcast sowing in the years immediately following the end of the War of 1812. The result was almost instant success.

After more than two decades of negligible experimentation, a flood of new implements was proposed and publicized in the ten-year span beginning with 1815. Of these, one quickly achieved prominence in the literature and acceptance on the farm. It offered a superior alternative that Tull had ignored. If broadcast sowing was intrinsically inefficient (as Tull was quick to emphasize), why not substitute a mechanical contrivance to replace the human hand, but continue to sow in the broadcast manner? This was what Bennett's machine was designed to do. As illustrated in Figure 4.13, the implement was a study in simplicity. A long, triangular-

Figure 4.13. Bennett's broadcast hand drill, 1816

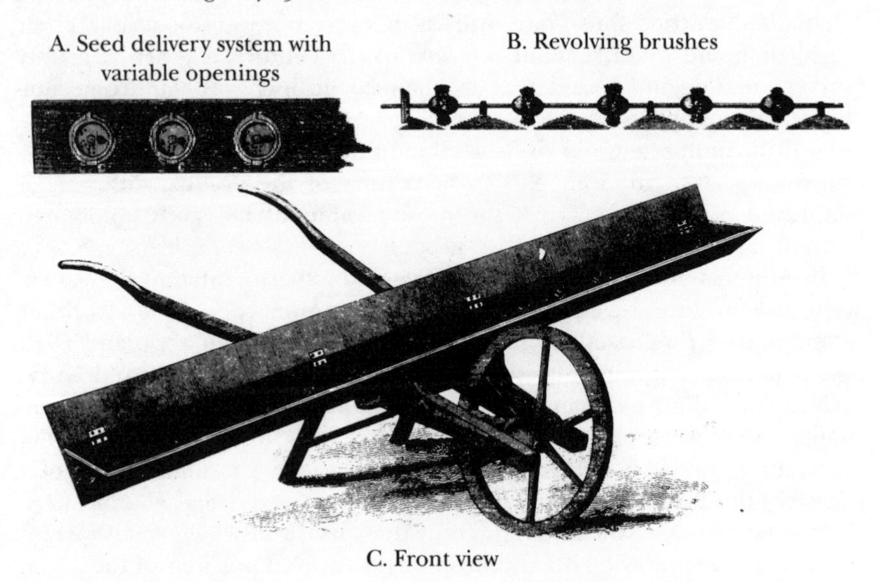

C. Front view

Sources: Philadelphia Society for Promoting Agriculture, *Memoirs*, 4 (1818), p. 45.

shaped seed box was mounted on a one-wheel vehicle that, in size and appearance, looked something like a wheelbarrow (Figure 4.13C). The seed box contained a set of revolving brushes (Figure 4.13B), which threw the seeds against openings at the back of the box (Figure 4.13A). The larger the seeds to be sown, the larger these openings had to be. Variability in both the size and number of the holes was achieved by having a single hole "covered with two perforated brass plates, the inmost one fixed, and the other so fitted as to be easily moved round." The end result was not unlike the flow options built into the top of a present-day box of salt or pepper. "The perforations being in groups of different numbers and sizes, the discharge of the seed is regulated to the desired quantity by causing the large or small, and many or few, perforations to coincide."[46]

Cheap to build and easy to maintain, "Bennett's Clover and Turnip Sowing Machine," as it was initially called, was repeatedly judged superior to the hand alternative. The early emphasis on turnip and clover seeds was to be expected. Both were exceptionally small, and for that reason extremely difficult to sow by hand. This new implement not only offered the prospect of a more even distribution, but promised that result "uninterrupted by wind or light rain,"[47] two natural forces notorious among farmers for disrupting broadcast sowing, although both were seldom mentioned in the writings of farming experts.

First publicized in 1816, Bennett's sowing device[48] was being commercially produced by 1819, and by 1823 could be purchased from distributors in New York City, Boston, and Baltimore. Its advertised capabilities soon expanded beyond sowing turnip and clover seed, and within a few years it was typically promoted as superior for "all kinds of grains" plus the spreading of plaster of Paris. Experts, both widely known and self-appointed, were quick to praise its reliability. Ordinary farmers repeatedly wrote to agrarian periodicals to record their satisfaction with its performance. Perhaps most impressive of all was its early prominence and lasting presence in advertisements of the 1820s for "new sowing drills."[49]

One obvious modification of this machine was to substitute horse power for hand power. A pioneer in this development was Robert Sinclair, a farm implement manufacturer of Baltimore who was among the first to produce Bennett's original sowing machine. As indicated in Figure 4.14, his innovation was nothing more than Bennett's original box (now twelve feet wide) mounted on a carriage. This horse-drawn version, according to Sinclair, was "particularly valuable on large estates." Like Bennett's original device, it appears to have had considerable success in the 1820s.[50] And like the hand-powered version, the horse-drawn variant remained popular for years to come, as illustrated in Figure 4.15.

In the years following 1815, American efforts to mechanize sowing took two main forms. Bennett's machine (and Sinclair's modifications) illustrate one.[51] The other was a simplified response to the old challenge spelled out by Tull in the 1730s: to devise a sowing implement that could carve a furrow and deliver seeds into that furrow in a regular and regulated flow. In the post-1815 world, the American search for solutions focused upon single-furrow machines that sowed one row at a time. The feed mechanism that dominated these experiments appears to have been

Figure 4.14. Sinclair's sowing machine, 1822

Source: American Farmer, 4 (July 5, 1822), p. 120.

Figure 4.15. Hatch's sowing machine, 1845

Source: Ohio Cultivator, 1 (June 15, 1845), p. 96.

a variation of the old cylinder-with-cavities device developed in the late eighteenth century.

One of the most esoteric and least successful was Conard's corn drill, depicted in Figure 4.16. The cylinder beneath the hopper (4.16A) is drawn with only one notch shown, but presumably the maker could vary size and number of notches. Novel in appearance—a "sled" rather than a wheeled implement (4.16B)—this was an inventive curiosity destined to be ignored.

Less esoteric and more successful was the 1821 "Corn and Seed Drill" depicted in Figure 4.17. Its manufacturer, Robert Sinclair of Baltimore, who was making a reputation as a producer of Bennett's seed drill, designed a single-furrow, seed-placing device that incorporated two ideas from the eighteenth century.[52] One was the notched cylinder revolving beneath a hopper to deliver seeds to a funnel and thence to a newly carved furrow.[53] The other, in evidence on Tull's original grain drill (see Figure 4.4), was to add two teeth at the rear to facilitate seed covering. Behind the teeth Sinclair positioned a small roller "to give some steadiness, whilst it also governs the depth of the coulter."[54] This machine appears to have achieved considerable popularity,[55] as did later variants. (See for example Figure 4.18.)

No good diagrams could be found of the feed mechanism of Sinclair's machine, or of other American variants. Here, too, British illustrations can be used to clarify how that mechanism worked. The seed drills of Figures 4.17 and 4.19, although separated in time by twenty years, nonetheless have a striking resemblance. As illustrated in Figure 4.19B, the

Figure 4.16. Conard's corn drill, 1819

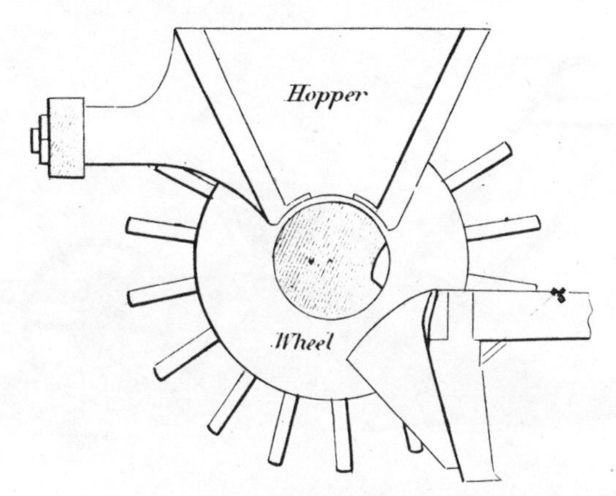

A. Detail of hopper and feed mechanism

B. Side view

Source: Philadelphia Society for Agriculture, *Memoirs,* 5 (1826), p. 100, Plate I.

British feed mechanism consisted of a cylinder (*C*) with a row of indentations (at K) revolving beneath a hopper. The "jack-chain" (F in Figure 4.19A) is designed to help cover the seeds through its dragging action. Sinclair's harrow teeth served the same purpose, but Willis's Improved Seeder (depicted in Figure 4.18) omitted these teeth and added a dragging chain.

One final American invention of a similar sort from the 1820s deserves notice. Francis Smith designed his "cotton planter," as the name implies, to appeal to the Southern market. (See Figure 4.20.) "A moving board (cc)" within the hopper was supposed "to keep the seed from clogging."

Figure 4.17. Sinclair's corn and garden seed drill, 1821

Source: American Farmer, 2 (January 26, 1821), p. 349.

But how the rotating cylinder plus what appears to be four rotating arms (at B) delivered a shower of seeds to the ground was never explained in Smith's written accounts of the machine. Although its initial advertisement included testimonials from satisfied customers, within a few years this particular seed drill disappeared from the agricultural literature.[56]

None of these one-furrow seed drills appears to have had outstanding success. The basic flaw was the lack of reliability of the feed mechanism. Willis's "Improved Seed Sower" of 1839, for example, claimed to have solved the clogging problems endemic to earlier machines,[57] but subsequent writers imply that claim was exaggerated. Three decades after Willis announced his solution, John J. Thomas reviewed the problem in his discussion of the comparatively new (as of 1869) force-feed mechanisms of the sort depicted in Figure 4.7 and characterized all previous feed mecha-

Figure 4.18. Willis's improved seed sower, 1839

Source: New England Farmer, 17 (January 9, 1839), Supplement, p. 6.

Figure 4.19. British turnip drill, 1801

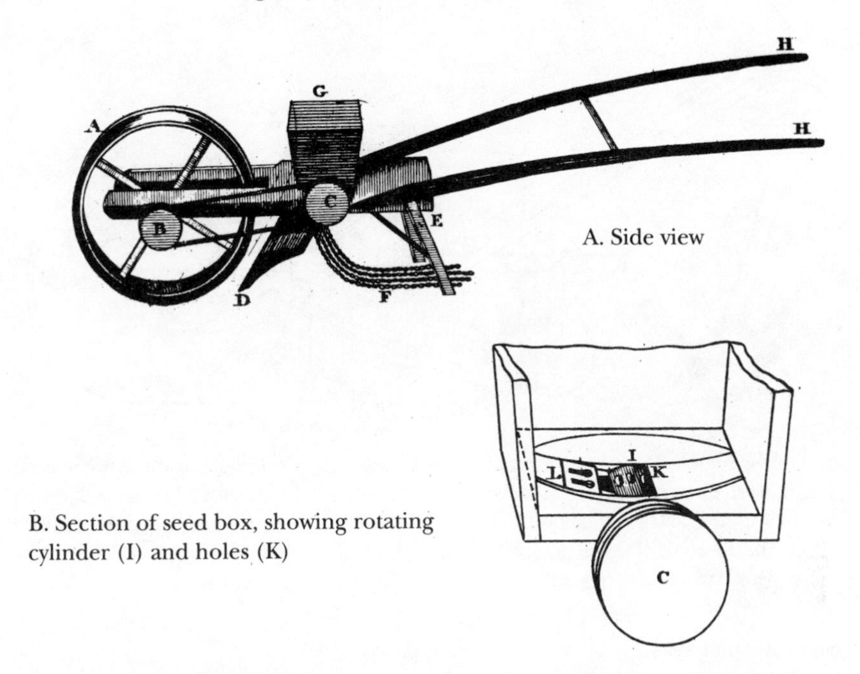

Source: A. F. M. Willich, *The Domestic Encyclopaedia* (Philadelphia: William Young Birch and Abraham Small, 1803), p. 205.

nisms as inherently defective.[58] Again the main defect he identified was the tendency of these machines to clog.[59]

One final point. Modern writers concerned with early nineteenth-century agricultural developments sometimes forget that, as of 1815, British and American experts generally considered two issues as yet unresolved: (a) whether "drill" culture was more efficient than broadcast sowing, and (b) how deep a machine should place seeds when drilling. The surge in sowing innovations in America after 1815 cannot therefore be viewed as inventors attempting to supply a well-established demand that reflected common acceptance of the superiority of drilling. A more accurate version of events would seem to be the gradual progress of, and interaction between, mechanical drilling devices, on one hand, and on the other, the knowledge of why drilling was preferable, and how it should be done.[60]

The Long View: 1733–1830

American efforts to replace broadcast sowing with mechanical implements seem to have occurred in two waves. The early attempts are per-

Figure 4.20. Smith's cotton planter, 1826

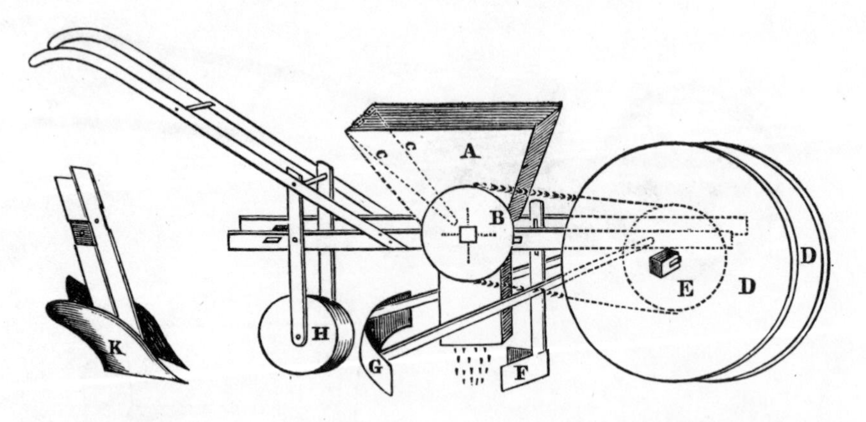

A. The hopper—B. a sheave attached to the wired axle for separating the seed. *c. c.* a dotted line, represents a moving board within the hopper, to keep the seed from clogging. D/D the wheels which give motion to the machine. E. another sheave between D/D over which the band passes to give motion to B. F. the coulter. G. the coverer. H. the roller. K. two small ploughs for cultivating the cotton, to be applied after A. F. G. & H. are removed.

Source: American Farmer, 7 (February 3, 1826), p. 364.

haps more appropriately viewed as occasional ripples, dispersed in time and with little force to modify the rock-ribbed allegiance of American farmers to ancient sowing practices. After the pioneering efforts of Jared Eliot about the middle of the eighteenth century, there follows almost three decades of inventive inactivity, during which no implement of note received even modest public attention. Immediately after the Revolution, a handful of inventors designed and publicized mechanical sowing implements, all within a short span of time that ended well before 1800. Two seed-dispensing systems dominated these designs. The one—whose purpose was the systematic placement of seeds into furrows—featured a share in front and a feed mechanism comprised of an indented cylinder rotating beneath a seed hopper. The other, less systematic in results, passed the seeds through holes in the bottom of a hopper by shaking the seed container and/or momentarily uncovering the holes by the back-and-forth action of a sliding cover. None of these much affected farming practices, and all had almost disappeared from the agricultural literature by 1810.

The second wave followed the conclusion of the War of 1812. A host of new farming devices burst forth from the literature and swept across the landscape, of which those concerned with sowing were but instances of a larger revolution. This wave of innovations, in turn, signaled the beginning of a major shift in attitude among American farmers that affected how they thought about, and went about, their agrarian tasks. Sowing innovations once again were of two main types, one old and one new. The novel one began with a problem Tull had ignored: how to replace the human hand with a machine but continue to sow in the broadcast manner. The implement devised to meet this challenge was cheap to build, easy to maintain, and by most accounts, successful in achieving its modest goal. Predictably, it soon achieved commercial success and remained a presence on American farms for years to come. The other thrust of inventive effort was directed to finding the simplest possible solution to Tull's original problem. These newly designed implements had long-established features, including a share to carve a furrow and a feed mechanism (rotating cavity-filled cylinder plus a funnel) to place the seed into those furrows, but with the modest goal of sowing only one row at a time. This second type of sowing implement proved less successful than the first, the major cause of its undistinguished record being the well-established propensity of feed mechanisms to clog. The American solution to this latter problem—the force-feed seed drill—did not appear for several decades.

In the fifteen-year period ending in 1830, attempts to design a better seed drill were by no means confined to the machines discussed thus far. Time and again the agricultural literature of the period featured advertisements, endorsements, and accounts (favorable and otherwise) of "turnip drills" and "hand seed sowers," "corn and bean droppers" and "drill barrows," "rutabaga drills" and "drill machines made to order," with perhaps first prize for simplicity belonging to the inventor whose sowing device consisted of a tube fixed to a garden hoe.[61]

The decade of the 1820s is often cited as a turning point in American politics, unambiguously signaled by the election of Andrew Jackson in 1828. Certainly by the time of the seventh president's inauguration the democratic forces unleashed in the new republic had taken on a new appearance. (Following the inauguration ceremonies, Jackson had to leave the White House by a back window to escape the crush of supporters with bronzed faces and muddy boots coming in the front.) But if a new spirit was animating some of the electorate, new attitudes of a different sort were beginning to be manifest throughout the rural landscape. A conservative approach to agrarian procedures was giving way to a new inquisitiveness, and novel implements, previously regarded as prima facie useless, were now regarded as prima facie worth exploring for possible adoption. This clearly was not true for all of the farmers all of the time, but it was true for some of the farmers part of the time. And that was a new development in American agrarian history. By the time that Jacksonian revelers were spilling punch and breaking glasses as part of a less decorous and more rustic inauguration style, American farmers had discovered the path to modernization.

[5]

Harrowing

Harrow Development to 1790

Throughout much of Western history, harrows have had two main functions, one well defined and the other not. To cover seeds that have just been sown seems to have been their primary function from the beginning. But inadequacies of other agricultural implements (notably the plow) resulted in harrows being used for a variety of field-tilling assignments. The plow's wedging action, for example, tended to compress soil, particularly wet clays. Subsequent use of the harrow helped to counteract this compression. In addition, harrowing a field after plowing (and before seeding) served to pulverize remaining clods and clean off weeds and roots uprooted by the passage of a plow. A further benefit was the flattening of ridges and filling in of gullies, resulting in a smoother surface. The flatter the tilled field, the more easily the standing crop could be reaped, particularly if the reaping implement was some variant of a scythe working close to the ground. For all these reasons, from the apogee of Imperial Rome to the upheavals of the French Revolution, harrows served a variety of purposes on the fields of Europe. Whenever they were used, the preferred draught animal by a wide margin was the horse. Speed of passage was crucial to stir and pulverize the soil, and oxen were commonly considered to be much too slow for the job. Century after century, the implements dragged were of two main types: bush or brush harrows, and wooden frames with wooden or occasionally iron teeth, ideally all uniform in length.

Brush or bush harrows, as the name suggests, were nothing more than branches or bundles of branches, often clamped into a wooden frame. Cheap to make and easy to assemble, they did not change much in appearance from the time of their first use in the Roman Empire until well

Figure 5.1. Bush harrows

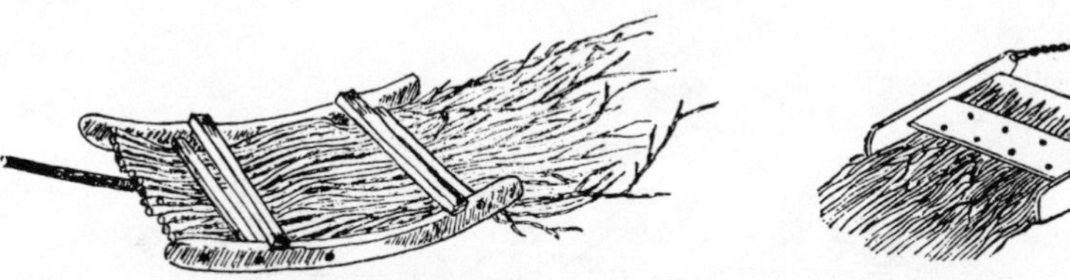

A. From Roman times

B. American, mid-nineteenth century

Sources: K. D. White, *Agricultural Implements of the Roman World* (Cambridge, England: Cambridge University Press, 1967), p. 146, © Cambridge University Press 1967; reprinted with the permission of Cambridge University Press; John J. Thomas, *Farm Implements and Farm Machinery* . . . (New York: Orange Judd, 1869), p. 142.

into the nineteenth century. (Compare for example Figures 5.1A and 5.1B, which are separated in time by almost two millenniums.) Because they were both lightweight and irregularly shaped—larger branches invariably projecting to score the soil with deeper furrows—these relatively inefficient harrows were used primarily to cover grass or other small seeds which, after sowing, required a minimum of topsoil thrown over them. Even the best of farmers commonly employed such implements despite their irregular structure and erratic performance. In America, such pioneers in good practice as Thomas Jefferson and Landon Carter frequently used them, as did many of their contemporaries on the other side of the Atlantic.[1]

The origin of the wooden frame harrow remains obscure. The Bible does mention "toothed harrows of iron," but only as instruments of torture and never as implements of agriculture. Not much is known about the processes whereby newly sown seeds were covered in most of the ancient civilizations of the Middle East. What little evidence has survived indicates that at least in some instances livestock, such as sheep or donkeys, were driven across newly sown fields to tread in the seeds.[2]

The first clear references to the harrow as an agricultural implement appear to date from Roman times. The bush harrow, as previously noted, was used mainly to administer a light covering to small seeds. Those consisting of wooden frames and spiked teeth offered the advantage of greater strength and greater uniformity in tilling points in contact with the soil. They served, then and for centuries to come, to pulverize, smooth and clean up what had been previously plowed, as well as to cover newly sown seeds. Their shape cannot be inferred from surviving relics—not surprising for a wooden implement—but literary evidence suggests a triangular form of the sort depicted in Figure 5.2.[3]

From the fall of Rome to the Norman conquest, the history of the har-

Figure 5.2. Roman harrow (irpex)

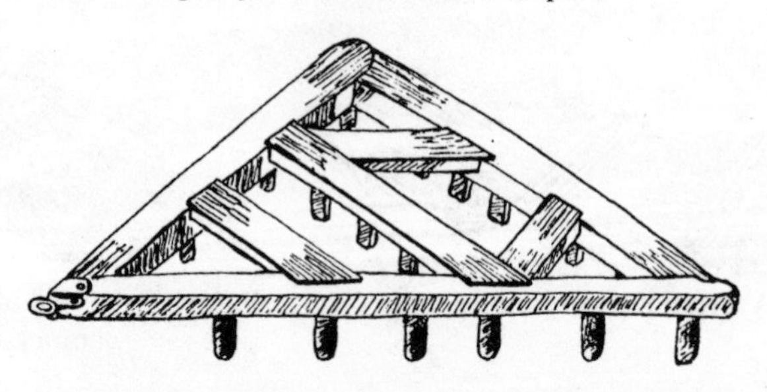

Source: K. D. White, *Agricultural Implements of the Roman World* (Cambridge, England: Cambridge University Press, 1967), p. 148, © Cambridge University Press 1967; reprinted with the permission of Cambridge University Press.

row remains something of a mystery. The first depictions of this implement date from the twelfth century, and such drawings as have survived from this and the next two centuries are hardly triumphs of pictorial clarity.[4] What is clear is that (a) by 1100 A.D. square or rectangular harrows were in use in Northern Europe, (b) by the fifteenth century the triangular shape was also employed, and (c) these two geometric shapes (with minor variations) continued to dominate wooden frame harrows on both sides of the Atlantic until the end of the eighteenth century.

The continued reliance upon triangles and rectangles was not surprising. Whether the triangular shape persisted from Roman times or (as some have argued) had to be reinvented in the thirteenth century, both geometric shapes served distinct functions in a primitive agricultural setting, and therefore once devised, were likely to persist. The triangular shape was better suited to rough ground littered with stumps, roots, and boulders. When the teeth of a triangular harrow hit such obstacles, precisely because of the triangular shape of the frame, the harrow was more likely to bounce to one side or the other and continue forward. A square or rectangular frame could hold more teeth, and for that reason was more efficient. (Consider, for example, the teeth-carrying capacity of the harrow in Figure 5.3D, compared with any triangle of similar dimensions at the base.) But if a rectangular harrow was used on a field littered with roots and rocks, some of which were immovable, the results could be disastrous. An American expert of the 1820s explains why. "When the teeth come in contact with a fast root or stone, the whole power or strain comes on the teeth and draws nearly crossways the timbers on which they stand, which occasions them to split or break and be rendered useless."[5] The general rule for harrow shapes in the premodern world was accordingly the same in Europe and America: triangles for rough ground and rectangles for well-tilled fields comparatively free of debris.[6]

The Doldrums: 1790 to 1814

As of 1790 the harrows employed in the fields of the newly independent United States looked much like those in use for centuries. Animals dragged brush harrows of various shapes (depending upon the shape of the brush assembled) across furrows in which were scattered grass or other small seeds in need of nothing more than a light covering. Heavier harrows were made of wooden frames that were triangular or rectangular in shape. Their teeth were made of either wood or iron, although most agrarian commentators suggest that wood still dominated in the post-Revolutionary world. In 1775 one New England observer complained that "the harrows are . . . of a weak and poor construction; for I have more than once seen them with only wooden teeth." Fifteen years later in his landmark study of New England agriculture Samuel Deane chastised the farmers of that region for neglecting to use teeth "steeled at the points" despite their low cost and superior efficiency.[7]

Nothing much changed for a quarter of a century. Harrows in use at the close of the War of 1812 were essentially the same as those in use at the close of the Revolution, and there is almost no evidence of experimentation in the intervening years with the shape of the frame or the setting of the teeth. Samuel Deane's second review of New England agriculture in 1797 was as devoid of details on possible new designs as was his initial survey of 1790. Most observers in this period found nothing worthy of remark under the heading of new harrow designs. One notable exception was Dr. James Mease, who, in his 1803 review of new designs in the British publication *The Domestic Encyclopaedia*, still questioned their relevance for American farmers. One novel design, for example, he claimed "consists of too many parts to be generally adopted by farmers, who require simple, strong, and cheap utensils." Accordingly, in Mease's version of the British book (first published in Philadelphia in 1803), the British author's written account of this particular harrow was deliberately omitted.[8]

Revolution in Designs: 1815–1825

In 1819 George Jeffreys of North Carolina enjoyed sufficient renown to receive a questionnaire from *The American Farmer* about agricultural practices in his region. On the subject of harrow design, he found little worthy of remark.

Q. What kind of harrows used?

A. There has been but little improvement on the harrow—the square harrow is the most common.[9]

Figure 5.3. European harrows

A. British harrow, circa 1340

B. Flemish harrow, late fifteenth century

C. French harrow, fifteenth century

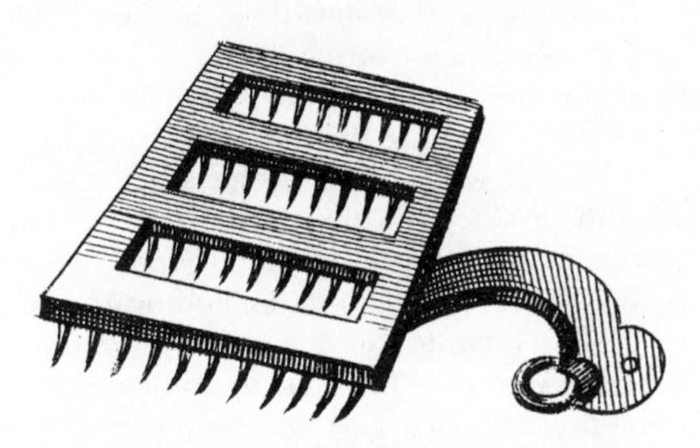

D. "Common" British harrow, 1725

Sources: Eric George Millar, *The Luttrell Psalter* (London: British Museum, 1932), f. 171; W. O. Hassall, ed., *The Holkham Library* (Oxford: Privately printed, 1970), Plate 159, by permission of the Earl of Leicester; Jean Longnon and Raymond Cazelles, eds., *The Très Riches Heures of Jean, Duke of Berry* (New York: George Braziller, 1969), Figure 10v, by permission of the Musée Condé, Chantilly; R. Bradley, *A Survey of the Ancient Husbandry and Gardening . . .* (London: B. Motte), 1725, Plate III.

Jeffrey's judgment was undoubtedly an accurate description of developments in his locale. Further north, however, the design of harrows was being revolutionized. Part of that revolution—the less dramatic part—involved a change in the materials and placement of harrow teeth. Square teeth were now being mounted with a corner (rather than a flat side) facing forward, and materials used to make them were gradually shifting from wood to iron to steel-tipped (the latter having been advocated by Samuel Deane as early as 1790). But this was hardly the stuff of agricultural revolutions. The main problem in design, to this point seldom addressed and never resolved, involved the shape of the harrow frame. The wider apart the teeth were placed, the less thorough scouring of the soil. The closer the teeth, the more frequently they clogged with the refuse inevitably encountered even on well-tilled fields: weeds and roots and stones and clumps of soil, the latter a problem particularly when soils were wet. Every clogging required a momentary halt to clear away debris. Forced to choose between a less thorough scouring or repeated stops while harrowing, most farmers opted to use the more widely spaced teeth of a less efficient implement.[10] The unsurprising result, as summarized by the president of the Cumberland County (N.J.) Agricultural Society, was a harrow "in common use in this country . . . with the teeth too wide apart" that failed to "break the clods sufficient for producing *fine tilth.*"[11]

In retrospect, the solution to this problem seems obvious enough: angle the bars that carry the teeth. This restructuring of the harrow frame took two main forms. Starting with the common rectangular shape, the innovator could take all the arms attached to the forward horizontal bar and swing them sharply to one side, as illustrated in Figure 5.4A. This produced a rhomboid in which a set of parallel arms containing teeth had the same advantages as the single arm of triangular harrow of the sort illustrated in Figure 5.4B. Slanting the arms relative to the forward motion of the harrow caused teeth of any given spacing on a (previously rectangular) frame to produce much narrower spaces between the furrows these teeth were scratching out on the ground. The forward bar in Figure 5.4A was now made of iron (for "defending the rest of the harrow") and the teeth were mounted so that no tooth followed in the track of another.[12] A different variant of the same solution was achieved by starting with the common triangular frame and simply multiplying the number of arms, as in Figure 5.5. The rhomboid shape was devised (according to R. K. Meade of Virginia) by 1816. The triangular solution, according to its New England inventor, was first constructed in 1818, improved and "in use" by 1820, and for sale "at the Agricultural Establishment, No. 20, Merchant's Row, Boston" by 1824. In design, this latter type of harrow with multiple triangles clearly anticipated the geometry of the celebrated Geddes harrow of the 1840s. (Compare Figures 5.5 and 5.6.)[13]

During the decade following the conclusion of the War of 1812, the revolution in American harrow designs was by no means confined to the modifications discussed thus far. A new function for this agrarian implement began to be explored, and with that exploration came other proposed modifications in the shape of the frame and the placement of the teeth. One problem of long standing was the weeding of crops between

Figure 5.4. American harrows, 1816

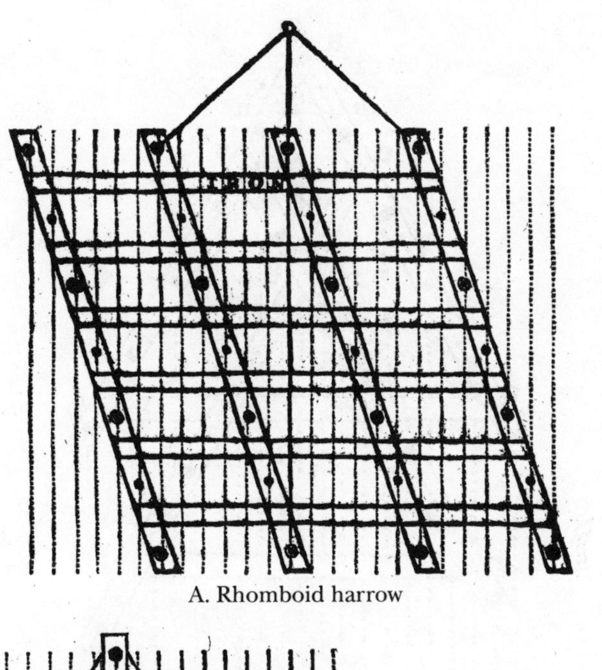

A. Rhomboid harrow

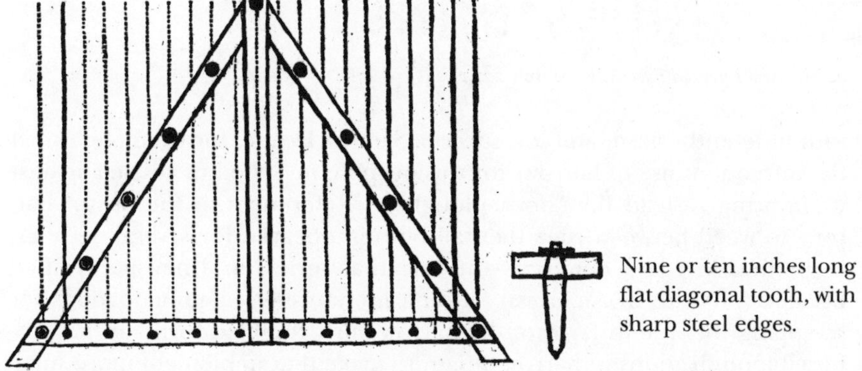

B. Triangular harrow

Source: R. K. Meade, "Mode of Cultivating Indian Corn: Harrows," Philadelphia Society for Promoting Agriculture, *Memoirs,* 4 (1818), pp. 186, 188.

first planting and final harvesting, particularly corn, potatoes, and other vegetables grown in rows. Commonly this required repeated hoeings by hand, a labor-intensive activity which had consumed the hours and strained the backs of colonial farmers from the time of the earliest settlements. The possibility of substituting a horse-drawn harrow for a hoe-wielding worker had been explored even in the late eighteenth century, but

Figure 5.5. American Abbot harrow, 1824

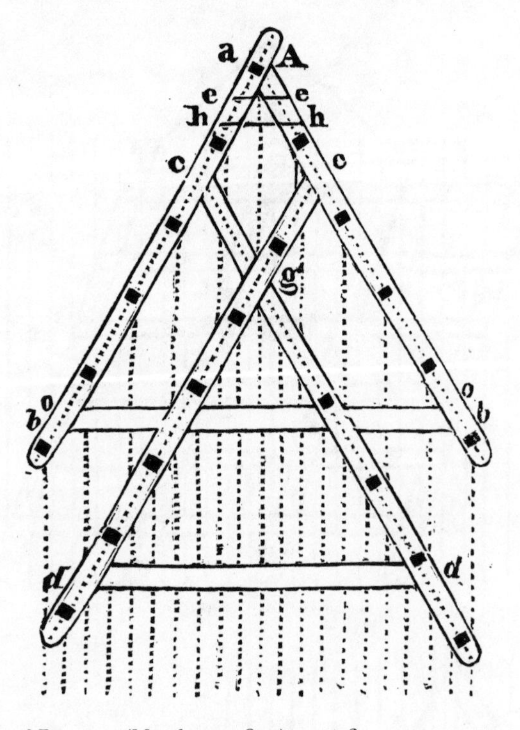

Source: New England Farmer, 2 (March 27, 1824), p. 276.

with little enthusiasm and less success. Samuel Deane, for example, noted the infrequent use of harrows for this purpose in 1790 but advised against it (favoring instead the "horse-plough").[14] References to the use of harrows to weed between rows then all but disappeared for several decades. Beginning in 1819, however—the year that agricultural newspapers first became a fixture in American farming literature—recommendations for scrapping the hoe in favor of the harrow fairly tumbled forth, as did proposed modifications in harrow design to make that implement more suitable for inter-row weeding. One of the first design modifications to appear was the double-harrow of Figure 5.7, evidently a Virginia invention "intended to run on each side of the corn," with "coupling bolts" in the middle to keep all ten teeth working, "however irregular the surface may be."[15] For cotton, a South Carolina planter recommended the horse-rake harrow of Figure 5.8, with two rows of staggered iron teeth, one behind the other, in a semicircular frame.[16] These are only two examples from a long list of implement suggestions from farmers and would-be inventors of how the old harrow design (with or without major modifications) could be accommodated to this relatively new function (at least for harrows) of

Figure 5.6. American Geddes harrow, 1845

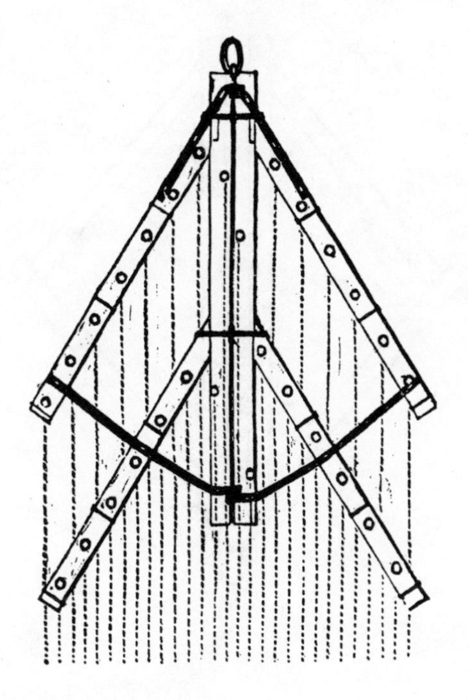

Source: American Agriculturist, 4 (June 1845), p. 187.

mechanized weed removal, particularly in the culture of potatoes and corn.[17]

Now that custom was in question and novelty was more in season, a multiplicity of other harrow designs were advocated, some more successful than others. These included "oblique" harrows (a rhomboid shape, but pulled from one of the rhomboid's corners), a "cotton seed coverer" shaped like the upper part of a wooden funnel cut in half, a "trowel hoe plow" (more closely resembling a cultivator than a harrow), multiple frame harrows and self-cleaning harrows, expandable harrows and self-sharpening harrows, and finally harrows with removable teeth combined with rollers.[18] (See Figure 5.9.)

The pace of progress in new designs can be traced in the advertisements of agricultural implement producers. In 1817, on Front-street in New York City, Richard Harrison opened what he claimed at that time to be, "the only establishment of the kind in the United States," selling every conceivable type of agricultural implement. Under "harrows" he offered for sale "common," "Scotch, with hinge," "expanding," and "Young's potatoe harrow made to order." (This spelling of "potato," made famous by Vice President Quayle in 1992, was quite common in the early nineteenth

Figure 5.7. American double-harrow, 1818

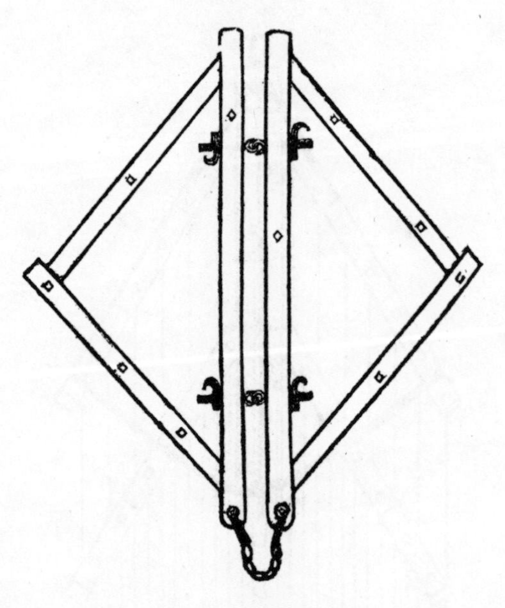

Source: American Farmer, 1 (July 2, 1819), p. 112.

century.) Whether or not Harrison was first, other competitors soon appeared and made their presence known by advertising in the agricultural press the availability of agricultural implements for sale in such urban centers as New York City, Baltimore, Philadelphia, and "Hagers-Town," Maryland. By 1825 the list of implements commonly advertised included

Figure 5.8. Horse-rake harrow, 1825

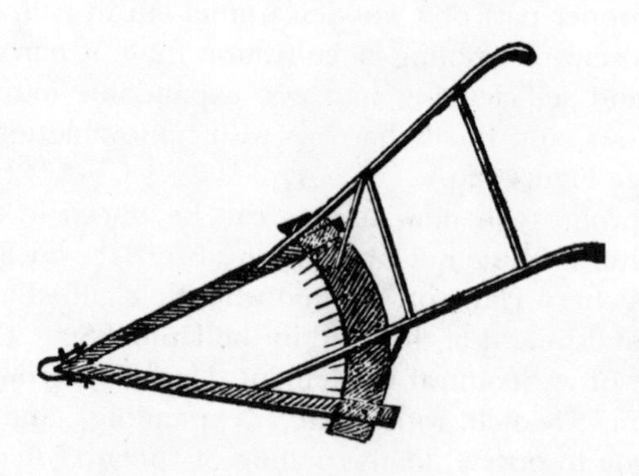

Source: American Farmer, 7 (August 12, 1825), p. 162.

Figure 5.9. American combined roller and harrow, 1820

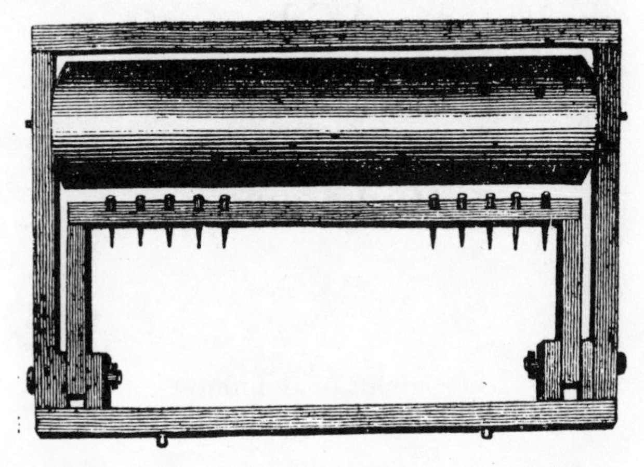

Source: American Farmer, 2 (December 15, 1820), p. 307.

expanding and hinge harrows, diamond harrows, common corn harrows, harrows "made to order," and "harrow teeth, all sizes."[19]

As noted at the outset of this section, as late as 1819 George Jeffreys of North Carolina reported that in his locale "there had been but little improvement on the harrow." But two years earlier Jeffreys himself had been urged by a friend in Pittsfield, Massachusetts, "If you have no well constructed ploughs and harrows, it will be well to send to Philadelphia or New York, and obtain the improved models."[20] By 1825, up and down the east coast of America such advice was a commonplace, because a host of new designs were commonly available, avidly endorsed, and widely used.

[6]

Cultivating

Problems of Definition

"Cultivator" is a word with an irregular history. By the middle of the eighteenth century, the meaning seemed clear enough. Johnson's famous dictionary of 1755, for example, defined "cultivator" as a person (not an implement): "One who improves, promotes, or meliorates; or endeavors to forward any vegetable product, or any thing else capable of improvement."[1] Clarity of meaning was undermined, however, as new agricultural implements were designed, some of which were also called "cultivators." Adding to the mounting confusion was the tendency to use the same word to designate a different kind of implement in Britain and America. In both countries, identical synonyms for "cultivator" evolved, notably "scuffler" and "scarifier." But what was a "cultivator"?

Two kinds of innovations in farm implements were involved. One is the primary topic of this chapter. After the seed has been sown and harrowed in, for most crops the remaining task of paramount importance between first sprouting and final harvest is weed removal. If the crop is grown in rows, weeds might possibly be eradicated, not by workers armed with hoes, but by some horse-drawn implement dragged between the rows. The working surface of that implement obviously must duplicate in some way the root-slashing action of the hand hoe. A second and quite different kind of implement evolved at roughly the same time to fulfill a different function: to stir and pulverize the soil (particularly "stiff" soil) to prepare the field for subsequent planting. This new implement was designed to complement the plow and the harrow. Unlike the plow, it lacked a moldboard, having instead a set of sharp teeth, usually inclined slightly forward to rip into the soil, but not invert it. Unlike the harrow which merely scratched the surface, this new implement had sharp metal teeth designed to dig deep and rip hard at clods and other obstructions, all with the objective of getting acreage into "fine tilth." In short, the second implement was designed for stirring and pulverizing prior to planting (and possibly for harrowing in seed in special circumstances). When the seed was sown and the resulting crop was growing in rows, the first implement could be used for inter-row weeding.

In Britain, the inter-row weeder was commonly called a "horse-hoe" (or "hoe-plough"), with the pulverizing implement usually referred to as a "cultivator." In America, the word "cultivator" was commonly applied to any inter-row weeder, and the pulverizing implement was little used until the 1820s. In that decade, American terminology became somewhat muddled, as a single word was used for two quite different implements. In subsequent years, "cultivator" would remain the preferred word for inter-row weeders, and adjectives would be appended to indicate the pulverizing implement, as in "claw-toothed cultivator" or "field cultivator" or "wheeled cultivator."[2]

British Beginnings

The history of American development of horse-drawn implements for inter-row weeding begins in Britain. The pioneer, as noted in Chapter 4, was Jethro Tull, who might have entitled his pathbreaking work *Seed Drill Husbandry* instead of *Horse-Hoeing Husbandry*. The two implements are intimately linked in Tull's approach to farming, the seed drill creating rows by mechanical means, and the horse-drawn "horse-hoe" subsequently used to weed between those rows.[3] If Tull's seed drill deserves high marks for inventive imagination, his "horse-hoe" or "hoe-plow" would seem to warrant little better than a minimum passing grade. As Figure 6.1 makes clear, Tull's horse-hoe was merely a modified common plow with a complex hitching device, as Tull himself conceded.[4] The weed-removal part of the implement was simply an iron share with an adjustable wing attached to one side. (The complex hitching device was designed to enable the trailing position of the horse-hoe to be varied to the left or right relative to the forward pull of the horse.)

This pioneering effort to devise a horse-drawn inter-row weeder had little impact upon established practice or upon designs developed by later inventors. Tull's ambition from the start was to keep "the irons among the roots."[5] But would this particular use of iron prove more efficient than the hand-hoeing alternative? The evidence suggests an answer in the negative. Roughly two decades after Tull's ideas were originally published, the Frenchman Duhamel, although a fan of Tull's, advised his readers to avoid Tull's "hoe-plough" and do their weeding by hand.[6] In 1766 the Englishman Hewitt, himself an inventor of farm implements, surveyed the state of horse-drawn hoe-plows with a jaundiced eye. "For the mechanism

Figure 6.1. Common plow versus hoe-plow, 1733

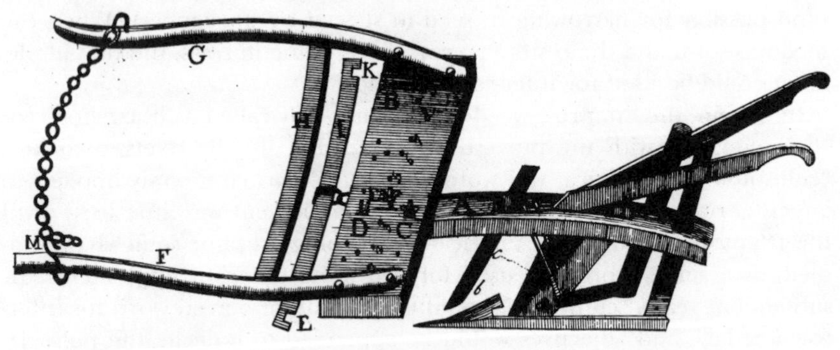

A. Jethro Tull's "hoe-plow"

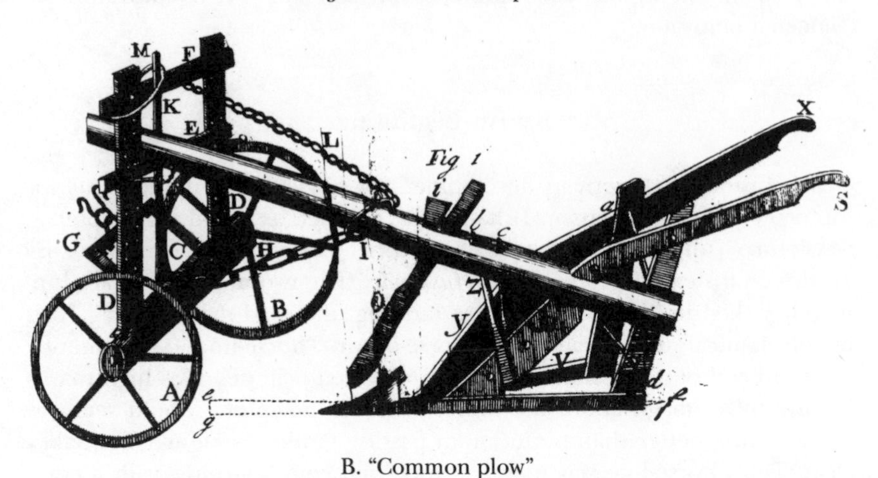

B. "Common plow"

Source: Jethro Tull, *The Horse-Hoeing Husbandry* (Dublin: A. Rhames, 1733), Plates 1 and 6.

of those devised hitherto, by ingenious lovers of agriculture, is of so perplexed, and complicated a nature, that it will no ways answer the common purposes of husbandry: but, being perpetually out of order, will throw the poor ploughman into despondency; . . . as neither he, nor the country plough-wright, can comprehend how to rectify any defects, or accidents, except with extreme difficulty."[7]

If the problem was complexity, the solution was simplicity, or so Hewitt believed, and tried to incorporate that principle into the horse-hoe that he designed. (See Figure 6.2.) The wheel at the front would become a standard feature of later inter-row weeding innovations (although the single broad, flat cutting blade would not).[8] The modest pace of horse-hoe development for the remainder of the eighteenth century is suggested by

Figure 6.2. British horse-hoe, 1766

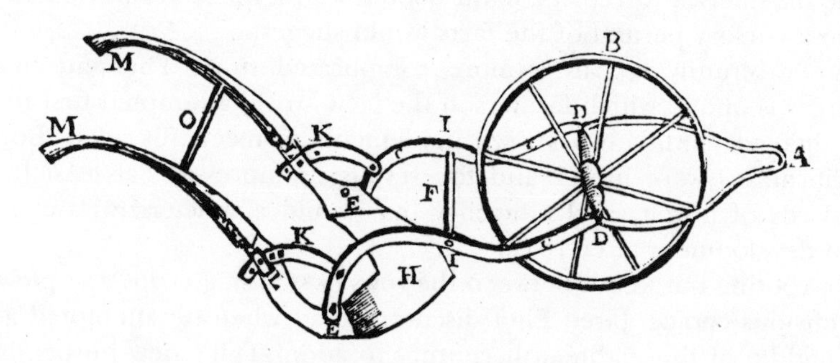

Source: Museum Rusticum et Commerciale, 6 (1766), p. 402.

the 1793 design reproduced in Figure 6.3. This simple man-powered (rather than horse-powered) implement received a prize of twenty guineas from the London Society for the Encouragement of Arts, Manufactures, and Commerce. At the turn of the century, according to one farming expert of the day, it could rightfully claim "the attention of all agriculturists."[9] This was not a pattern of implement development well calculated to revolutionize farming practice on the other side of the Atlantic, as American pioneers in agrarian technology were quick to realize.

American Beginnings: Experiments to 1815

If necessity serves as a prod for inventive activity, the weeding assignments in the American colonies should have prompted the search for mechanical substitutes for hand hoeing. Labor was scarce, wages were high, and the hoeing of row crops, particularly corn and tobacco, was a major drain on the labor resources of farms and plantations throughout British North America. And yet from the publication of Tull's work in the

Figure 6.3. British improved hand-hoe, 1793

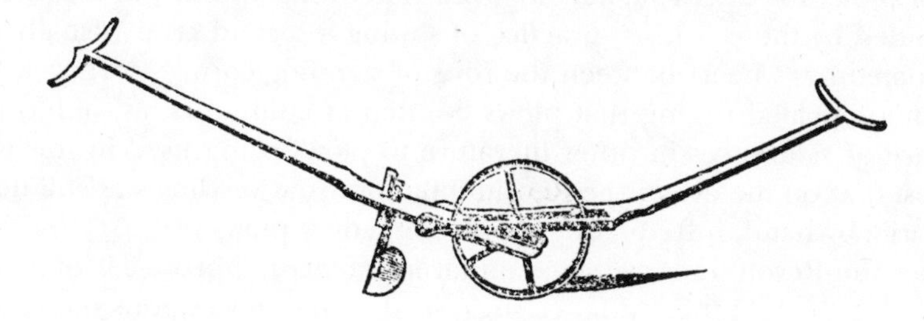

Source: A. F. M. Willich, *The Domestic Encyclopaedia* (Philadelphia: William Young Birch and Abraham Small, 1803), p. 294.

1730s to the outbreak of war in 1812, no effort of any consequence was made in America to replace hand hoeing with a horse-drawn implement. Or so a cursory perusal of the facts would suggest.

Closer scrutiny reveals a more complicated story. The pattern that emerges is one in which farmers in the New World attempted first to use, and then to modify, two known implements to meet this need. But the modifications were minor and the results unimpressive, at least by the standards of novelty and efficiency that would characterize farm implement development after 1815.

For rooting out weeds between the rows of standing crops, the plow was the obvious choice. Jared Eliot discovered this when he attempted about the middle of the eighteenth century to adopt Tull's new husbandry to New World conditions. As Eliot tells the story, when he showed Tull's pictures of a "Hoe-Plough" to the local "Plough-Wrights," they could not understand the diagrams and could not reproduce the implement.[10] Momentarily discouraged, Eliot soon realized that any inter-row weeding that a "horse-hoe" could do could also be done "well enough" by "a common Plough." (This insight seems hard to miss, given the role played by "a common plough" in Tull's original design.[11]) The desirable position of the moldboard depended upon the task to be accomplished. If hilling of the standing crop was required, the plow could be run with the moldboard next to the rows to turn the earth toward the plants. If hilling was not desirable, the plow could carve the same furrow, now running in the opposite direction to turn the soil away from the plants. As the plants grew and their roots spread, the plow necessarily had to pass further from the standing crop, leaving more to be done during follow-up weeding by hand.

Tentative evidence suggests that this use of the plow was slow to develop during the colonial period. The Virginian Landon Carter, for example, experimented with the plow as a means for cultivating corn and peas, but only in the early 1770s, and the results left Carter unconvinced that a plow pulled by oxen was more efficient than hoeing by hand.[12] The author of *American Husbandry*, published in 1775, reported the occasional use of plows for inter-row weeding, but noted this option was frequently precluded by the colonists' practice of sowing a second crop (usually rye but sometimes wheat) between the rows of standing corn.[13] Moreover, the author's repeated urging that plows be used as cultivators, as well as the absence of references in other literature to plows being used in this way, suggest that on the eve of the Revolution, inter-row weeding was still done primarily by hand, infrequently supplemented by plows.

After the Revolution, evidence on farm practices improves, if only to a minor degree, and that evidence suggests that in many regions the exception quickly became the rule. Samuel Deane's survey of New England farming practices, first published in 1790, indicates that at least for corn, inter-row weeding with moldboard plows had become relatively common.[14] Nine years later, John Beale Bordley in his *Essays and Notes on Husbandry and Rural Affairs* made similar observations about agrarian practices, particularly those common in Maryland and Pennsylvania.[15] Scattered references in other literature from the immediate post-Revolutionary period support the impression of increasing use of the plow by American farmers for the inter-row weeding of crops, particularly corn.[16]

As the plow became more popular in a use for which it was not originally designed, the question naturally arose whether modifications in its shape might enable it to perform this new function more efficiently. What is striking is how limited the resulting modifications were.

One innovation did nothing more than retain the common plow design while doubling its working surface. If weeding between two rows required two passes with a moldboard plow—once down and once back—a plow with two shares and a pair of moldboards facing opposite ways might accomplish the same task in a single pass. This "double-mouldboard plow," as it was called, was evidently well known by the time Samuel Deane published his two surveys of New England agriculture in 1790 and 1797. Indeed, in Deane's 1797 edition he defines a "cultivator" as "a plough with a double share and two mouldboards," although, in his opinion, the common plow passed back and forth "may answer just as well."[17] Although Deane gives no hint that this cultivating implement could be expanded and contracted to adjust to variations in the width between rows, this variability was a feature of some British versions of the same implement.[18] (See Figure 6.4.)

The second modification in plow design proved to be similarly undramatic (and arguably even less imaginative). For many weeding assignments, the moldboard's action of lifting and inverting soil needed to be replaced by a weed-ripping action minus the inversion. Farmers could have chosen to drop the moldboard and keep the share, but usually they elected to replace both (moldboard plus share) with a single cutting blade that ripped out weeds with a minimum of soil displacement. One of the more obvious blades to use could be found on virtually every American farm; namely, the blade of a shovel. The resulting implement, not surprisingly, was called a "shovel plow." Prior to the Revolution, adoption of this implement (plow frame plus shovel blade) had begun and by the turn of the century it had become relatively common, particularly in the South.[19] One indication of how common it had become by 1808 was the manner in which it was depicted in that year (see Figure 6.5). The originator of this diagram wanted to discuss the merits of a "ripper," a device with a plow frame and a three-inch blade for cultivating Indian corn "when quite young." Almost as an afterthought—and with no explanation whatsoever in the text—he added a separate picture of the head of a "shovel plow" (H in Figure 6.5), which could be substituted for the "rip-

Figure 6.4. British double-moldboard plow, 1803

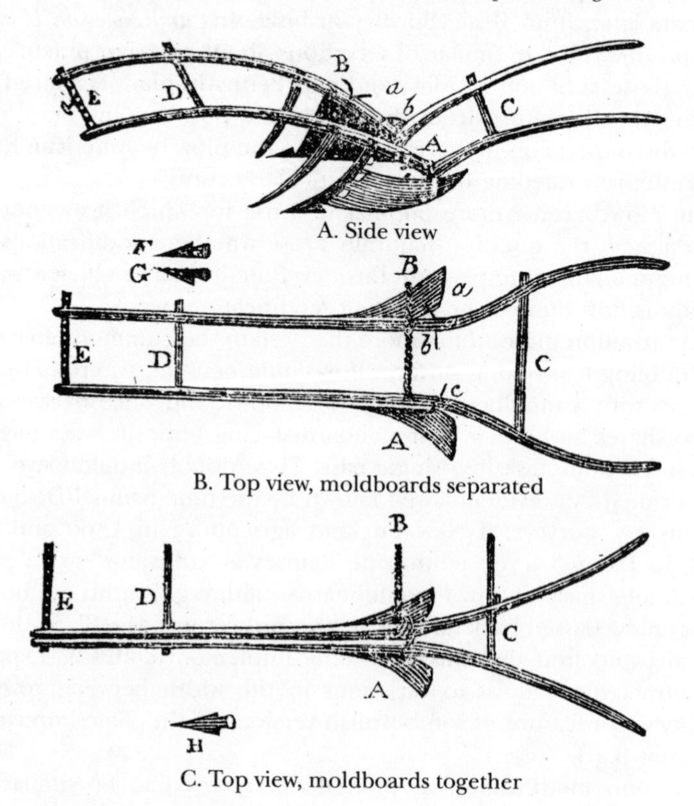

A. Side view

B. Top view, moldboards separated

C. Top view, moldboards together

Source: A. F. M. Willich, *The Domestic Encyclopaedia* (Philadelphia: William Young Birch and Abraham Small, 1803), p. 207.

per" blade on the same frame. The author evidently assumed his readers were so familiar with the resulting implement that no explanation was needed. The "ripper," to judge by its disappearance from the agrarian literature, was, by and large, a failure from the start. The shovel plow, however, remained an important implement on American farms and plantations for decades, particularly in the South. Figure 6.6 illustrates its appearance when fully assembled. Its blade, as shown here, usually sloped slightly to the rear. The effect on soil resembled the action of the ard of ancient civilizations described in Chapter 3. The shovel blade was symmetric (or roughly so) in appearance and performance: in moving through well-tilled soil, it tended to displace a similar amount to either side. Cheap to produce, simple to use,[20] and easy to maintain, the shovel plow had multiple uses, of which inter-row weeding was only one. But for that use, as previously indicated, it remained popular for decades to come.

Figure 6.5. American ripper, with shovel plow attachment, 1808

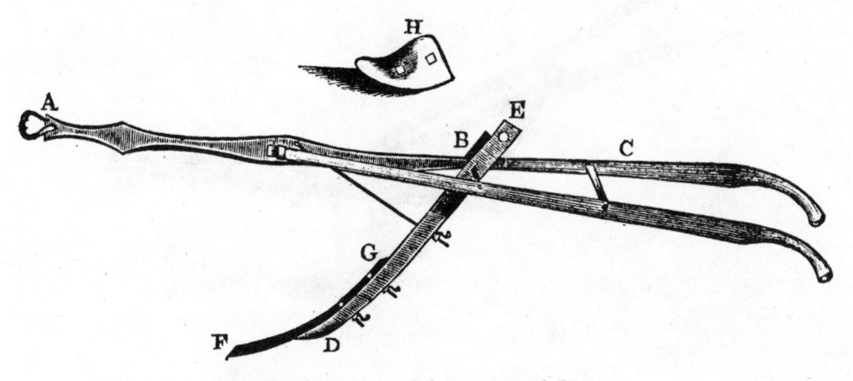

Explanation of the annexed Cut.

A B C, 5 feet 1 inch in length.
E to D, iron stud, 3 feet 4 inches long.
F to G, the ripper, 1 foot long, the iron 3 inches broad, screwed on to the stud.
H, the shovel plough.

Source: Caleb Kirk, "Substitute for Trench Ploughing, and new Mode of putting in Winter Grain, and on live Fences," Philadelphia Society for Promoting Agriculture, *Memoirs,* 1 (1808), pp. 88–89.

A third modification was to replace the shovel blade with a hoe-like blade designed to slash roots, and modify the frame as illustrated in Figure 6.7A. The resulting implement resembled a simplified version of English "horse-hoes" of the late eighteenth century, as suggested by Figures 6.7B and C. When the American version was first used remains obscure, although variations on this basic design clearly had achieved some popularity in the South by the early nineteenth century.[21]

Figure 6.6. American shovel plow, circa 1820

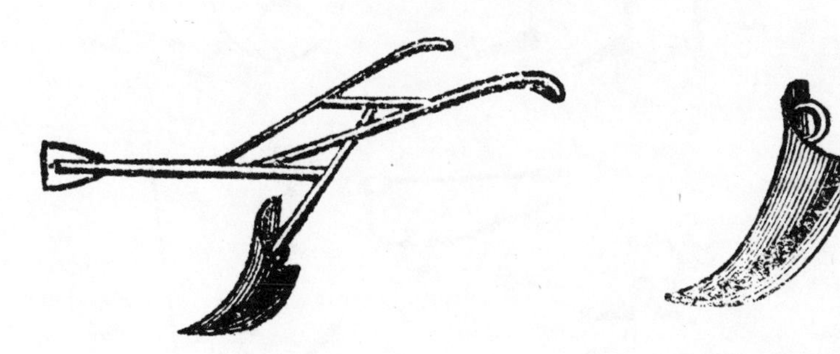

A. Side view of implement

B. Side view, blade only

Source: American Farmer, 2 (November 10, 1820), p. 262.

Figure 6.7. Early American and British devices for inter-row weeding

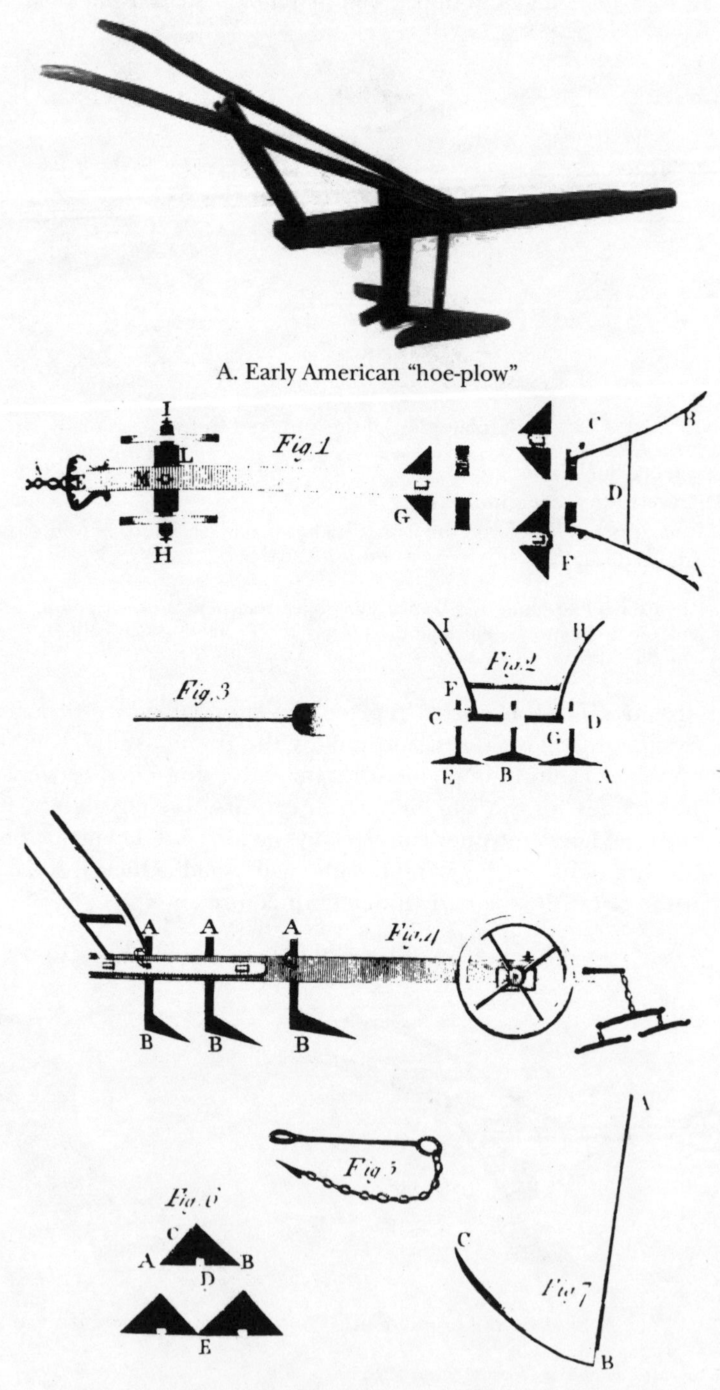

A. Early American "hoe-plow"

B. Robert Hinde's "New-invented Three-shared Plough," 1777

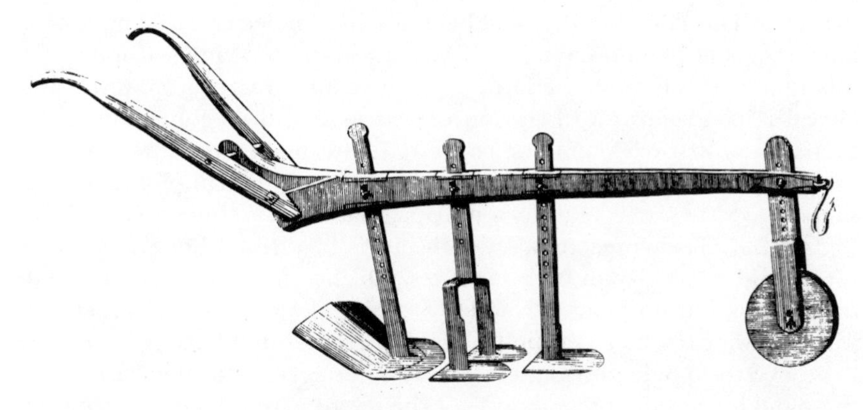

C. James Sharp's horse-hoe, 1777

Sources: North Carolina Division of Archives and History, Negative N72–8-258; used by permission (this museum artifact is recorded as "1840 handmade plow from N.C. Fairgrounds"); *The Farmer's Magazine and Useful Companion* (London), 2 (1777), plates opposite pp. 148 and 224.

If the plow (slightly modified) was one conspicuous candidate for American farmers in search of mechanical means for inter-row weeding, the other obvious candidate was the harrow. Both were common implements on most farms. Harrows offered the advantage of teeth to perform this function, but suffered two drawbacks. One was size. The common harrow was often too wide to fit between rows. This could easily be remedied (and was) by designing smaller harrows of a triangular shape with the required dimensions. The other difficulty was the teeth. These were designed primarily to stir the soil only lightly, in order to cover newly sown seeds. As such, they were not well suited for a subterranean ripping at the roots of weeds. One possible solution was to redesign the teeth. But as discussed in Chapter 5, this possibility remained largely unexplored prior to 1815. The net result was that from the Revolution to the War of 1812, the horse-drawn implement of choice for inter-row weeding (when an implement was used) was overwhelmingly some variant of the plow, with the harrow a distant second, infrequently used and seldom recommended.[22]

Almost as late as 1810, therefore, Americans had no novel farm implement deserving to be called a "cultivator" (although Samuel Deane had used that word in the 1790s to designate a double-moldboard plow). What they had for inter-row weeding were plows modified, at most, in minor ways and harrows modified in trivial ways. Many farmers continued to spurn both options in favor of weeding by hand with hoes. But as the decade ended and war with Britain loomed, a handful of new inter-row weeding implements were proposed that foreshadowed a burst of inventive activity to come. Cultivating inventions were not completely unknown in previous decades, but they had been few in number, received no lasting interest in the literature, and achieved no perceptible impact upon prevailing practices beyond the farm of the inventor.[23] In 1811, however, as a logical appendage to a discussion of a new seed drill, John Lorain described how that drill could be converted to a "hoe harrow" by replacing the shares with harrow teeth. (Lorain even proposed teeth of a triangular shape.)[24] In the same year, in a nine-page article on "Indian Corn," *The Agricultural Museum* recommended the use of "a trowel hoe plough and two mould boards," that bore some resemblance to the old double-moldboard plow, but now had a trowel-like cutting device and a "skimmer . . . with two wings."[25] In 1812 the Columbian Agricultural Society awarded a prize of $10 to Solomon Cassedy of Alexandria, D.C., for the "best weeding plough," but the press reports of the award gave no details about the implement's design.[26] These scattered references, few in number and with scanty details, suggest that at least a few Americans were beginning to ask of existing inter-row weeding practices: Is there a better way? In the decade following the conclusion of the War of 1812, this question would come to dominate the agricultural literature, and would generate a host of new farming implements, of which new cultivators were but a case in point. Their combined effect would be to change American farming practices in ways that were at once decisive and enduring.

Enter the Cultivator: 1815 and Beyond

In the decade following 1815, the development of inter-row weeders in America took four different directions. Two implements already in use for this purpose, the plow and the harrow, were modified in major ways. A third option tried decades earlier by John Beale Bordley—a single wide blade that Bordley called a "shim"—resurfaced and was improved. The fourth direction produced a new implement: a "cultivator" with a frame and teeth specifically designed for inter-row weeding. In every case the dominant priority in design was simplicity. The end result in most cases was a farm implement cheap to produce, easy to operate, and simple to repair.

Harrow improvement after 1815, as described in Chapter 5, involved primarily the redesigning of harrow frames and teeth (including interchangeable teeth) to make the implement more suitable for inter-row weeding.[27]

Plows, long used in several forms for cultivating purposes, were now further modified to make them more efficient for that task. The old designs continued to be employed, particularly the shovel plow and, to a lesser extent, the double-moldboard plow.[28] But now the design of the shovel plow was modified in two major ways, both of them in retrospect somewhat predictable. One was simply to double the working surface by adding a second blade, the resulting implement called, not surprisingly, a

Figure 6.8. American double shovel plow, circa 1820

Source: American Farmer, 2 (September 1, 1820), p. 183.

"double shovel plow." (See Figure 6.8.) The initial patents for this implement date from 1817. By 1820 it was no longer a curiosity, and could be purchased from farm implement dealers for roughly twice the price of a single shovel plow ($8 as opposed to $4.50).[29] The other design modification consisted of retaining a single blade, but radically narrowing the cutting surface to achieve an implement more suitable for cutting deeper, "running 'round corn at its first tending," or eradicating weeds among rows of a standing crop whose root development had progressed to the point where the use of a wider implement (such as a regular shovel plow) threatened root destruction. This implement appears to have had, at first, no name of its own. Usually it was referred to by the width of its blade—often six or seven inches, which was roughly half the width of a common shovel plow. But before long, no doubt because of the blade's resemblance to part of an animal common to most farms, this narrow blade version of the shovel plow became known as a "bull tongue plow." The one depicted in Figure 6.9 has a coulter in front of the tongue (labelled

Figure 6.9. Bull tongue plow

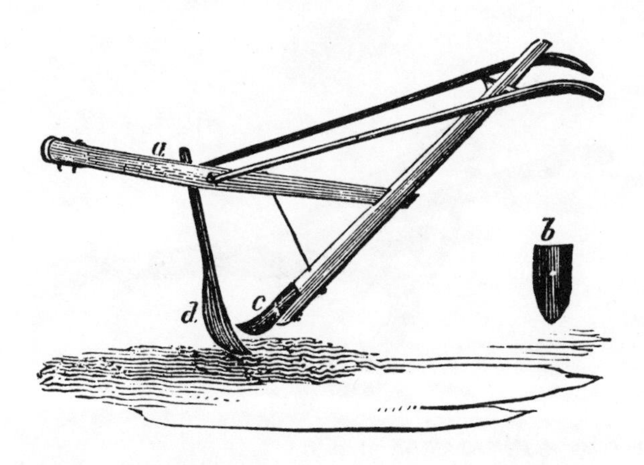

Source: The Agriculturist, 3 (October 1844), p. 305.

d) to facilitate a "jumping" action when it was used to break up stiff land full of obstacles.[30]

By 1822 the president of the South Carolina Farmers' Society was convinced that for most shovel plow tasks, a narrower blade gave superior results because the latter appears "to answer all the purposes of a large one, and with less exertion to the animal to effect the desired end."[31] Others were less sure. Thus in subsequent decades both implements remained popular, as did the twin-bladed version of the conventional shovel plow.

Not all experiments with blade design were similarly successful. Two that appeared in the early 1820s—of little note and not long remembered—are depicted in Figure 6.10. The shovel shape of the blade was replaced by a circular blade with a sharp edge, which permitted rotation of the surface as the forward cutting edge became dulled from use. The result was either called an "adjusting shovel plow" (top diagram) or, with additional reinforcement of the frame behind the blade, a "substratum plow" (bottom diagram). According to their manufacturer, Gideon Davis of Georgetown, D.C., the modified shovel plow, because of its concave

Figure 6.10. American modified shovel plows, 1823

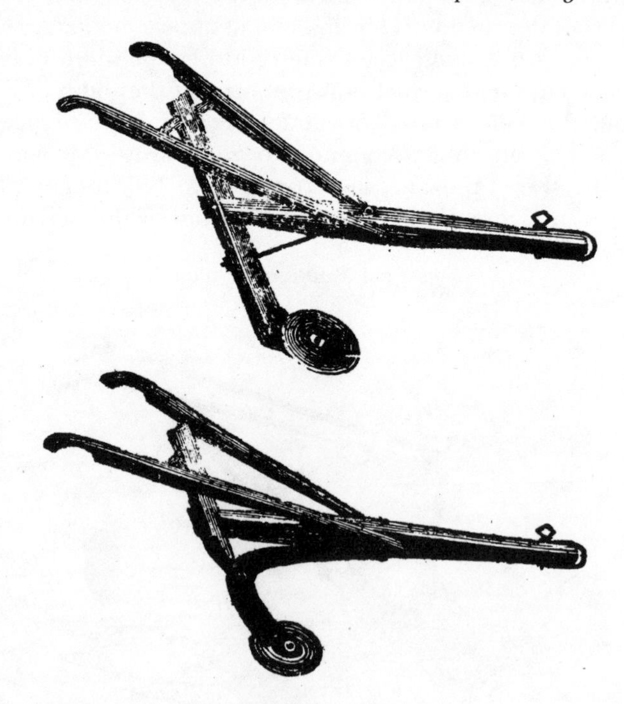

Source: American Farmer, 5 (March 28, 1823), p. 1.

circular blade, could "work deeper, pulverize the ground finer, and leave it much evener on the surface, than the common shovel plough does."[32] Evidently few believed the advertisements or used the implements. Neither version shows any evidence in the literature of receiving widespread praise or achieving widespread use.

A third type of inter-row weeder developed after 1815 was merely a design extension of an old idea. In the eighteenth century, one possible cutting surface explored was a single wide blade, as illustrated in Figures 6.2 and 6.3. The same idea, as previously noted, was the central feature of Bordley's "shim" developed in the 1780s. In the general outpouring of new design ideas for farm implements, those preoccupied with novel possibilities for inter-row weeding revived this idea, primarily in two shapes. One mimicked the straight horizontal blade of earlier designs, producing an American implement not unlike the European wheel hoe depicted in Figure 6.11. (No pictures of the comparable American implement could be found.) The popular American name for this device was "sweep-all," and it was described as "nothing more than a broad thin bar of iron, 3 or 4 inches wide, and about 30 inches or 3 feet long; connected with a wooden frame at each end, by two iron uprights in such a manner, that the front edge of the bar which is made sharp, and well steeled, has nearly the same inclination to the surface of the earth, that a drawing-knife in operation, has to the board on which it acts. This edge then cuts every thing before it, running from 2 to 3 inches below the surface."[33]

If a single wide blade was to do the weeding, one possible shape was a broad, narrow rectangle. Another was a triangle with the point mounted forward. One variant of this latter possibility is depicted in Figure 6.12.

Figure 6.11. European wheel hoe, circa 1823

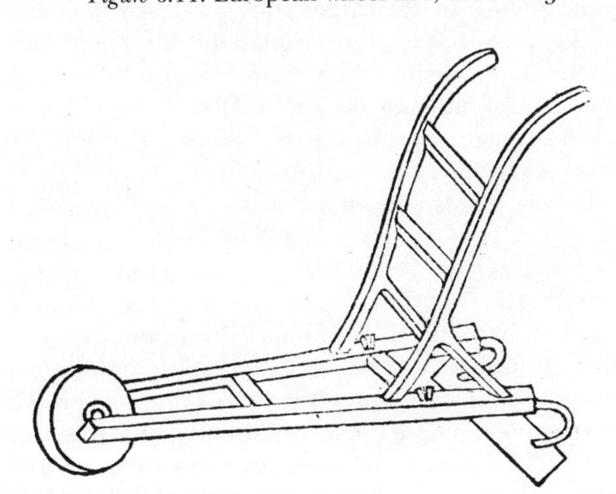

Source: New England Farmer, 2 (December 13, 1823), p. 154.

Figure 6.12. American skimmer, circa 1820

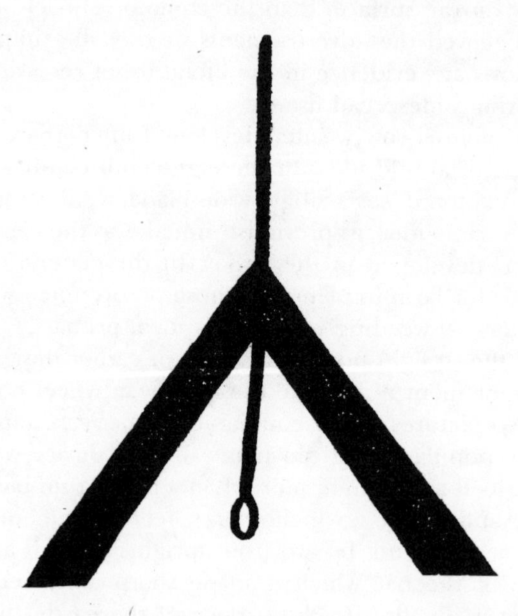

Source: American Farmer, 2 (September 15, 1820), p. 195.

This particular "skimmer," recommended for weeding corn or cotton, consisted of two wings, twenty inches long by five inches wide, forming a triangle twenty-six inches across at the base. According to John Taylor, a planter of some stature in the 1820s, this particular skimmer was "far preferable to the harrow as a weeding plough" and could be "drawn by one horse or one ox."[34] Again, others were less sure. John Beale Bordley spelled out one crucial difficulty decades earlier. A single blade might be "an excellent instrument against young weeds," but was "insufficient where grass and weeds have obtained strength."[35]

Such places where weeds were dense or soil stiff demanded a different blade design. After 1815 a novel weeding implement appeared that quickly earned the name of "cultivator" and revolutionized inter-row weeding on many American farms. The reasons for its blade design are easily grasped when its features are compared with those of modern cultivators. If the force applied to weeding is limited to human muscles, the cultivator can feature multiple triangular blades that are relatively thin, as depicted in the hand-hoeing device of Figure 6.13 (top diagram). But for inter-row weeding by draft animals (or modern tractors), where blades are to be drawn through the soil at high speeds with considerable force, their thickness must be increased, as illustrated by the "high-speed sweeps" of

Figure 6.13. Twentieth-century cultivators

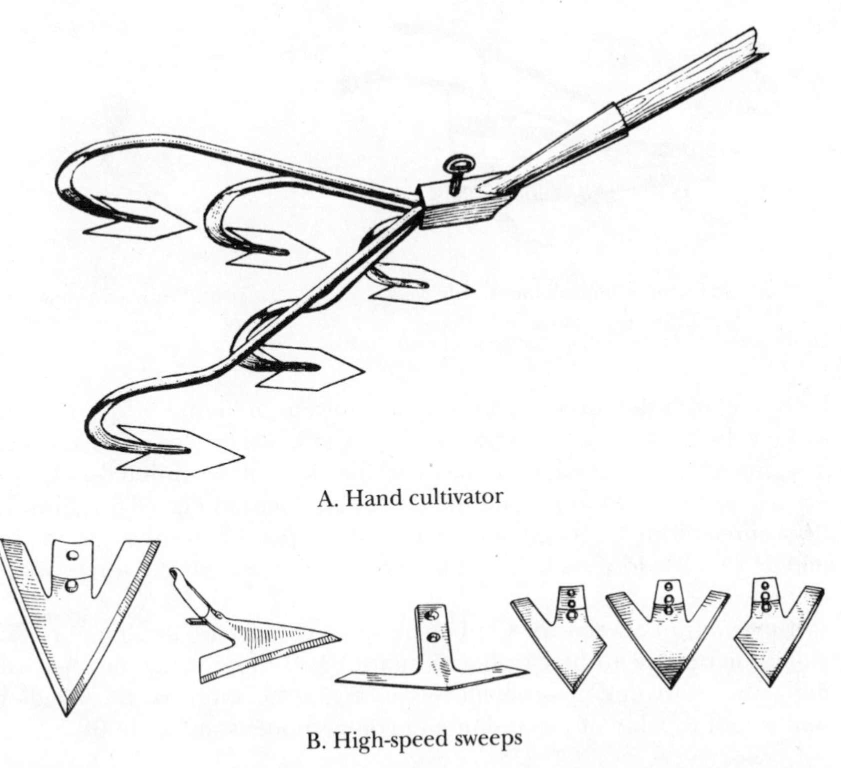

A. Hand cultivator

B. High-speed sweeps

Sources: H. J. Hopfen, *Farm Implements for Arid and Tropical Regions* (Rome: Food and Agriculture Organization of the United Nations, 1960), p. 64; H. P. Smith, *Farm Machinery and Equipment* (New York: McGraw-Hill, 1955), p. 231, reproduced with the permission of The McGraw-Hill Companies.

Figure 6.13.[36] A further refinement, evident in the sweeps depicted, is to trim back the cutting surface of that blade destined to pass closest to the standing crop (while preserving the triangular shape on all the others). Progress toward this ideal proceeded in several steps during the 1820s, all of them achieved with notable speed.

Four design problems had to be solved.

First came the shape of the frame. A triangular design had long been preferred for harrows used in inter-row cultivation. The dominance of triangular shapes among cultivators of the sort depicted in Figure 6.14 did therefore not require a sharp departure from previous traditions.

The second problem was where to place the teeth. Here, too, developments in harrow design suggested several possibilities. Most obvious of all, perhaps, was simply to place the teeth along the two wings of the triangu-

Figure 6.14. American cultivator, circa 1820

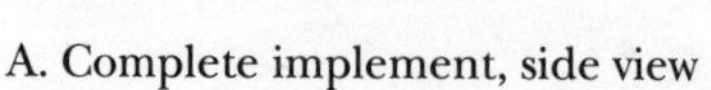

A. Complete implement, side view

B. Tooth only, side view

Source: American Farmer, 2 (November 10, 1820), p. 262.

lar frame, with the foremost tooth at the apex as in Figure 6.14A. But the further the teeth could be separated, front to back, the lower the odds of clogging when the implement was working the soil, as outlined in Chapter 5. This may help to explain the teeth placement in Figure 6.15, where the center tooth has been shifted from the apex of the triangle to the middle of a bar located behind the lead pair of teeth on the wings of the cultivator.

If the width between rows to be cultivated could vary, the frame of the cultivator needed to be capable of adjusting to those variations. Accordingly, the third design problem was to make the width of the weeding implement capable of expansion and contraction, as in Figure 6.16.

Figure 6.15. American cultivator, 1821

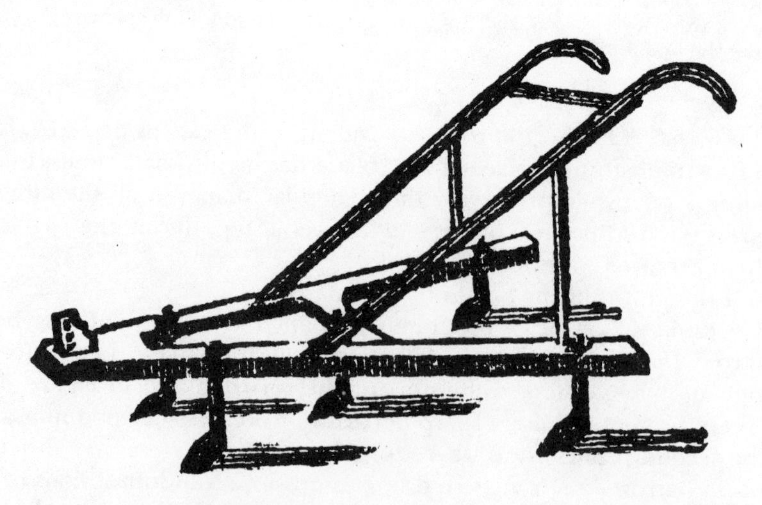

Source: American Farmer, 3 (December 21, 1821), p. 312.

Figure 6.16. Seaver's expanding cultivator, 1834

Source: New England Farmer, 12 (June 25, 1834), p. 401.

The final problem—possibly the most important and complex to solve—concerned the optimal shape of the teeth. The difficulty was that optimality depended upon several variables, including whether the soil was wet or dry, stiff or friable; whether the crop required hilling in its early stages or not, and what kinds of root systems developed in later stages of growth; and whether the weeds were to be "cut over" (slashed) or rooted out. The solution was to design teeth of various shapes for various tasks and make them readily interchangeable on the cultivator frame. (The requirement for all shapes was a capacity to rip through weeds.) One of the earliest tooth designs, depicted in Figure 6.14B, resembled the blade of a shovel plow in miniature. Or as the inventor put it, the teeth were "curved . . . and brought to a point, like a pointed spade or shovel."[37] Later models offered greater variability, as illustrated in Figure 6.17. For regular inter-row weeding, a triangular shape (*C*) was used. But "should circumstances require," these could be replaced with "coulters" (*G*), "which will open and pulverize the soil as well as rake out the weeds." If the cultivation process required the taking of "earth from the plants, as in the case with the ruta baga when quite small," teeth in the shape of a "half share" (*D*) could be used. "In hard clay," however, often the only teeth that "could be made to enter the soil" was a combination of a triangular hoe (*C*) at the front and four coulters (*G*) on the wings.[38] By the 1830s, manufactures produced various tooth designs, which they promoted in language that sometimes stressed the self-evident: "it is most convenient to have sets of different kinds, and the cost is trifling, that they may be shifted at pleasure."[39]

Figure 6.17. Bement's cultivator, 1837

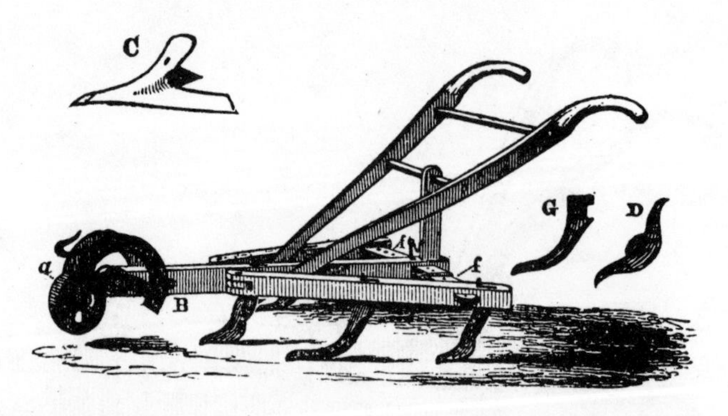

Source: The Cultivator, 3 (January 1837), p. 180.

One final design refinement, evident in Figure 6.17, became a common feature of later cultivators. The wheel at the front served both to steady the implement and control the depth at which the cultivator's teeth worked.[40]

The word "cultivator"—once used in America for a variety of implements including a double-moldboard plow—after 1815 came to be applied to the triangular weeding devices described above. The terminology, however, soon became confused again, because a second kind of implement, a soil pulverizer, also became known as a cultivator.

Some of these were imported from Great Britain. Among the most popular was Beatson's "scarifier," depicted in Figure 6.18A. Although this name suggests a clear distinction between pulverizers and inter-row weeders, in fact Americans typically used "scarifier" as a synonym for "cultivator." The details of Beatson's model are not particularly clear in his sketch, but the general nature of the design can be inferred from the British cultivator depicted in Figure 6.18B. Its popularity in this country derived partly from its simplicity, for it was little more than a conventional triangular frame with sharp teeth angled forward. The American price (roughly $20), while significantly above the cost of single or double shovel plows ($4.50 and $8, respectively), was far from prohibitive.[41]

Not all pulverizers were imported. Americans developed and marketed several of them, in part through experimentation with inter-row cultivators. Once the shapes of teeth became a subject of inquiry, the question naturally arose whether a triangular frame might be combined with teeth for other purposes, such as pulverizing soil (particularly when implements were being imported from Britain expressly for that purpose). One of the earliest shapes proposed for breaking up clods and stiff soil was a tooth "about the shape and size of a colt's foot, the blade inclined like the coulter of a plow, sharpened and steeled and fastened over the top of the

Figure 6.18. British soil pulverizers

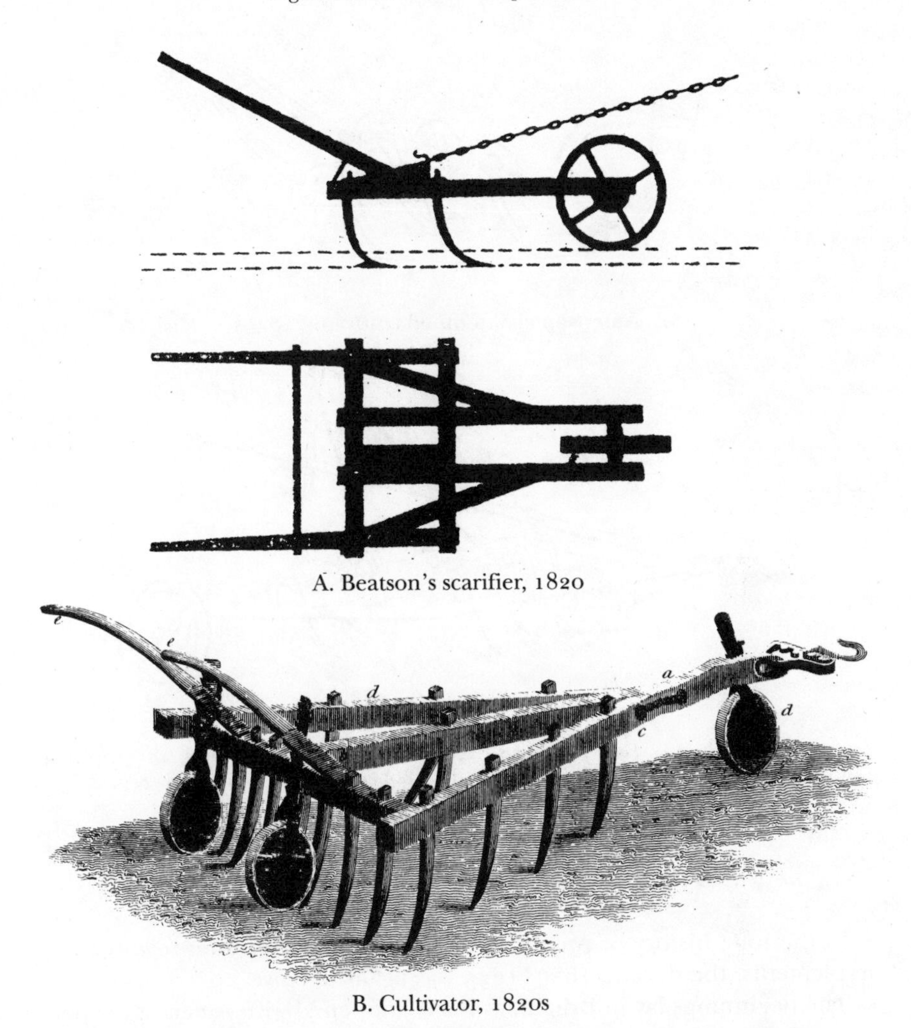

A. Beatson's scarifier, 1820

B. Cultivator, 1820s

Sources: Alexander Beatson, *A New System of Cultivation* (London: W. Bulmer and W. Nicol, 1820), Plate I; Abraham Rees, *The Cyclopedia, or Universal Dictionary of Arts, Sciences, and Literature* (Philadelphia: Samuel F. Bradford and Muray Fairman, 1810–1824), Plate X.

frame by means of a screw cut on the top of each tooth."[42] Between the 1820s and the 1850s, on both sides of the Atlantic a variety of pulverizers were developed to supplement or complement the work of plows and harrows in the preparation of a field for planting. In general, American models tended to be simpler than those of the British, although by the 1850s many of the implements used in both countries were strikingly similar in appearance. (Compare the British "grubber" with the American "claw-toothed cultivator" in Figure 6.19.)

Figure 6.19. Soil pulverizers

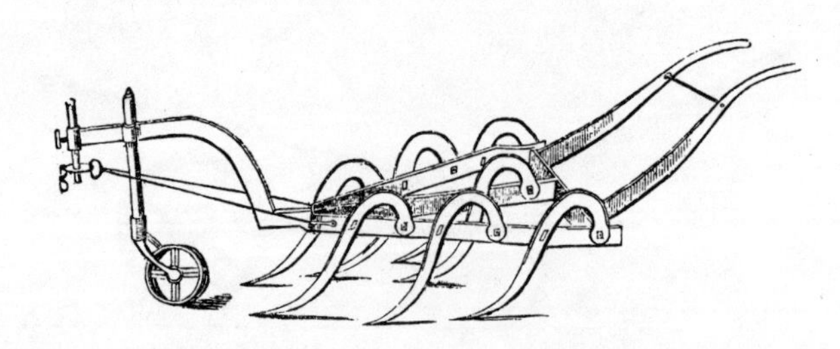

A. American claw-toothed cultivator, 1850s

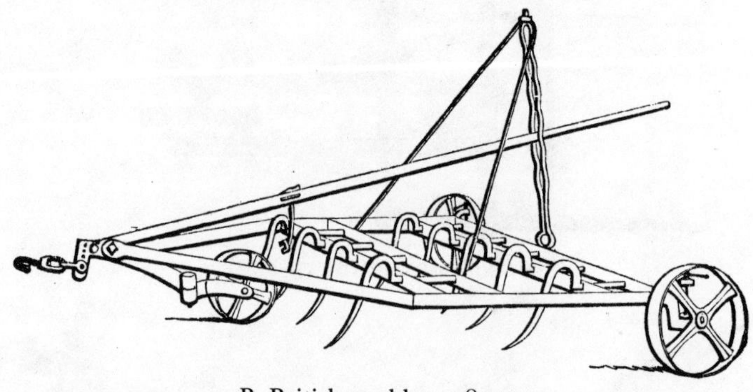

B. British grubber, 1850s

Sources: John J. Thomas, *Farm Implements* (New York: Harper and Brothers, 1855), p. 144; Cuthbert W. Johnson, *British Husbandry* (London: Robert Baldwin, 1848), Vol. 2, p. 27.

The Long View: Colonial Times to 1840

In the long history of American attempts to develop inter-row weeding implements, the decade 1815–1825 was again decisive.

The beginnings lay in Britain, as often happened with American experimental efforts. To Jethro Tull belongs the credit for popularizing the idea of using a horse-drawn implement to weed between rows. But the implement Tull designed for this task—his "horse-hoe"—was little more than a common plow with a complex hitching device. Americans responded by jettisoning complexity while exploring the option of using a plow instead of the hand hoe for inter-row weeding. The other possible option readily available was the harrow. To make these implements more suitable for weeding tasks, the plow was modified in minor ways and the harrow in trivial ways. The working surface of a common plow was doubled in the double-moldboard plow, and the resulting implement used where weed removal was usefully complemented by the inverting action of a moldboard. Where weeding proceeded better without inversion, the frame of the plow was altered to accommodate a different working surface: a blade of shovel shape that left symmetric results (or roughly so) as it churned through soil between the rows. The design of harrows was little modified for weeding purposes beyond reducing the size of the traditional triangular frame to enable the implement to pass between the rows.

The use of both plows and harrows for inter-row weeding apparently became progressively more common in America as the eighteenth century drew to a close, with the plow the preferred choice by a substantial margin. By 1810 both the double-moldboard plow and the shovel plow were widely known and often used.

Between the Revolution and 1810 examples can be found of other innovations designed to weed between rows with the aid of horse power—Bordley's "shim" of the post-Revolutionary era, or Kirk's "ripper" of 1808—but these were few in number and seem to have had no impact of any consequence upon established practice. As the second decade of the nineteenth century began, however, fragmentary evidence suggests that experiments with implement design began to rise: the awarding of a prize by an agricultural society for the "best weeding plow"; Lorain's discussion of how to convert a novel seed drill into yet another novelty by substituting weeding teeth for seed drill shares; the mention of a "skimmer" for weeding corn in *The Agricultural Museum*, a periodical whose appearance (if short-lived) indicated a growing interest in agricultural topics. The question raised, and still in doubt when the War of 1812 began, was whether these were merely isolated and ephemeral departures from the status quo, or whether they heralded the beginning of a fundamental shift in attitude toward the design and application of new farm implements in America.

When peace returned, the answer came in a rush. Within a decade, the literature and the landscape confirmed that a revolution in attitude had become accomplished fact. Developments in inter-row weeding design merely serve as one more example of this more comprehensive transformation.

Almost all the innovations described above had been developed and applied by 1825. Harrow frames and teeth had been improved to make the resulting implement more effective for weeding tasks. The shovel plow had been modified into both the bull tongue plow and the double-shovel plow. The array of possible weeding implements also now included "skimmers" and "sweep-alls": horse-drawn weeders featuring a single, broad cutting blade, either triangular or rectangular in shape. Finally, and arguably most important of all, a triangular, multi-toothed implement had been designed exclusively for inter-row weeding. Commonly called a "cultivator," it was available by 1825 with teeth that varied in number and placement on the frame, and with frames that in some instances were expandable.[43] Although a beginning had been made in designing different weeding teeth for different weeding tasks,[44] a reasonably full range of options does not appear to have been developed until the 1830s. In the early 1820s, however, designs (both domestic and imported) did become available for a pulverizing implement with a frame resembling the cultivator but with teeth specifically devised not for weeding but for breaking up soil prior to planting.

After many decades during which little change occurred in implement design—and the few novelties proposed, with a handful of exceptions, were generally ignored—the rate of transformation was astonishing. As late as 1820, the merits of inter-row weeding by horse-drawn implements remained a topic of debate in the agricultural press.[45] By 1825 the debate had virtually disappeared, and in its stead could be found endorsements and discussions of preferable designs. Rising demand for new implements prompted retailers of farm tools to include in their advertisements "cultivators of all sizes and construction" ranging in price from $4 to $12.[46] By 1827 one Baltimore implement manufacturer could boast of having "agents . . . where gentlemen can leave their orders" in seven major East Coast cities.[47]

Small wonder that agrarian expert Jesse Buel, surveying the state of cultivator design in the mid-1820s, found much to praise (although the options available in teeth design struck him as somewhat limited).[48] Improvements were soon to include more teeth designs, as Buel desired,[49] and later refinements would feature riding cultivators, straddle-row cultivators, self-sharpening cultivators, and disk cultivators.[50] But much had been accomplished during the ten years following the War of 1812. Indeed, by the closing years of the 1820s one might characterize American development of inter-row weeding implements with language not unlike that used by Churchill after the victory at El Alamein: it was not the end; it was not even the beginning of the end; but it was—quite unmistakably—the end of the beginning.

[7]

Reaping

The Puzzle

Harvest-home, harvest-home,
We have plowed, we have sowed,
We have reaped, we have mowed,
We have brought home every load,
Hip, hip, hip, harvest-home.

Thus ran the chant in bygone days of English reapers marching in processions to mark the end of the harvest season,[1] their words a mixture of elation and relief, and rightly so. In the premodern world, harvesting small grains was a high stakes game. The available time to move the crop from field to barn was seldom more than ten to fourteen days, and often less depending upon the weather.[2] Delay could result in overripe grain prone to shatter or lodge in wind and rain and sure to shed when cut by any reaping implement, leaving much of the crop on the ground. A downpour during harvesting was a special problem, particularly for wheat and barley, which, if cut and stored wet, tended to sprout. For those with low incomes and limited savings—which is to say, for most farmers of the premodern world—the economic implications of a major loss at harvest time were at best, serious, and at worst, catastrophic.

All these difficulties were evident in the farming communities of both colonial America and the young republic. The need for haste was paramount, and the reaping tools, by modern standards, were primitive in the extreme. A visitor to the United States in 1800 would have found four harvesting implements in use. Sickles and scythes were common in every major farming region, and grain cradles much in evidence in the Southern and Mid-Atlantic states. Last, and least in terms of geographic dispersion, was the Flemish or Hainault scythe, whose use was confined almost exclusively to Dutch and German settlements of the Hudson, Mohawk, and Schoharie Valleys. The puzzle is why such variety? If speed at harvest time was of the essence, and some implements harvested more quickly than others—the cradle faster than the Flemish scythe, or so some claimed, and both of these much faster than "the tardy, piddling sickle,"[3] as was self-evident to all—why should the use of all four implements persist well into the nineteenth century? The answer involves a variety of costs and benefits associated with each harvesting tool, plus the operation of a market mechanism historians have generally ignored.

The Premodern Implements

The oldest of the four by a wide margin is the sickle. Archeological evidence suggests that in the Western World small grains were initially cut down by an implement with a straight blade made of flint that, over time, evolved into a curved blade made of metal. Which implement was first used where, and the pace of design modifications from straight blade to curved are puzzles to which, predictably, there are few established answers and a variety of hypotheses based upon fragmentary findings. Reaping knives with straight blades appear to have been used some 8,000 years ago, but by the time the Egyptians were building pyramids, grain was more commonly harvested with a curved implement.[4] The progress in curvature design is itself a protracted story. One of the suspected evolutionary paths from early crescent shape to "balanced" sickle is portrayed in Figure 7.1.[5] What is striking is not the diversity of designs in ancient implements, but the similarity in the blade shape that came to dominate Western agriculture. All evolutionary paths seemed to converge upon some variant of the shape of the American sickle depicted in Figure 7.2. (Compare, for example, this American implement with implement 7 in Figure 7.1.) Nor is this surprising. Consider the strain on the wrist that would result from a full day in the field reaping grain with any sickle similar in shape to implements 3 through 6 in Figure 7.1. The problem is the "unbalanced" design of the blade. Almost all of its weight is distributed to the left of the handle during a cutting stroke. Sickles of this shape therefore invariably tended to rotate toward the ground, a tendency the reaper would have to combat from dawn to dusk on a harvest day. The solution to this tendency was to bend that part of the blade closest to the handle backward prior to curving it forward in a crescent shape.[6] This shifted the center of gravity of the cutting surface to a point directly in line with the axis of the handle (or approximately so), and thereby re-

Figure 7.1. Sickle development: Northwest Europe

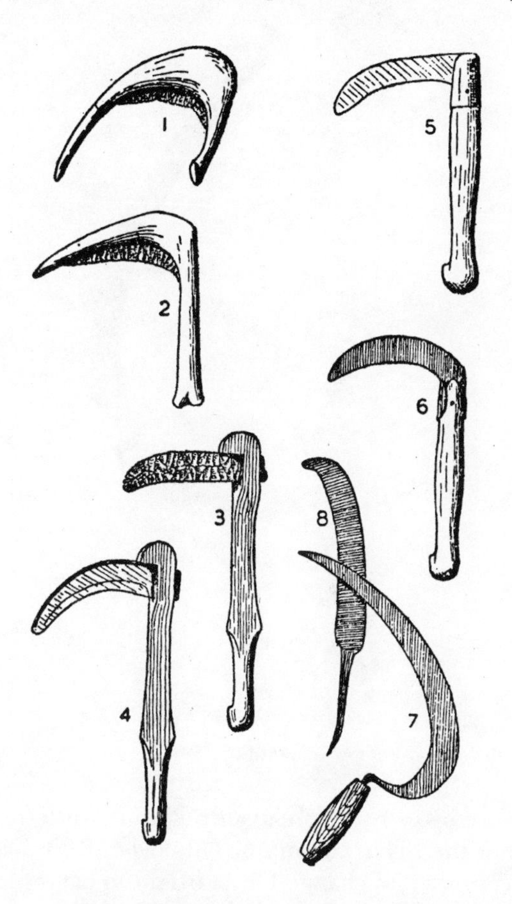

1. Composite crescent of three flints in original wooden handle; Kahun, Egypt, 12th dynasty. 2. Crescentic flint sickle of Scandinavian type; mounted as No.1. 3. Flint sickle of British type; one horn of the Scandinavian crescent has become a butt or tang. 4. Flat bronze sickle, mounted as No.3. 5. Socketed bronze sickle, British type. 6. Iron sickle with wing socket; Iron Age; the Caburn, Sussex. 7. Balanced iron sickle; medieval Danish. 8. Leaf knife; Viking period; Norwegian.

Source: E. C. Curwen and Gudmund Hatt, *Plough and Pasture: The Early History of Farming* (New York: Henry Schuman, 1953), p. 109.

moved any pronounced tendency of the implement to rotate in the hand while the blade was being swung parallel to the ground.

Just when sickles first became balanced—or when the balanced (versus unbalanced) shape first became popular—is far from clear. A few depictions of balanced shapes can be found on Egyptian tombs. During Roman times this form of sickle blade became more common, and thereafter the

Figure 7.2. American balanced sickle, probably nineteenth century

Source: Smithsonian Institution Photo No. 44796.

balanced shape seems to have spread throughout Western Europe.[7] The implication is that the sickle commonly employed on American farms in the eighteenth century had changed little from the harvesting implement used in ancient Rome.

Early American depictions of balanced sickles in use are difficult to find. One of the earliest is reproduced in Figure 7.3. Unfortunately the artist, Alexander Anderson, had little farming experience, which helps to explain why his reapers have the wrong body position and are wielding implements with blades that are remarkable for length and curvature.[8] For harvesting small grains, the common body position for a reaper with a sickle is depicted in Figure 7.4. Crouching low, the worker grasps a handful of grain with the left hand and then slices the stems with a right-hand pulling motion that includes a twist of the wrist. When well performed, this harvesting technique produced a field of stubble that was more or less uniform in height.[9]

No doubt this stressful position long ago raised for many reapers the question whether harvesting might be accomplished standing up. Ultimately it could, with a scythe roughly of the sort depicted in Figure 7.5.

Figure 7.3. American depiction of harvesting with sickles, 1820s (incorrect body position)

Source: Alexander Anderson Scrapbooks, Vol. 1, p. 53, Print Collection, Miriam and Ira D. Wallach Division of Arts, Prints and Photographs, The New York Public Library, Astor, Lenox and Tilden Foundations.

Figure 7.4. British depiction of reaping with sickles, 1850s (correct body position)

Source: Henry Stephens, *The Book of the Farm* (London: William Blackwood, 1855), Vol. 2, p. 332.

Figure 7.5. American straight-handled scythe, eighteenth century

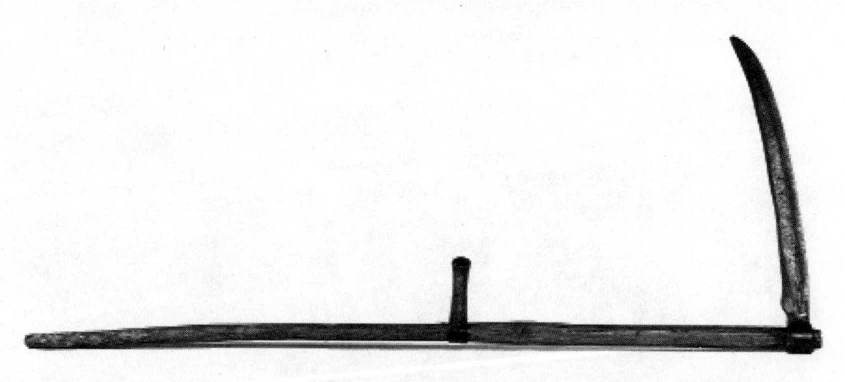

Source: Smithsonian Institution Photo No. 78-19142.

At first glance its advantages relative to the sickle seem overwhelming. The operator remains more upright, the blade is longer, and a single stroke cuts a larger swath closer to the ground, thereby harvesting more stalk and leaving shorter stubble in the field. One drawback for the harvesting of small grain is the lack of order in the cut stalks so important for subsequent binding. Another is the tendency of the scythe's slashing motion to shatter grain, particularly if the crop is dead ripe, of which more in the next section. But this prospective loss disappears if the crop to be cut is hay for livestock instead of small grains. And therein evidently lies the key to the scythe's beginnings, at least as most archaeologists hypothesize about those beginnings.

Toward the close of the Bronze Age and the start of the Iron Age, in Northwest Europe the climate cooled and the use of scythes began. One popular hypothesis links these two developments, arguing that the drop in temperature necessitated the provision of fodder over winter months for livestock that previously had been able to forage throughout the year. To use a sickle to gather hay for winter feeding would have been enormously time consuming. The resulting need for an implement to cut faster and closer to the ground, runs the argument, led to the development of the first scythes.[10] These early implements differed somewhat from the American scythe depicted in Figure 7.5. The blades were shorter, as were the handles, which lacked the grips or "nibs" of the sort seen half-way down the handle of the American version.

About the subsequent evolution in design from the early Iron Age to the late eighteenth century, few experts would contest the summary judgment of Sian E. Rees: "The development of the scythe . . . is by no means understood."[11] Pictorial evidence nevertheless documents several decisive steps in the process. Long-handled scythes without nibs or grips apparently prevailed in the Roman World.[12] When they became common

Figure 7.6. Reapers with straight-handled scythes, with and without nibs, eleventh and twelfth century

Source: James Carson Webster, *The Labors of the Months in Antique and Mediaeval Art* (Evanston: Northwestern University Press, 1938), Plates XX, XXIV, LX.

throughout the rest of Europe—or even Northwest Europe—is far from clear. But by the eleventh century such implements were much in evidence in the agrarian scenes included in depictions of labors of the months.[13] (See Figure 7.6.)

Two modifications in design were of the first importance. A scythe with a straight handle and no nibs forced reapers to keep their left hands well above the right, and the resulting implement could only be used with a short, chopping motion. Even the addition of grips did not necessarily change the hand positions, to judge by the reapers portrayed in Figures 7.6 and 7.7. A less awkward and more effective stroke, designed to cut more and tire the body less, was a broad sweep with arms extended. To produce this, both hands had to work at roughly the same height as they swung through an arc parallel with the ground, a stroke facilitated by either adding long nibs to a straight handle or short nibs to a curved handle. (The relationship between the two options is illustrated in Figure 7.8.)[14] Scythes of this design were in use in America at least by the late eighteenth century (and possibly much earlier). Tailored to the height of the reaper—a taller man required a longer implement—their handles or snaths were light and yet strong (Americans favored white ash), with a stiffness that minimized "trembling" in the course of a stroke.[15] An expert

Figure 7.7. Reapers with straight-handled scythes with nibs and bows, sixteenth century

Source: Francis M. Kelly, *The Seasons of the Year* (London: B. T. Batsford, 1936), Plate V.

Figure 7.8. Scythe: curved versus straight handle

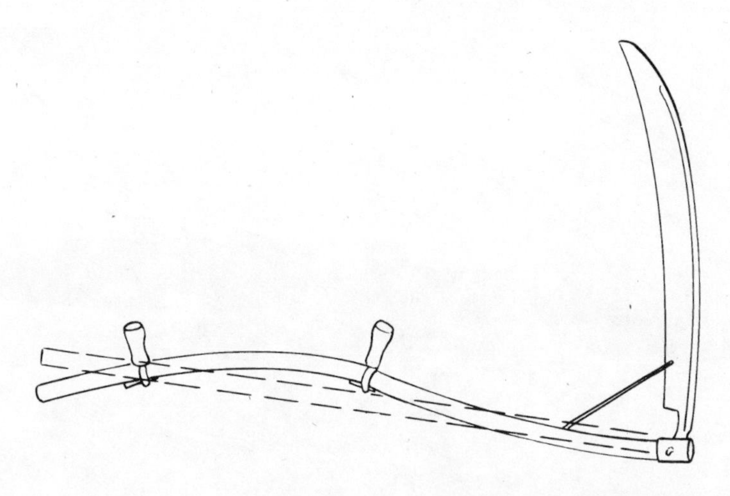

Source: Percy W. Blandford, *Old Farm Tools and Machinery: An Illustrated History* (Fort Lauderdale: Gale Research, 1976), p. 114, copyright © 1976, Gale Research, Inc. Reproduced by permission. All rights reserved.

could swing such a scythe easily and smoothly through a broad semicircle just above the ground, with devastating effect upon grasses and small grains. One of the earliest American depictions of the scythe in action appears in Figure 7.9. Again the artist is Alexander Anderson, and again he portrays the reapers' body position incorrectly (and omits the nibs). The correct position of the hands and torso is illustrated in Figure 7.10.[16]

Figure 7.9. American depiction of harvesting with scythes, 1820s (incorrect handles and hand positions)

Source: Alexander Anderson Scrapbook, Vol. 1, p. 54, Print Collection, Miriam and Ira D. Wallach Division of Arts, Prints and Photographs, The New York Public Library, Astor, Lenox and Tilden Foundations.

Figure 7.10. Harvesting with scythes

Source: Smithsonian Institution Photo No. 86-4128.

After reapers with scythes passed through a field, the remaining task was to gather the fallen crop and take it to the barn. The question that one would expect to arise sooner or later was whether the reaping implement could be modified to assist with the gathering process.

One possible modification appears in Figure 7.11. A "bow" of light wood was added at the base of the handle, supported by a brace. This modified harvesting implement usually had a blade somewhat shorter than the common scythe. The cutting stroke also changed. Instead of a full swing in a broad arc through, and then away from, the standing grain, the reaper now had a shorter stroke, mowing "inward, bearing the [grain] he cuts on his scythe [plus bow], until it comes to that which is standing, against which it gently leans."[17] Each reaper therefore had to be closely followed by a "gatherer" who would use a "hook or stick about two feet long" to pull the cut grain from its upright position and deposit it in a bundle on the ground.[18]

Just when this scythe-plus-bow was first used in Europe is not known. It is noticeably absent in eleventh- and twelfth-century pictures of the sort appearing in Figure 7.6, and much in evidence by the sixteenth century, as illustrated in Figures 7.7 and 7.12.[19] What is strikingly apparent is its

Figure 7.11. American curved-handled scythe with bow, 1828

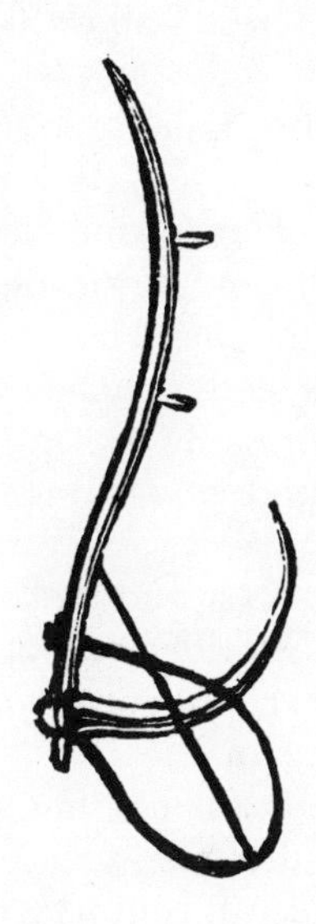

Source: New England Farmer, 7 (August 22, 1828), p. 33.

Figure 7.12. European scythe with bow, sixteenth century

Source: Detail from Pieter Brueghel's "Summer," reproduced in E. M. Jope, "Agricultural Implements," in Charles Singer, E. J. Holmyard, and A. R. Hall, eds., *A History of Technology* (Oxford: Clarendon, 1956), Vol. 2, p. 96.

lack of popularity in the New World. Almost never is this reaping implement referred to in American literature. On those rare occasions when it is, the harvesting method described above is usually referred to as "the English manner of mowing."[20]

Americans preferred a more dramatic solution for the same problem: the implement depicted in Figure 7.13. Its purpose is evident at a glance. The mowing part consisted of an ordinary scythe with curved handle, supplemented not by a small bow at the neck of the scythe but by an entire tier of teeth. Roughly the diameter of a man's finger (hence their common name "fingers"), the teeth matched the curvature of the scythe's blade. Light in weight but not easily broken, these gathering teeth were usually made of ash, "that wood being tough, and yet yielding gently to pressure."[21] The result was a massive harvesting device called a cradle,[22] poorly illustrated in Figure 7.14 and well illustrated in Figure 7.15. (Again, Alexander Anderson provides one of the earliest American illustrations, this one also of unexceptional accuracy.)[23] To wield this implement required "a great swing of the body,"[24] preferably with the wind blowing in the same direction as the cradling motion. The resulting distribution of grain on the ground can be seen in Figure 7.16. As an American reported in the late eighteenth century: "this gathers the wheat as it is cut, and helps [the cradlers] to throw it very evenly in a row, with the heads all one way. A raker, with very little trouble, may gather it into sheaves."[25] The cutting technique with a cradle was somewhat different from that employed with the scythe-plus-bow. The latter tended to be used in heavy stands, and left the cut grain leaning against uncut stalks. A

Figure 7.13. American grain cradle, 1839

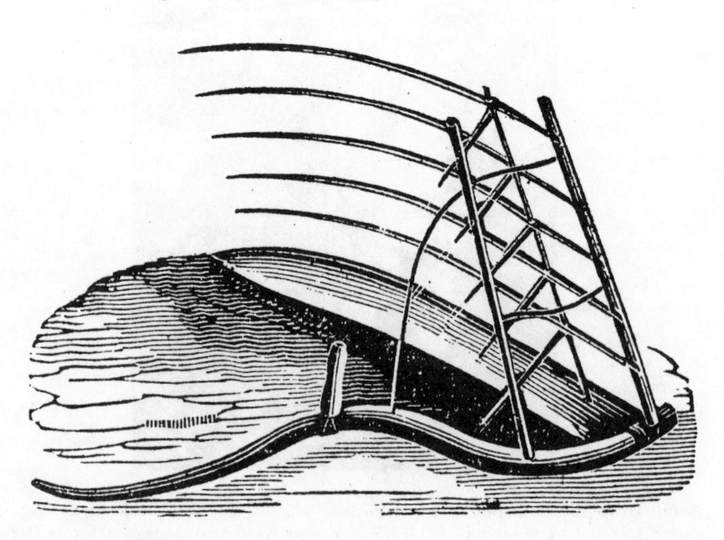

Source: New England Farmer, 18 (July 16, 1839), p. 11.

Figure 7.14. American depiction of harvesting with grain cradles, probably 1820s

Source: Alexander Anderson Scrapbooks, Vol. 1, p. 58, Print Collection, Miriam and Ira D. Wallach Division of Arts, Prints and Photographs, The New York Public Library, Astor, Lenox and Tilden Foundations.

cradler mowed "away from the crop" (not "to the standing crop"), leaving windrows of grain laid out on the ground "with admirable precision."[26]

As previously noted, by the sixteenth century the scythe-plus-bow was often featured in European illustrations. Pictures of grain cradles are difficult to find prior to the eighteenth century. Whether the second implement was an outgrowth of the first[27] and how the cradle spread through Europe are topics not well understood. By the early eighteenth century, cradles appear to have been used for harvesting in various parts of Europe, and by the end of the century the implement was well known both in Britain and on the Continent.[28]

Its American origins are similarly obscure. Historians still debate which regions pioneered its use and the subsequent process of diffusion. With no systematic data on ownership or use and few anecdotes from the pre-Revolutionary period, uncertainty about the cradle's early history in the colonies would seem unavoidable.[29] Early in the eighteenth century "cradles" appear in farm inventories; by the 1750s, they appear in farm account books; by the 1760s such pioneers in agricultural improvements as George Washington and Landon Carter were using the implement to harvest wheat.[30] Immediately following the Revolution, references to the cradle increase significantly.[31] In his landmark study of New England agricul-

Figure 7.15. Harvesting with American grain cradle

Source: Smithsonian Institution Photo No. 86-11242.

Figure 7.16. British depiction of reaping with cradles, 1850s

Source: Henry Stephens, *The Book of the Farm* (London: William Blackwood, 1855), Vol. 2, p. 341.

ture published in 1790 Samuel Deane reported that the cradle was more often used for cutting oats and rye than wheat.[32] By 1799 John Beale Bordley was so confident that knowledge of the implement was common that he made no effort to explain it when making a passing references to cradles in his *Essays and Notes on Husbandry and Rural Affairs.*[33] By 1803, in an editorial commentary on the subject of European cradles, the American James Mease observed: "The American tool *it is well known,* has five teeth; and the handle is somewhat crooked."[34]

A harvesting technique not well known to most Americans, either in 1803 or during the decades that followed, involved a kind of modified scythe whose use was confined largely to Dutch and German settlements in the Hudson, Mohawk, and Schoharie Valleys. As illustrated in Figures 7.17 and 7.18, the person reaping used not one implement but two: a cutting device ("sith" in America, "pik" or "zicht" in the Low Countries) which at first glance looks like an abbreviated scythe, and a hook ("mat-hook" in America, "pikhaak" in the Low Countries) to gather and hold the grain to be cut.[35] The hook consisted of a wooden handle some three to four feet long plus an iron crook at the end. The Flemish or Hainault scythe, as it was commonly known in America, had a curved blade somewhat shorter than that of a common scythe, and a handle long enough to permit the reaper to slash at the base of standing grain without bending over. (Note the upright body position in Figure 7.18.) At the top of the scythe's handle was a curved flange (*a* in Figure 7.17) that fitted over the

Figure 7.17. Flemish or Hainault scythe and mathook, 1829

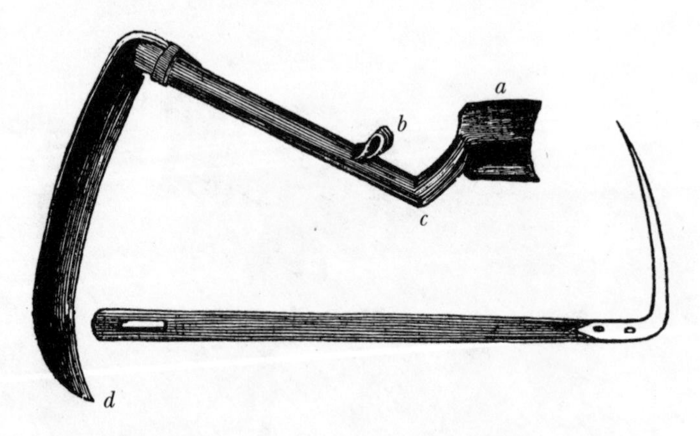

Source: Committee of the Highland Society of Scotland, "Report in regard to Experiments made in Scotland, in Harvest 1825, with the Flemish Scythe," Highland Society of Scotland *Transactions,* 7 (1829), p. 249.

user's wrist. Below the handle was a small leather loop (at *b*) into which the reaper's forefinger was inserted to help prevent the implement from slipping from his grasp. As a cutting device, the Flemish scythe had exceptional balance. This was achieved partly by having the reaper's right hand directly over the center of the blade when the blade rested on the ground, and partly by having the tip of the blade in line with the uppermost part of the handle. (In Figure 7.17, this would require point *d* to be in the path of a line extended from *ac.*)[36]

Figure 7.18. Reaping with Flemish or Hainault scythe, 1850s

Source: Henry Stephens, *The Book of the Farm* (London: William Blackwood, 1855), Vol. 2, p. 338.

The way these two implements were used to reap small grains is not easily described. The early part of the stroke was simple enough. The hook acted as a kind of substitute for the sickle reaper's hand that grasped the grain prior to cutting. As the mathook, held about the middle of its long handle with the left hand, pressed the standing crop slightly down and to the left, separating the grain for cutting from other stalks, it cleared the way for the blade to follow, which happened almost immediately. With the mathook exerting pressure about half-way up the stalks (the hook in Figure 7.18 is too low), the Flemish scythe, wielded primarily by the wrist, cut the stalks within a few inches of the ground. The conclusion of the stroke brought the severed grain to rest against the standing crop (as with the scythe-plus-bow). The complex part of the procedure, difficult to describe and difficult to execute, converted severed grain, then upright, to a neat bundle on the ground. A British expert described the process as follows. "The cut [grain] leans against that which is standing; and when as much has been cut as will make half a sheaf, the workman turns half round, and hooking up part of what is cut with as much of what is standing, he cuts and rolls up the whole in the form of a sheaf, using his leg and foot to keep it in the bend of the blade: the legs are protected by pieces of strong leather over the shins. Thus it is laid down for the binders."[37] Small wonder that this reaping technique was not easily learned or commonly endorsed, even by those already expert with the sickle, scythe, or cradle.

The Flemish scythe, as the name suggests, appears to have originated in Belgium some time during the Middle Ages, possibly as early as the thirteenth century.[38] By the sixteenth century it often appeared in pictures depicting agrarian activity.[39] By the early nineteenth century, although it was not used in England, English writers characterized the implement as "the well-known Hainault or Flemish scythe."[40] Early settlers from the Low Countries may well have carried the implement to the New World. Inventories of the seventeenth century sometimes include "Dutch" or "Flemish" scythes, and by the early eighteenth century New York newspapers occasionally advertised "Dutch scythes."[41] Two features of American use are striking. The first is the failure of the implement to diffuse much beyond Dutch and German settlements of the Hudson, Mohawk, and Schoharie Valleys. The second is the sustained popularity of the implement within that region well into the nineteenth century. Thus a New York farming expert noted as late as 1826: "the Dutch Scythe . . . was introduced by the first settlers along the banks of the Hudson, . . . is still in use by their descendants, and is sold in our shops."[42]

The puzzle is why this limited diffusion if, as many claimed, the Flemish scythe was much superior to the sickle as a reaping implement. For that matter, if the grain cradle was superior to either sickle or scythe, as many others claimed, why should the use of all four implements persist on both sides of the Atlantic well into the nineteenth century?

The Determinants of Adoption

The process of adoption—which reaping device came to be used where and for which crops—was determined by a simple market mechanism, plus a number of costs and benefits peculiar to each implement.

The nature of the mechanism is easily understood if a host of factual details are momentarily ignored. Suppose farmers are trying to decide whether to harvest a given crop (say, wheat) by hiring either sickle reapers or cradlers. Assume further (economists are always assuming further) that the only difference between the two reaping devices is speed of operation, the cradle being twice as fast as the sickle. This suggests that only cradlers will be employed. But as obvious as that choice may seem, it ignores the operation of the labor market that determines the wages of harvest workers. If all farm owners demand cradlers, the wages of cradlers should rise.[43] Indeed, the wage of cradlers should continue to rise until their higher wages roughly offset their higher productivity. At any wage differential less than this—say, compared with sickle reapers, cradlers harvest twice as fast but cost less than twice as much—all those hiring harvest workers will want cradlers, and that demand should drive up the wages of these particular workers. Over time, the supply of cradlers should increase as more agrarian workers learn to use this implement in order to earn the higher wage. An increase in supply may lower the cradler's wage, but—and this is the central point—as long as both kinds of reapers are hired in markets with any pretensions to functioning as markets should, the forces of supply and demand should create a strong tendency for the wage differential between the two roughly to equal the productivity differential between the two.

An example of this market mechanism at work is provided by a Pennsylvania farmer in 1829.[44] He begins by noting:

(1) because cradlers only cradle, a farmer must also hire a raker and a binder,
(2) sickle reapers "reap through, bind back and shock in the evening" (and thus do not require the hiring of ancillary workers), and
(3) "a cradler will not do much more than three [sickle] reapers, and never as clean harvested."

The costs of reaping three acres by these two different methods are accordingly estimated as follows:

		Wages
by cradling:	1 cradler	\$0.75–\$0.80
	1 raker	\$0.37½–\$0.40
	1 binder	\$0.37½–\$0.40
	Total	\$1.50–\$1.60
by reaping with sickles:		
	3 reapers at 50¢ each	\$1.50

The near equivalence of the two estimates prompts this farm analyst to conclude that "there is but little difference in the expense of [sickle] reaping and cradling."

Not all such calculations show a similar equality. But often when an inequality is discovered—when a "clear advantage" is demonstrated in favor of one harvesting implement over another—that conclusion is suspect because the wages of the reapers have often been incorrectly estimated. One of the most celebrated examples of a flawed calculation is provided by the Committee of the Highland Society of Scotland.[45] Seeking to introduce the Flemish scythe to Scottish farmers, they brought to Scotland in 1825 "two respectable intelligent young farmers of Flanders" to demonstrate the use of that implement at a series of harvest exhibitions. These exhibitions suggested that one reaper with a Flemish scythe and mathook could cut as much as "two good reapers with the sickle." After adjusting for differences in ancillary workers needed, the committee concluded the Flemish harvesting technique led to a saving of 25 percent. But in making this calculation the committee assumed that both kinds of reapers would receive the same wage. Had many Scottish workers learned the Flemish technique (they did not, for reasons discussed below), presumably the wages of these workers would have risen relative to those of sickle reapers to reflect their superior productivity.

The actual functioning of labor markets for reapers in different countries at different times was less smooth and more complex than indicated thus far. Much more than speed of operation was involved in deciding which reaping implement to use for which crop, and the many associated costs and benefits were not easily estimated even by those inclined to arithmetic calculations.[46]

Sometimes the decision was easy because conditions in the field precluded the use of any scythe-like implement. Crops that were twisted and beaten down by the elements usually had to be harvested with the sickle,[47] as did those growing on soil littered with stumps and rocks.

Where this was not the case, the farmer had to compare, for reaping with sickles as against any implement with a scythe blade, a variety of costs and benefits, of which three were particularly crucial.

The first concerned the desirability of long or short stubble. Sickles tended to cut higher, scythe-like implements to cut lower. Longer stubble could be used for grazing or alternatively plowed under to add to the fertility of the soil.[48] Shorter stubble meant more straw taken to the barn for subsequent use as winter fodder, livestock bedding, or for thatching roofs. Which option was preferable depended upon whether the lower part of grain stalks had more value to the farmer as residue in the field or as part of the harvested crop.

The second consideration was loss due to shattering. Harvesting with a sickle with a serrated blade (the common form)[49] required a cutting motion far less violent than the blows delivered by a reaper armed with a scythe or grain cradle or Flemish scythe. In the premodern era, most small grains were harvested when almost ripe, or, on occasion, "dead ripe." The heavier the stand and the riper the grain, the greater the probable loss from shattering if any implement other than a sickle was used.[50]

Third and last, but far from least, was the matter of the strength and skill required to wield each implement effectively. The use of a sickle was quickly learned and required little strength (but considerable dexterity for repeated stooping). Accordingly, those harvesting with this implement frequently included women and occasionally children, with some observers even suggesting that women were superior to men.[51] Wielding any of the scythe-like implements was considered "a man's work." Strength was important, as were skill and stamina, the great sweeping motion of the scythe or cradle twisting the upper torso and not infrequently creating cramps in the lumbar region of the back.[52] The more difficult the implement was to learn and to use, the higher the relative wage received by those who could reap with it.

The list of relevant costs and benefits does not end here.[53] But even if it did, consider again the market mechanism outlined at the beginning of this section. The economic argument is simple, but the associated calculations obviously were not. Productivity differentials should be reflected in wage differentials among reapers who were proficient with different harvesting implements. But estimating the many associated costs and benefits was seldom easy and no doubt often not attempted. And yet however primitive labor markets were, and however imperfect the decision making of those who sought to hire reapers, two features should be observed over time as past experience conditioned present knowledge about the relative merits of each implement for different crops. Where one implement was demonstrably superior, over time that implement should have tended to displace inferior alternatives as the supply of reapers expert with the superior reaping device gradually increased. Alternatively, where two or more reaping implements were used in the same farming region to harvest the same crop over a sustained period of time, observed wage differentials should have tended to reflect productivity differentials, the latter being a net calculation of relevant benefits and costs. These two tendencies may help to explain a number of historical puzzles, although the following are appropriately viewed as tentative hypotheses.

(1) The initial development of the scythe, as previously noted, was closely linked in certain areas (particularly Northwest Europe) with the need for winter fodder. The rapid displacement of the sickle for mowing purposes undoubtedly reflected the superior speed of the scythe, but also the fact that one prospective cost of scythe harvesting—increased shattering—was irrelevant, given the nature of most hay crops.

(2) The displacement of the sickle by scythe-like implements in the harvesting of small grains (where shattering loss was important) was a gradual and complex process in both Western Europe and America, and one not well understood. Few historians note the curious contrast between England and the Continent, particularly the Low Countries. The use of scythe-like implements for harvesting small grains was well established in the Low Countries by the fifteenth century, but was not much in evidence in Britain until the end of the eighteenth century.[54] When sickles were displaced, the implements of choice in the Low Countries were predominantly the Flemish scythe and the scythe-plus-bow. Both gathered the grain as it was cut, leaning the gathered stalks against the standing crop, thereby reducing loss from shattering. While the scythe-plus-bow was also used in England, much of the sickle displacement that took place in small grain harvesting from the late eighteenth century onward involved the "naked" or "common" scythe (that is, a scythe without a bow or withy) and the grain cradle. The first did not gather the cut stalks at all, and the second required a cutting motion away from (rather than up to) the unsevered stalks of the standing crop. Shattering losses for both implements, therefore, should have increased. This difference in implement performance may help to explain the lag in the British move away from the sickle as a reaping implement relative to a similar move on the Continent,[55] but it does not explain why the British choice of implements to displace the sickle were so different. One is struck by the appropriateness of Low Country implements for reaping in heavy stands, where leaning severed stalks against the uncut crop was feasible. The British choice may well reflect, or partially reflect, a difference in the relative density of crops in the two areas.[56]

(3) American reaping implements also differed somewhat from those popular in Britain, and again the relative density of crops to be harvested may have been an important determining factor. When sickles were displaced by scythe-like implements for harvesting small grains in America, the scythe-plus-bow (appropriate to heavy stands) was almost never used. American grain cradles also tended to have longer teeth, and more of them, than the comparable British implement. Both choices may reflect the high cost of labor relative to land in land-abundant America. As British observers never tired of pointing out, farming techniques in the New World were "wasteful" in the sense that few techniques popular in Britain to enhance fertility were much used in America, particularly crop rotation and extensive use of fertilizers. A farming style that was "extensive" rather than "intensive"—designed to husband labor more than land—would naturally tend to create thinner stands of small grains, and therefore require reaping implements appropriate to that lesser density. On both sides of the Atlantic, experts agreed that heavy stands were difficult to harvest with any cradle, and clearly the massive American implement of the sort depicted in Figure 7.15 was well suited to thinner crops.[57]

(4) As previously indicated, the pattern of use of the Flemish scythe in both the Old World and the New is something of a puzzle. Once established in certain regions, such as the Low Countries in Europe and parts of New York State in America, its use persisted for generations, and yet it failed to diffuse in obvious directions—across the English Channel into Britain, or beyond the confines of three valleys in America. One explanation often offered for this absence of diffusion is that, while superior to the sickle, the Flemish scythe was inferior to grain cradles then in use in both America and Britain.[58] But were that true, diffusion should have gone the other way, the grain cradle gradually displacing the Flemish scythe. The most likely explanation is that the net benefits of each type of implement were roughly comparable. Or as one British analyst put the matter (using the royal "we"), "we suspect [the Flemish scythe] has, if any, no great advantage" over grain cradles already in use in Britain.[59]

American Reaping Innovations: From First Settlement to 1815

In the two centuries between the initial settlement at Jamestown and the War of 1812, American efforts to improve reaping implements took several directions, of which two were strikingly successful.

Neither involved the sickle or the Flemish scythe. Their design when the United States made its final peace with Britain on December 24, 1814, appears to have changed little since migrating settlers first brought these implements to the New World.[60] This record of unchanging form was sharply at variance with the American history of the scythe. In 1646 Joseph Jenks of Massachusetts obtained the exclusive privilege "for making scythes and other edge tools" at the first iron works in that colony. Nine years later he acquired a patent for an improved scythe "for the more speedy cutting of grass," the improvements in question consisting of (a) a longer and thinner blade, and (b) an iron bar welded to the back of the blade to strengthen what was intrinsically a more delicate implement. The resulting design change was a triumph. More than two centuries later an expert on American manufacturing history conceded that since Jenks's original patent, "no radical change has been made in the form of the implement."[61] Such continuity from one perspective is a tribute to the genius of Jenks, but from another is illustrative of a static agrarian technology, at least during colonial times. Within New England, from Jenks's scythe design of 1655 until the American Revolution only one other mechanical device received a patent (a surveying instrument).[62]

Such an absence of patents should not obscure the other major achievement in redesigning reaping implements during the colonial era. The American grain cradle, as previously noted, could claim to be the great reaping novelty of the eighteenth century, at least in the New World. How cradle designs of the Old World influenced the size and shape of the first American implements remains something of a mystery (as does the initial development and subsequent diffusion of the grain cradle in Europe). But probable European origins should not obscure American achievements in adapting this implement to a frontier setting. By the beginning of the nineteenth century, compared with its British counterpart the American cradle tended to have longer teeth, and more of them[63]—both changes making for a harvesting device well suited for cutting and gathering crops in stands of lower density.

These two changes in harvesting implements—an improvement in scythe design in the seventeenth century and the adoption and modification of the European grain cradle in the eighteenth century—exhaust the list of major American reaping innovations prior to 1815, if "major" means having a significant impact upon established practice. By the beginning of the nineteenth century, admittedly Americans with agrarian expertise and an eye for innovation had recognized the obvious: that a horse-drawn mechanical reaper for small grains was the Holy Grail of farm technology, promising huge rewards for the inventor and massive benefits to the farming community. To judge by patents filed, the search was clearly under way by 1803, when the first American patent was granted for a machine to cut grain. Between that year and 1815, a handful of further patents were awarded for machines with a similar purpose.[64] But all these early efforts are nothing more than historical curiosities, having, as they did, no impact whatsoever upon the literature or established practice. The reasons for this undistinguished record are not difficult to identify. The mechanical problems to be solved were many and complex. Equally important, the mechanical solution had to feature an exceptionally reliable performance in the field, because any breakdown during the few days available for harvesting could have dire consequences for those relying upon horse-powered machines to reap their grain. A mechanical reaper promising such reliability would not be available for decades. Why this was so, and how it was ultimately achieved are secondary themes of the next section. The primary theme is inventive efforts that were successful in the fifteen years following the Treaty of Ghent.

American Reaping Innovations: 1815 to 1830

In 1823, in his address to the assembled members, the president of the Pennsylvania Agricultural Society risked laboring the obvious when he noted: "Nothing is more wanted than the application of animal power in the cutting of grain," adding "such an invention can be no easy task, or the ingenuity of our fellow citizens would, ere this, have effected it."[65] Twelve years later, one of those Americans of ingenuity was not disheartened by the unsuccessful field trial of his harvesting machine at Flower Field, Michigan. "I see the shore far off," Hiram Moore insisted, "and it will take me a long time to get there, but I will succeed in time."[66] He spoke for a generation of American inventors who sought to mechanize the reaping of small grains with horse-drawn implements. Most would fail, but two realized a stunning success, although, in Moore's words, it took them a long time to get there.

The development of a successful reaper by Cyrus McCormick and Obed Hussey is a thrice-told tale, beginning with field trials (McCormick in 1831, Hussey in 1833) and patents (the first for Hussey in 1833, for McCormick in 1834), and ending with a workable machine that would revolutionize American farming. Our concern is not with their various experiments and modifications in design, but with the mechanism at the core of their success, and why incorporating such a mechanism into a commercially successful reaper was so difficult.

The key to a successful reaper was, understandably, the devising of a reliable cutting mechanism, and both men hit upon a solution that would come to dominate American reapers: the back-and-forth motion of multiple teeth in a long bar mounted close to the ground, the motion of the bar provided by a gearing mechanism linked to a wheel in contact with the ground. Hussey's original cutting mechanism was slightly different from McCormick's. As illustrated in Figure 7.19, Hussey's device, in his own words, consisted of "a row of strong iron spikes about the size of small harrow-teeth. These spikes are formed of two pieces of iron, one above and one below, leaving a horizontal slit in each spike for the cutting-blades to play in. These blades are formed like lancet-points, being sharp on both edges, and several inches long. They are fastened side by side on an iron rod, as many blades as there are spikes on the platform. . . . As the machine is drawn ahead in the grain, the . . . straw is received between the spikes, while the vibrating blades cut it off as it enters; the straw being held by the spike both above and below the edge of the blade, while the blade passes into the spike—thus the cutting is made sure."[67] McCormick's cutting surface also moved back and forth between spikes, but to cut the stalks he used, not pointed blades, but "a mere saw-edge, of a waved outline."[68] (Again, see Figure 7.19.) The Hussey system worked better, as McCormick eventually conceded and revised the design of his machine accordingly.[69] The early McCormick reaper had one superior feature missing in the Hussey, a revolving reel that held the grain against the reciprocating blades and helped to lay the severed stalks upon the platform behind the cutter-bar.[70]

The early performance of both machines made clear that commercial success, however clearly visualized, lay far in the future. As McCormick later admitted, his reapers "were not of much practical value until the improvement of my second patent in 1845."[71] From that date until the

Figure 7.19. American mechanical reapers, 1850s

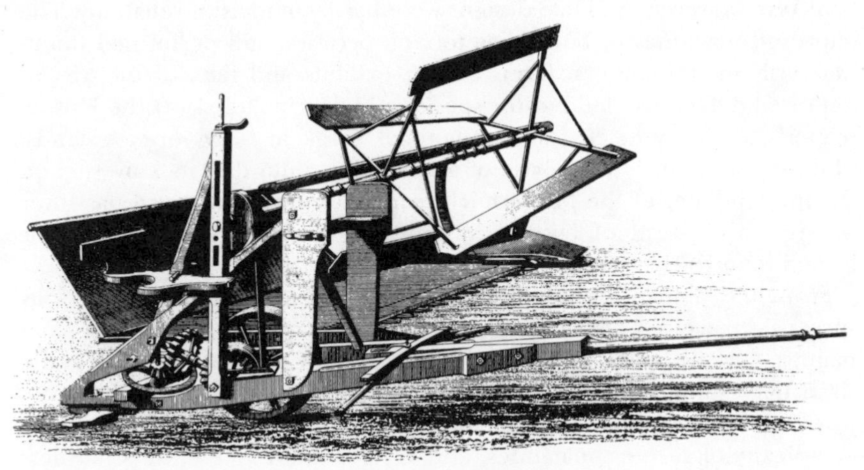

A. McCormick's reaper

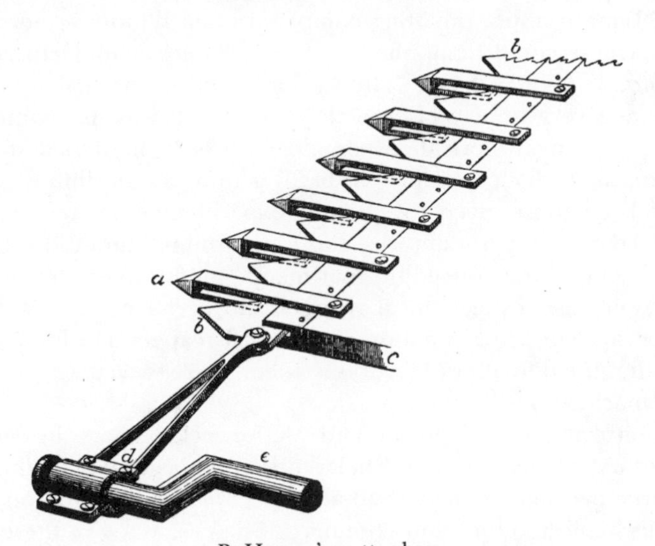

B. Hussey's cutter-bar.

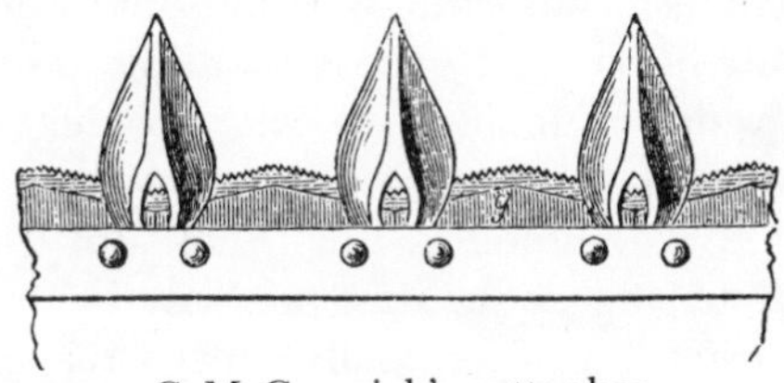

C. McCormick's cutter-bar

Source: John C. Morton, *A Cyclopedia of Agriculture* (London: Blackie & Son, 1856), Vol. 2, Plate XXXV, pp. 743, 744.

Civil War, however, available designs were hardly models of reliability. The improved machines of both inventors on occasion still performed unsatisfactorily on purchasers' farms and at exhibits and fairs of the 1850s. Reported defects singled out for special mention included (a) the Hussey reaper's tendency to clog and sometimes break in heavy or wet stands, while soil adhering to the base of grain stalks could dull its knives,[72] and (b) the tendency of the McCormick reaper "to press down and pass over the [grain] instead of cutting it, unless it stood perfectly upright or leaned towards the machine."[73]

From the pioneering experiments of the 1830s to the commercially successful machines of the 1850s, four problems generated the most complaints: (1) the need for exceptional reliability in the field, given the short harvesting season of most crops, (2) the primitive state of both the machine-tool and metallurgical industries in America, (3) the intrinsic complexity of reaping machines, and (4) the ignorance of most farmers about the operation and repair of such a complicated mechanism. As Clarence Danhof notes, the 1853 complaint of an Illinois farmer about a Hussey machine could be applied to most complex manufactured products of the pre-Civil War era: "The workmanship in the making of these machines was very imperfect, and they were built with bad timber and brittle iron castings where there should have been hardened steel, and faults concealed which I could not discover until the machines were put in use."[74] Even those given to agrarian experimentation were therefore often reluctant to experiment with this new farming innovation. Edmund Ruffin, for one, while conceding that many "good farmers" in his neighborhood were using mechanical reapers, still preferred "the scythe and cradle" because of the reputation mechanical reapers had for "[getting] out of order, the difficulties of working them, and especially my own ignorance of machinery."[75]

If the histories of the Hussey and McCormick reapers are stories of protracted experimentation and delayed success beginning in the 1830s, in the three decades prior to 1830 attempts at reaper innovation are stories of mechanical failure and commercial fiascos. Many of these experimental designs tried, in one way or another, to reproduce the cutting motion of scythes witnessed in the field. Almost every imaginable kind of whirling blade mechanism was employed by some would-be inventor in England or the United States.[76] Figure 7.20 illustrates one of the better-known American inventions, featuring a rotary blade.[77] It was not the first (the first American patent for a mechanical reaper was issued in 1803),[78] nor was it the only one designed in the 1820s. But it was the one most publicized. What is striking about that publicity is how quickly it disappeared. The first illustration of Jeremiah Bailey's reaper appeared in the agricultural press in 1823, with its inventor's assurance that it had been "extensively used and approved of during the last season, in the neigh-

Figure 7.20. Bailey's mowing machine, 1823

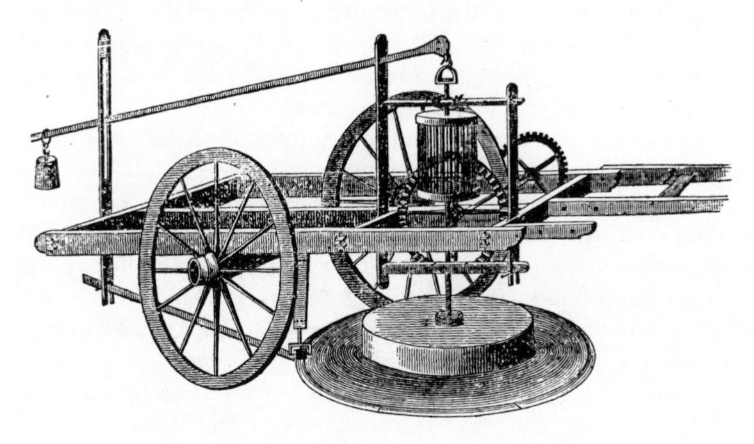

Source: American Farmer, 5 (December 12, 1823), p. 199.

bourhood of the patentee [Chester County, Pennsylvania], and promises to be of great public utility."[79] Twenty-seven months later in the same periodical Bailey published the written endorsements of sixteen farmers from his region, of whom only two had actually used his reaper. (The rest had seen it in operation and judged "that it fully answers the purposes intended, both for grass and grain.")[80] But no publicity of any consequence ever appeared again. Such evidence of quick demise is not surprising. Given the complex mechanical challenges to be solved as illustrated by the histories of the Hussey and McCormick reapers—both of which began with the cutting mechanism (reciprocating, not rotary) that would ultimately dominate reaper design—the commercial failure of these earlier harvesting inventions would seem a norm to be expected, not an aberration requiring explanation.[81] Such was the fate of all American mechanical reapers patented prior to 1830, including that of Jeremiah Bailey.[82]

But if those who sought to improve the reaping process in the period 1815–1830 invariably failed if they tried to realize the grand ambition of a reliable horse-drawn reaper, many others succeeded in small ways to make the harvesting of hay and small grains more efficient.

Improvements in the scythe and grain cradle are cases in point. During the Napoleonic era, the progressive strangulation of foreign trade stimulated demand for most domestic manufactures. With the restoration of peace, many of these local industries continued to flourish. Scythe production, for example, was well established by 1810, particularly in Massachusetts, Connecticut, and New York. By 1816 production had spread to Cincinnati; by 1825 firms employing ten to twelve hands and using water power were producing "upwards of 700 dozen scythes" per year.[83] Fragmentary evidence suggests that Americans were producing not only more scythtes, but better scythes. In 1823 the president of the Pennsylvania Agricultural Society declared "the scythe is improving from year to year." ("Not so with the sickle," he added.)[84] By 1826, German steel was being imported to make "cast steel" scythes, because American steel was still considered to be inferior for such purposes.[85]

As scythe blade quality improved, presumably so did the quality of cradle blades.[86] But the latter implement was also being redesigned in minor ways. About 1820 Charles Vaughn of Hallowell, Maine, developed a lighter cradle.[87] By 1825 the gathering function of the cradle had also been improved by sewing "a piece of coarse linen" on the first two fingers "from the heel of the scythe to within four or five inches of the points of the fingers," thereby enabling the cradler to cut off "the heads of the clover and [throw] them into small heaps."[88]

Other innovations had the same objective of speeding up the gathering of clover heads, but the mechanism employed had been anticipated by the Romans centuries earlier.

These Roman devices were to gather grain (not clover). As illustrated in Figure 7.21, the gathering process consisted of stripping the heads with

Figure 7.21. Roman reaping devices

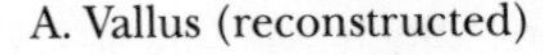

A. Vallus (reconstructed)

B. Carpentum (reconstructed)

Source: K. D. White, *Agricultural Implements of the Roman World* (Cambridge: Cambridge University Press, 1967), p. 157, © Cambridge University Press 1967; reprinted with the permission of Cambridge University Press.

a comb-like device attached to the front of a wagon propelled by animal power from the rear. These "headers" were evidently used on large estates located on level plains in the provinces of Gaul. Although such reaping carts saved labor at harvest time, they generated enormous wastage. Not only did they leave most of the straw left uncut in the field, but such cumbersome stripping mechanisms must have left large quantities of shattered grain on the ground. Their continued use for several centuries suggests that the benefits probably outweighed the costs. But that could only result where some or all of the following prevailed: (a) harvested straw had little value, (b) labor was extremely costly relative to the price of grain, and (c) additional workers of sufficient numbers to work large estates at harvest were difficult, if not impossible, to find.[89] However much these cumbersome head-strippers made economic sense on large estates in Roman times, they quickly disappeared with the collapse of the Roman Empire, never to resurface as a significant implement in the agriculture of the same Gallic regions through all the centuries of the Dark Ages and the Middle Ages.

An implement strikingly similar in appearance did surface in America in the second decade of the nineteenth century.[90] In 1819, a cart bearing a marked resemblance to the Roman reaping device was illustrated in the *American Farmer.* (See Figure 7.22.) The American version was pulled from the front, not pushed from the rear (in this instance, Americans evidently had an aversion to putting the cart before the horse), and given its objective of gathering clover heads (not grain), the stripping teeth had to be set closer to the ground. An implement "so simple that any carpenter may make it," it was to be drawn "over the cleanest parts of clover so as to avoid the seeds of weeds."[91] Who first designed this mechanism and where it was first used remain unknown. One farmer claimed to have used such a clover header for "six or seven years" prior to 1819, and another believed it was a New York invention, but the bulk of the evidence indicates its early use was concentrated in the southern states.[92]

A year after this horse-drawn implement first appeared in the agricultural press, hand-drawn implements of similar design for the same purpose were also publicized. The "iron rake" (1 in Figure 7.22B) required "a strong man" to pull it, but the "wooden rake" (2 in Figure 7.22B) could be used "by a child of 9 or 10 years."[93] The popularity of these clover headers, particularly the horse-drawn variant, is indicated by the endorsements they received in the 1820s from such authorities as the editor of the *American Farmer* and the New-York Agricultural Society, as well as by the appearance and endorsement of virtually identical clover-gathering mechanisms in British agricultural literature during the 1830s.[94]

At the same time that these new devices for gathering clover heads were first publicized, two other innovations for gathering hay received considerable attention in the American press.

Figure 7.22. American clover reaping devices

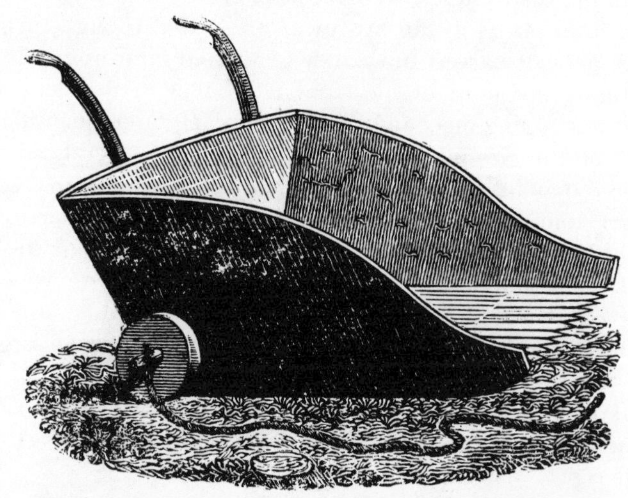

A. Horse-drawn device for gathering clover heads, 1819

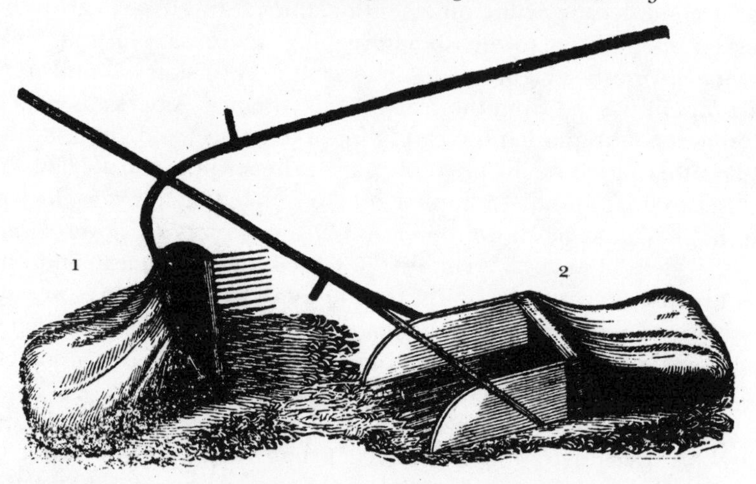

B. Clover head "rakes," 1820

Sources: American Farmer, 1 (September 5, 1819), p. 254; *The Plough Boy*, 1 (February 19, 1820), p. 301.

The simpler of the two is illustrated in Figure 7.23. Once the hay was cut, usually by the common scythe, the fallen stalks had to be raked, usually by hand. One horse-drawn implement occasionally employed in America for this task prior to 1820 resembled the British bean stubble rake illustrated in Figure 7.23. (The rake's tines may seem at first glance to curve in the wrong direction, but the handles shown were for the

Figure 7.23. Horse-drawn implements for gathering crops

A. British bean stubble rake, 1807

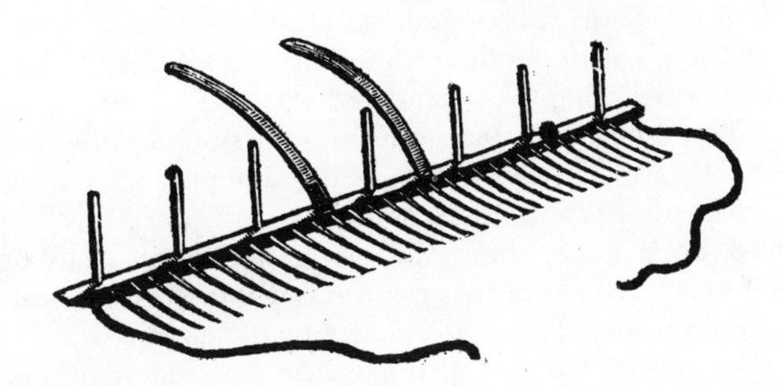

B. American "improved" horse rake, 1821

Sources: American Farmer, 3 (June 20, 1821), p. 135; Arthur Young, *General View of the Agriculture of the County of Essex* (London: R. Phillips, 1807), Plate 35, p. 152 (Photo: Division of Rare and Manuscript Collections, Carl A. Kroch Library, Cornell University).

worker, not the horse, as the latter pulled the mechanism from hitching devices at 5 and 6.) The American hay rake of the post-Revolutionary era was "dragged over the ground like a harrow," and had the disadvantage that "it would drag much of the soil up, and mix with the hay in loose ground."[95] The latter difficulty was neatly solved by rotating the collecting teeth 90 degrees, and making them longer so they would "run horizontally under the hay."[96] Such "improved" hay rakes may have been in use as early as 1812, and were certainly well known by 1820, particularly in Massachusetts and Pennsylvania.[97] Where adopted, this horse-drawn device for gathering hay received enthusiastic praise,[98] but the areas where it was used remained limited throughout the 1820s. The topographical requirements of the implement explain this seeming contradiction. A hay-collecting mechanism featuring long, protruding teeth running close to the ground usually did not perform well on uneven fields, and even less well on uneven fields littered with boulders and stumps that could catch and often break the wooden teeth.[99]

A second hay-gathering innovation that looked much like the first had the same limited success during this period, and for similar reasons. The revolving horse-rake, as its name implies, was just a variant of the "improved" horse rake, but with the added capability of flipping over when full. How it was operated is not easily inferred from illustrations of the 1820s, but can be readily grasped with the aid of a diagram from the 1850s. As indicated in Figure 7.24B, the rake is pulled by a horse from *a*, and guided by the handles *b* at the rear. When the teeth became full, the operator, using the handles, could tip the implement up and forward. This caused the entire device to rotate about the bar *c*, the front teeth swinging to the back (dropping the gathered hay en route), with this backward rotation checked when the teeth encountered the bar *e*. (The strut *de* served to lift this horizontal bar clear of the rear teeth as they swung up and forward to continue the gathering process.) The obvious advantage of the revolving rake compared with the nonrevolving version was the ability to dump gathered grain without stopping the horse's forward progress. Figure 7.25 illustrates how this dumping was arranged on a mown field, with long, continuous windrows at measured intervals. Like the improved horse rake, the revolving version appears to be a distinctively American invention, first reported in the second decade of the nineteenth century, and clearly in evidence by the early 1820s.[100] And like the simpler version, the revolving horse rake was widely praised where used, and not used very widely. The problem, once again, was the implement's need for a smooth and unobstructed surface to function well, and the reluctance of most farmers whose fields were not naturally smooth to invest extensive resources to groom them.[101]

But for farmers on the margin—those whose fields could be made relatively flat and unobstructed with a minor investment of resources—the requisite implement was a roller. Such needs led to increased demand for rollers,[102] and partly in response to those demands, rollers of the 1820s also were improved in modest ways.

The roller typically used at the beginning of the decade tended to resemble the "common wooden roller" of Figure 7.26. Usually made of a single solid piece of wood, these implements had several shortcomings that innovations of the 1820s sought to remedy. Any farming implement

Figure 7.24. American horse rakes

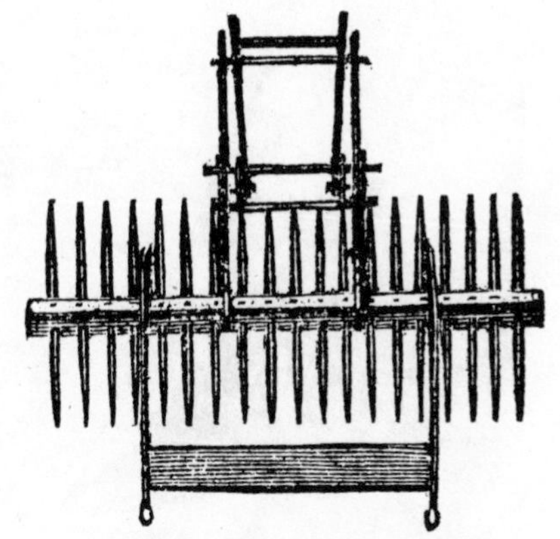

A. "Hay and grain rake," 1822

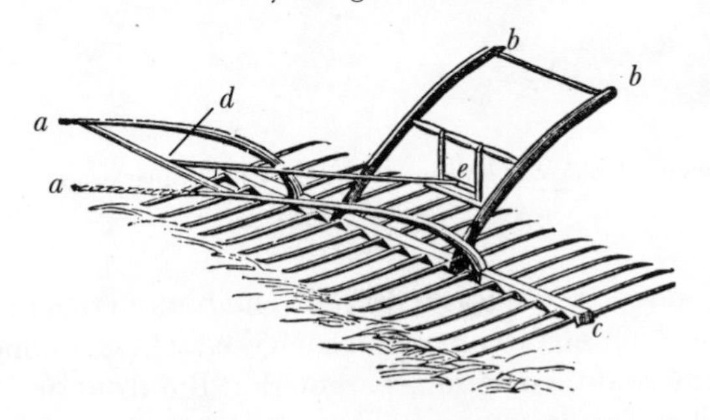

B. Revolving horse rake, 1850s

Sources: American Farmer, 4 (June 28, 1822), p. 112; John J. Thomas, *Farm Implements and Farm Machinery* (New York: Orange Judd, 1869), p. 168.

this wide was difficult to turn, but could be turned much more readily if divided into segments in the manner illustrated by the bottom diagram in Figure 7.26. Segmented rollers of the 1820s probably did not have quite so many divisions, but the merits of division were clearly known at that time. In 1822, for example, the Massachusetts Society for the Promotion of Agriculture singled out for commendation "an improvement on the common roller" consisting "of two [cast-iron] rollers moving upon one axis, [which] thereby turns easier and makes less ridges."[103] Efforts to remedy the deficient weight of common rollers took several directions. One, as previously intimated, was to make the implement of iron. (Implement dealers advertised cast-iron rollers as early as 1820.)[104] A second possibility, although laborious to construct, was to make the roller of stone.[105] A third option, often used irrespective of the material of the roller, involved mounting a box directly above the roller into which stones could be

Figure 7.25. Hay rake at work, 1850s

Source: Henry Stephens, *The Book of the Farm* (London: William Blackwood and Sons, 1855), Vol. 2, p. 238.

dumped when added weight was required. The bottom diagram of Figure 7.26 illustrates a common placement for such boxes. Less common were the twin boxes of the middle diagram, mounted at the front and rear of the 1830 barrel roller. This latter implement increased the surface in contact with the ground. The common roller with its relatively small circumference had the disadvantage that "stones and grass roots" tended to be "carried forward and displaced, rather than pressed down in a perpendicular direction."[106]

A cautionary note should perhaps be added at this point. None of these developments signaled a revolution in roller use. American farmers had long known of that implement and commonly neglected it,[107] and that neglect in most cases probably reflected a rational choice to invest few resources in grooming fields in a land-abundant country. This reluctance to use rollers was not significantly altered by improvements in the implement or by improvements in hay raking best applied to flattened fields free of obstacles. But these improvements did indicate a new willingness to ask of every kind of agricultural activity: Is there a better way?

Innovations of 1815–1830 in Historical Perspective

In the long history of American improvements in harvesting devices from the first primitive sickles to the modern combine, the era 1815–

Figure 7.26. Ninteenth-century American rollers

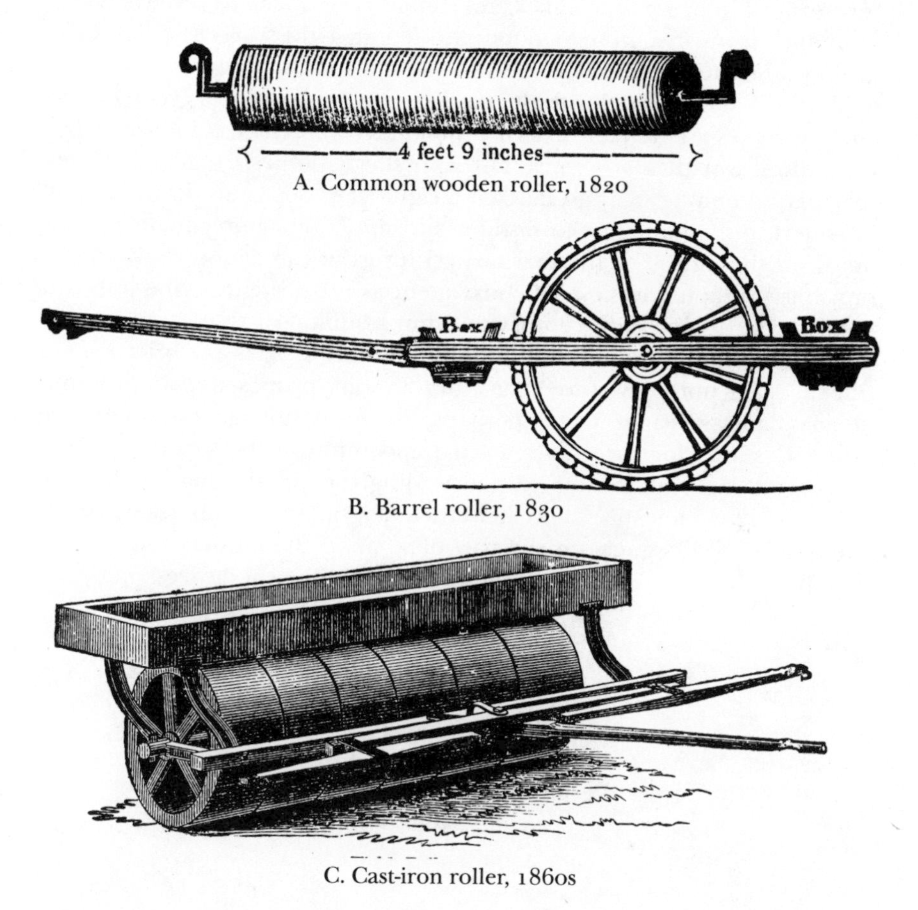

A. Common wooden roller, 1820

B. Barrel roller, 1830

C. Cast-iron roller, 1860s

Sources: American Farmer, 2 (August 4, 1820), p. 150; *New England Farmer*, 8 (April 16, 1830), p. 306; John J. Thomas, *Farm Implements and Farm Machinery* (New York: Orange Judd, 1869), p. 152.

1830 at first glance appears noteworthy for what it did not accomplish. The high drama and protracted development of complex reapers by Hussey and McCormick had not yet even begun. Nor could any of the pre-1830 innovative efforts be fairly viewed as mechanical precursors of the new devices destined to revolutionize American farming. The main implements used at harvest time in 1830 were essentially the same as those used in 1800: the sickle and common scythe, the grain cradle and Flemish scythe.

Closer scrutiny of the decade and a half beginning in 1815 nevertheless reveals a number of innovations of a modest sort designed to improve the cutting and gathering of small grains and hay. Better scythe blades and grain cradles modified to gather clover, horse-drawn and hand-drawn clover headers, improved hay rakes and similar rakes made to revolve, rollers fashioned from cast iron in multiple sections—these were all advances distinguished for simplicity of design and cheapness of construction, particularly when compared with the costly and complex reapers yet to come. Such novelties are unlikely to strike the modern observer as indicative of a revolution. But they are consistent with the hypothesis that the fifteen-year period ending in 1830 marked a significant departure from the past in Americans' willingness to fashion and apply new implements for use down on the farm. For the first two centuries of the nation's history, the only harvesting innovations of consequence were a single scythe improvement in the seventeenth century and the adaption of the European cradle to American needs in the eighteenth century. Now in the brief span of fifteen years not only were many innovations proposed that promptly failed (Bailey's reaper being perhaps the most notable example), but others destined for success were tried and found to be superior to past practice, recognized as such, and accordingly adopted. That adoption in most cases was indicative of greater efficiency, but more important, indicative of a new willingness on the part of some of the farmers some of the time to evaluate and subsequently use new mechanical devices.

[8]

Threshing

The Old Order

The Biblical expression about "separating the wheat from the chaff" oversimplifies the threshing problem. A stalk of wheat newly cut lying in the field consists of a kernel (which is highly valued) encased in a hull (which is not valued) atop a long stem that can be used for such diverse purposes as feeding livestock or thatching roofs. The problem is to separate these three different parts. The solution from remote antiquity to the beginning of the modern era was to beat or thresh (or "thrash") the grain after spreading it out on a hard surface.

Throughout the colonial period Americans achieved this separation primarily in one of two ways: one used wood to beat the grain, the other livestock to trample the grain. The first, arduous and slow, threatened the laborers with respiratory problems, while many, including George Washington, considered the second an "execrable" practice,[1] in part because of what not infrequently dropped from the livestock to the threshing floor. The puzzle is why two markedly different techniques should be used to perform the same task.

The wooden device commonly used for beating was the flail,[2] an implement whose design changed little from King John's signing of the Magna Carta in 1215 to the ratification of the American Constitution in the late eighteenth century. (See Figure 8.1.) To a long wooden handle (or "handstaff") was joined a shorter wooden beater ("swingel" or "fringel") by some combination of thong plus swivel so the beater could rotate through 360 degrees.[3] Although commonly depicted in American and European art as being wielded by a full swing over the head, the flail could just as effectively be applied to grain on a barn floor by a looping motion of the arm about waist high, combined with a flick of the wrist.[4]

This threshing implement appears to have been unknown in ancient

Figure 8.1. Western flails through eight centuries

A. European, twelfth century

B. European, fourteenth century

C. British, eighteenth century

D. American, nineteenth century

Sources: James C. Webster, *The Labors of the Months* . . . (Evanston, Ill.: Northwestern University, 1938), Plate xxxi; Eric G. Millar, *The Luttrell Psalter* (London: Bernard Quaritch, 1932), Plate f.74b; R. Bradley, *A Survey of . . . Husbandry and Gardening* (London: B. Motte, 1725), Plate III; John J. Thomas, *Farm Implements and Machinery* (New York: Orange Judd, 1869), p. 187.

Egypt, where unjointed sticks or clubs were used.[5] The flail probably originated during Roman times and, being superior to unjointed alternatives, remained in use after the collapse of the Empire. Certainly by the time farming implements first appeared in European art, workers wielding flails were a common theme.[6]

The other threshing method dating from antiquity and employed in the colonies involved not beating with wood, but crushing with hoofs. As of 1800 livestock had been used to trample out grain in the Western world for more than four thousand years.[7] Settlers in the New World used both oxen and horses for this purpose, but by the late eighteenth century the horse appears to have been the preferred choice by a wide margin.[8]

Evidence concerning which threshing technique dominated in which geographic area of America is extremely sparse. The historical literature on the subject tends to feature two regional assertions and one uncertainty. The flail was "commonly used" in New England (or so the evidence suggests), and treading out grain with livestock was the dominant threshing technique in the South, at least on the eastern seaboard.[9] What is not clear is which technique was dominant throughout the broad area in between.

Given the scarcity of factual accounts from the areas in question, this puzzle can perhaps more usefully be approached by examining the costs and benefits associated with each technique. To hear contemporaries tell it, both were dreadful. Wielding a flail was slow and arduous work, usually undertaken in a dust-filled room, with the blows sometimes bruising the grain so badly it would not germinate if used for seed in a subsequent planting. Treading with livestock tended to injure the animals and soil the grain. Dung was always a problem, and if the treading was done in the field on poorly prepared ground (as happend not infrequently), dirt became mixed in with the harvested crop, lowering its market value.[10] The driving of horses in ranks, as depicted in Figure 8.2B, as of the late eighteenth century, was an ideal advocated by a farming pioneer, not accepted practice in the agricultural community. What is labelled "Bordley's plans" are merely two sketches by that author of circular threshing floors, one with a barn in the center (as depicted in Figure 8.2B) and one without. The common practice was to drive the horses "promiscuously and loose" in an enclosed area, causing each animal to struggle "to be foremost to get fresh air, jostling, biting, and kicking the others."[11]

But the farmers, as the existentialists might put it, were not free not to choose. The grain had to be threshed. The question yet to be addressed is why one method was chosen over another. One possible starting point in pursuit of the answer is with the reputed determinants of choice in a geographic region where both techniques were used extensively. Delaware provides a case in point. In response to "Queries on the present state of Husbandry and Agriculture in the Delaware State" in the 1780s, one resi-

Figure 8.2. Examples of treading grain in America

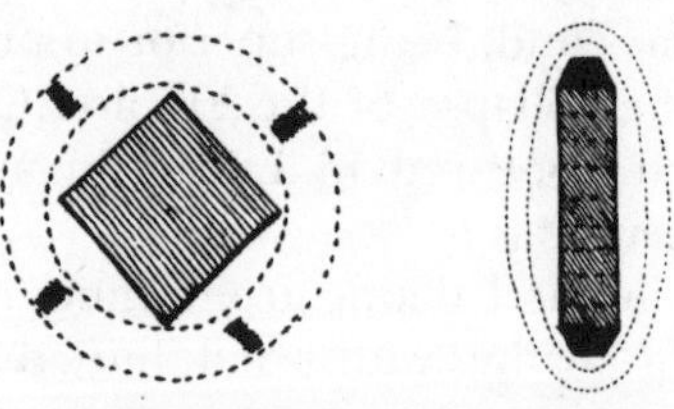

A. Bordley's treading floor plans, 1799

B. Outdoor treading, circa 1800

C. Indoor treading, nineteenth century

Sources: John Beale Bordley, *Essays and Notes on Husbandry . . .* (Philadelphia: Budd and Bartram, 1799), Plate III; Ralph Henry Gabriel, ed., *The Pageant of America* (New Haven: Yale University, 1926), Vol. 3, p. 99; Benjamin Butterworth, *The Growth of Industrial Art* (Washington, D.C.: U.S. Patent Office, 1892), "Thrashing and Cleaning Grain."

dent reported the major crops, such as wheat and barley, were "generally trodden out with horses," but "smaller crops, such as rye, oats, and buckwheat" were generally threshed with a flail.[12] The implication is that choice of technique was crucially dependent upon the size of the crop to be threshed. Small farms should accordingly prefer the flail, while large farms should use treading for "major" crops and the flail for "minor" crops.[13]

Reinforcing this general tendency were three factors, of different importance in different regions.

The first was capital availability. Eighteenth-century American capital markets were notoriously imperfect, and livestock constituted a significant capital investment, as did threshing floors if they were well made and sheltered from the elements. Small farmers had few horses and limited access to borrowed capital, whereas owners of large farms had greater access to both.[14] Second, the opportunity cost of farm labor in northern states declined markedly after the harvest was in and temperatures fell. Threshing with flails was often one of the few available tasks to keep the men, including hired workers, busy during the late fall and early winter.[15] Third, the flail produced sharply diminishing returns per blow in the later stages of threshing a given quantity of grain.[16] This could lead to significant wastage if workers were hired and paid according to the amount threshed. A lack of diligence (or from another perspective, the presence of rational calculation on the part of hired laborers) caused a significant portion of residual grain to remain unthreshed. If this was true for hired workers who lacked incentives to be thorough, it was even more true for slaves who had no stake whatsoever in either the productivity of threshing techniques used or the value of the final output. The implication for choices made on large plantations would seem obvious. Consider again the threshing with livestock depicted in Figure 8.2. The owner of a large plantation was sure to be well supplied with horses and slaves, and might invest a little or a lot in preparing a threshing floor. With these factor supplies in place, the incentive to thresh with livestock must have been almost overwhelming. Notice the nature of the labor input. All that was required was a few slaves with minimal skills to guide the horses[17] and remove the horse droppings—both activities easily monitored by an overseer. To shift from treading with livestock to threshing with flails would have required the hiring of a considerable number of temporary workers for any crop of any magnitude. Even if such workers were available (and were skillful and diligent), their wages would be an incremental cost, compared with the minor outlays required for treading grain with horses and slaves already owned and an overseer already on the payroll.[18]

In the first instance, then, the choice of threshing technique depended upon the size of the farm, with small farms using the more labor-intensive technique of threshing with flails and large farms using the more capital-intensive technique of livestock treading for major crops (and possibly flails for minor crops, the labor in question being applied during slack times to small amounts of grain to get a cleaner product for market). These tendencies (and in the mid-Atlantic region quite possibly they were no more than that) were powerfully reinforced by the presence of more severe winters in the North and slaves in the South. In short, the conventional wisdom about dominant technique by geographic region in America—flails in the North and treading in the South—seems unsurprising, as is the uncertainty about which technique prevailed in the various farming areas of the middle colonies.[19]

Problems of Mechanization

Early efforts to mechanize threshing in Britain and America in the main relied upon either a "beating" or a "rubbing" mechanism to separate the grain. The latter type would come to dominate successful American machines, but that dominance was slow to be established. The reason for the delay was the intrinsic complexity of the mechanical processes of any threshing machine. In a world of imperfect metallurgy and primitive machine tool expertise, complex requirements resulted in flawed machines. Thus the question of which method was intrinsically superior—beating or rubbing—was difficult to resolve when all threshing implements were prone to defects.

Early inventors of beating machines often tried to duplicate the flailing motion witnessed in the barn. Figure 8.3 reproduces two of these duplication efforts. The operating procedures of the British machine[20] are self-evident, as are its intrinsic limitations. To each corner of a triangle was attached a beater (not unlike the swingel of a flail), and a hand crank (L) spun all triangles simultaneously. The American machine depicted in Figure 8.3 used pegs on a rotating drum to activate the beaters, although whether such a device was ever constructed is far from clear.[21] All machines based upon the beating principle suffered from the same problem. The violence of the mechanical action used to separate the grain soon produced broken parts and an implement in disrepair.[22]

"Rubbing" machines did not require such mechanical ferocity, although how they worked is somewhat difficult to describe. A useful starting point is with the mechanical principle of an old-fashioned clothes wringer. Its two rollers, placed close together, rotate in opposite directions and thereby pull the clothes through and squeeze out excess water. A similar principle is evident in Figure 8.4, but now the rollers are a bit further apart, and each one has embedded in its surface pieces of iron placed in such a manner that they narrowly miss one another in the course of roller rotation. This machine was to shell Indian corn, not thresh wheat, but it

Figure 8.3. Experimental threshing machines, late eighteenth century

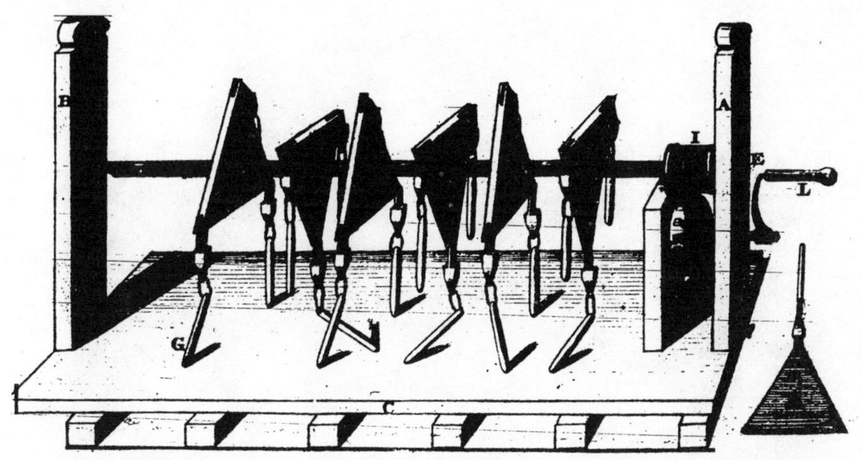

A. British flailing machine

B. American flailing machine

Sources: Pennsylvania Magazine, 1 (February, 1775), Plate opposite p. 70; Benjamin Butterworth, *The Growth of Industrial Art* (Washington, D.C.: U.S. Patent Office, 1892), "Thrashing and Cleaning Grain."

illustrates two features of the more complex machine of Figure 8.5, namely, rotating drums and iron protrusions positioned to pass close to one another during rotation. In Figure 8.5, the outer drum (with iron pegs facing inward) was stationary, whereas the inner drum (with iron pegs facing outward) rotated. Just as in the case of the corn shelling machine, the placement of the iron protrusions was crucial: close enough to assure that "rubbing" pressure would be applied to grain inserted into the thresher, but not so close as to risk collision. The resulting mechanism

Figure 8.4. Haven's corn shelling machine, 1813

Source: Charles W. Peale, "Account of a Corn Shelling Machine," Philadelphia Society for Promoting Agriculture, *Memoirs,* 3 (1814), p. 250.

could be roughly summarized as "an iron cylinder armed with teeth whirling in a toothed concave."[23] The rubbing action, as explained by an American in the 1830s, was produced in most American machines "by a revolving cylinder, with more or less cogs, spikes, or teeth, of different shapes and lengths, which pass through corresponding ones, placed in a concave, or bed-piece, which is stationary. The number of spikes vary [sic] from 100 to 3000."[24]

Whether based upon the beating or rubbing principle, most of these threshing implements of the late eighteenth and early nineteenth century—and certainly all of those with any claim to sustained success—were at the time among the most complicated and expensive of all agricultural implements in America and Britain.[25] High price and frequent breakdowns did not bode well for rapid adoption when "the least derangement . . . is death to the whole machine."[26] These problems persisted for decades, as indicated by the following list of American complaints from the early 1830s, by which time hundreds of patents had been granted and hundreds of different threshing machines were commercially available.

Figure 8.5. British peg drum threshing machine, 1850s

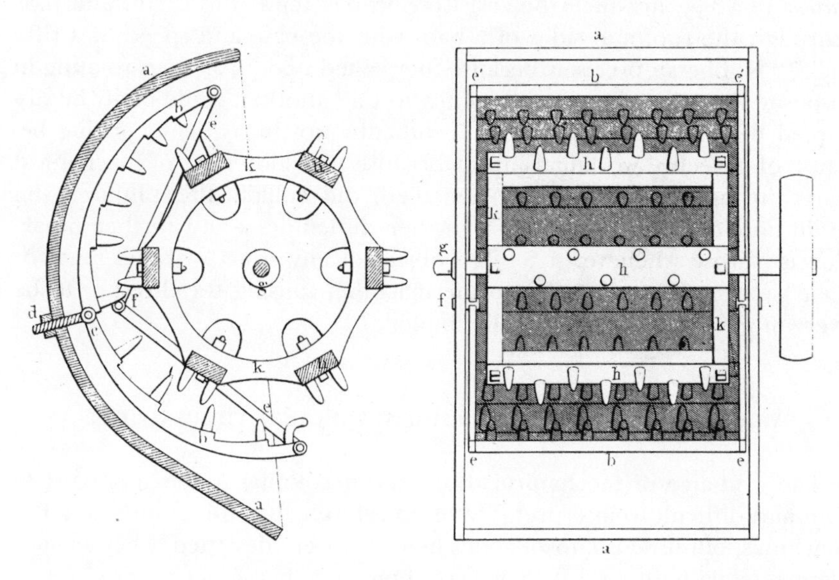

Source: John C. Morton, *A Cyclopedia of Agriculture* (London: Blackie & Son, 1856), Vol. 2, Plate XLVIII.

> In the first place, before we come to remark upon the merits of those now in use, we would observe, that one of the great and principal causes of failures, in many kinds of machines, is the flimsy, cheap, and do-for-the-present manner in which they are made. They are not unfrequently constructed by carpenters, or rather by those who are only an apology for a good one, and who could hardly construct a button to a barn-door, or as is frequently the case, by patentees, or their agents, a set of speculating cut-throats, alike ignorant of mechanical powers, of motion or matter, whose only talents consist in the *rigamarole* with which they recommend their wonderful discoveries.
>
> By the operation of these causes the farmer often gets an ill-constructed, weak, and rickety machine, which needs wedging, nailing, and bracing, at every revolution, and as much power to drive it as would operate a run of mill stones; and I have known many tolerably good machines condemned, from having connected with at a weak, ill-contrived, and ineffective horse power.[27]

The specific frailties of a given machine tended to reflect the threshing principle it employed. "Beaters" had to be run at considerable speed, and the resulting frenzy of mechanical action invariably strained key working parts. Breakdowns on occasion could imperil life and limb. "We know of one kind [of thresher]," a correspondent complained to the *Genesee Farmer* in 1833, in which the cast iron beaters tended to fly off and pass "through the roof and sides of a barn with the ease and speed of a rifle ball."[28] "Rubbers," precisely because they relied upon iron pegs rotating in opposite directions in close proximity to one another, could easily be disrupted by a slight misalignment, a difficulty not infrequently arising because of defective workmanship or because of a loosening of mechanical parts through repeated use. When teeth did collide, they "injured the grain and ruined themselves."[29] Further difficulties—and further breakdowns—arose whenever a foreign object of any size ("a stone, limb, or root") was accidentally fed into the machine, causing the thresher to be "very much injured, if not totally spoiled."[30]

American Mechanization Efforts: Colonial Times to 1815

The evidence of mechanical threshers in colonial America is hard to find and difficult to interpret. Scattered references can be found to a few machines, but little is known about how they were designed, whether they were actually built, and if they were, how well they functioned. Cases in point are Peter Oliver's announced efforts to contrive "a method to thresh grain" in 1762, John Clayton's machine approved for construction by the Pennsylvania Assembly in 1764, and John Hobday's "simple machine for separating wheat from straw" which the inventor insisted in 1772 "could be carried into execution by any tolerable carpenter" using models that he would "lodge . . . in different parts of the country" once "sufficient funds" had been raised.[31] The one certainty is that none of these efforts produced a commercially successful threshing machine, or even one deemed worthy of sustained commentary in the American literature. The first depiction of a machine to be published on this side of the Atlantic is the British machine pictured in Figure 8.3.[32] Its limitations, discussed above, assured a universal indifference among local farmers to its importation or replication. The colonial era therefore ended with American allegiance to ancient threshing techniques virtually unscathed by the handful of attempts in Britain and America to mechanize that process.

Compared with the indifference and inactivity of colonial years, the period between the Revolution and the War of 1812 seems at first glance to be one of innovative bustle. Leading American agriculturalists inspected machines, imported machines, and designed their own machines, sometimes using as a guide British models imported for that purpose. John Beale Bordley sent the plans of a British machine to the Philadelphia Society for Promoting Agriculture; George Washington corresponded with Arthur Young about importing British threshers in the

1780s and built his own machine in 1797; Thomas Jefferson inspected an American-made thresher with Washington in 1791, and ordered a model from Britain in 1792 which served as a guide for the machine he completed in 1796.[33] Novel threshing devices were reported in the press and inspected by agricultural societies.[34] Some were British made, some American, and beginning in 1791, a number of the latter were patented as indicated in the following table.

Table 8.1. American patents for threshing machines, 1791–1810

Year	Number	Year	Number
1791	2	1800	—
1792	—	1801	1
1793	—	1802	2
1794	2	1803	1
		1804	3
1795	—	1805	—
1796	—	1806	—
1797	2	1807	2
1798	1	1808	1
1799	—	1809	4
		1810	6

Source: U.S. Bureau of the Census, *Preliminary Report on the Eighth Census* (Washington, D.C.: U.S. Government Printing Office, 1862), pp. 96–97.

But this period of unprecedented American experimentation with threshing devices, when compared with the tornado of inventive energy unleashed in the second and third decade of the nineteenth century, reduces to little more than the first notable stirrings of inquisitiveness by the select few, producing no machines of widespread interest and none of lasting consequence. Prior to 1815, two problems were particularly evident, both traceable to the intrinsic complexity of all threshing machines with any prospect of demonstrating sustained superiority over established threshing practices. The first was cost. The second was the difficulty of maintaining mechanisms notorious for breaking down. American experts accordingly tended to dismiss British imports as "either . . . so very expensive, that few could use them, or so complicated that they soon were put out of order."[35] American-made machines were similarly suspect. With a reputation for being "soon put out of order" or abandoned as inefficient, none of them ever achieved commercial success, or even sustained popularity among farmers living in close proximity to the inventor. Washington's machine, for example, built using the plans of William Booker soon proved unsatisfactory; a Maryland machine constructed with the aid of Colonel Anderson's plans developed a warped wheel and was abandoned; in 1802 an immigrant from Edinburgh introduced into the Mid-Atlantic states "six or seven" based upon "the Scotch principle," but they soon developed problems and "common workmen" could not repair them; the machine patented in 1803 by Jedediah Turner, based "upon entirely new, and very plain principles" was never commercially produced. (Instead, Turner offered to sell the patent right "on the most reasonable terms."[36]) Washington perhaps spoke for a number of would-be agrarian pioneers when he voiced his frustration with this prospective new technology in 1793: "I have seen so much of beginning and ending of new inventions, that I have almost resolved to go on in the old way of treading."[37]

One indication of the limited importance of this early experimental period is what was missing in surveys made and prizes offered and essays written by distinguished agricultural pioneers. The list of questions sent by Washington to "the best informed" of agriculturists in the early 1790s included nothing about threshing techniques or threshing machines.[38] The essay written by John Beale Bordley in 1786, and reprinted substantially unchanged in 1799, was animated by Bordley's dissatisfaction with the manner in which livestock were used to tread out grain in the South. But his recommendations for improvement were confined to modifying existing practice by building circular threshing floors (of the sort depicted in Figure 8.2) and harnessing the horses in ranks (also depicted in Figure 8.2). His only reference to machines as a possible alternative is confined to a footnote, in which he notes that such devices are "not used in America that I know of."[39] Interest in, and endorsements of, threshing machines are similarly absent in the literature of contemporary agricultural societies. Nothing is asked about such implements in the fifty questions compiled in 1800 by the Massachusetts Society for Promoting Agriculture "with a view to collect the most accurate information on the principal Branches of Agriculture, as now practiced, and thus be enabled to propagate the knowledge of whatever shall be found useful; and to open more wide the way for future improvements."[40] The same list of queries—with the same omission—was used to gather information more than a decade later.[41] Nor did this society offer to reward inventors of threshing devices, although it offered premiums for such diverse improvements as "an effectual and cheap method of destroying the canker-worm," and "the best and cheapest method of introducing fine grass, fit for hay or pasture" into meadows "now producing coarse grass, or bushes."[42] The same disinterest is evident in the premiums offered by the Philadelphia Society for Promoting Agriculture, the oldest and arguably the most prestigious organization of its kind in the country. Their initial list of awards published in 1785 offered premiums for "the best experiment" in trench plowing, and sowing wheat "with a machine," but nothing was promised for threshing wheat with a machine.[43] Twenty-one years later, a revised list of premiums featured awards designed "to introduce a practice, found very beneficial where it has been fairly tried," but the practices of interest to the society involved manure use, crop rotation, and plowing, and once again no mention was made of threshing.[44]

Consider finally the number of patents listed in Table 8.1. These indicate some patent activity beginning in 1791 that showed signs of accelerating as the first decade of the nineteenth century drew to a close. But if the patents recorded for the last two years (1809 and 1810) are ignored, the average for the period 1791 to 1808 is slightly less than one a year or ten a decade. For the quarter century beginning in 1810, patents for threshing machines would average ten a year, or one hundred a decade.[45]

American Mechanization Efforts, 1815–1835

In 1815, for the first time in its twenty-three-year history, the Massachusetts Society for Promoting Agriculture offered a premium for a "machine for threshing and separating grain, adapted to a farm of medium size."[46] The initial response was not encouraging. None of the 1817 entries was deemed worthy of any prize.[47] A year later, with considerable reluctance the society awarded a premium to Elihu Hochkiss of Brattleborough for a threshing machine which, in their view, was much in need of improvement.[48] Nor were British imports judged to be an attractive alternative. In 1819 New York agrarian expert Jesse Buel complained that foreign-made threshers were so costly they appealed only to "rich amateurs," although he remained hopeful "that the mechanical genius of our own country (which is not inferior to that of any other) may be the first to combine *power* and *cheapness* in this machine."[49] That optimism was echoed in 1820 by the vice-president of the Pennsylvania Society for Promoting Agriculture, although he was quick to concede that "a machine effectual for the threshing of wheat, and not too high-priced," was still needed and not available.[50] All this soon changed. By the early 1830s, more than thirty new machines were being patented every year, and about seven hundred different mechanical threshers were in use (often a dozen different kinds in a single town) reflecting a "rage for these machines" among the American farming community.[51] The puzzle is what happened during the intervening decade.

None of the existing agricultural histories provides a satisfactory answer. Many are content to echo William Brewer's judgment: "About 1820 thrashing-machines ceased to be a curiosity in the better grain-growing regions of the middle states, but by far the most of the grain was thrashed by the old methods until after 1825. Between that date and 1835 their use spread with great rapidity."[52] Leo Rogin challenged this pattern of accelerated adoption after 1825 as being off by half a decade. "The beginning of rapid extension of the thresher," he asserts, began in 1830.[53] But the only evidence offered by Rogin in support of that contention is a single observation in 1829 by a committee of the Maryland Agricultural Society that "The machines made for threshing wheat hitherto have been at a price vastly beyond their value, and not within the means of the great mass of farmers."[54]

For Rogin to be right about accelerated adoption beginning in 1830, given that by 1833 those with the previously noted "rage for threshing machines" could choose from over seven hundred models, the associated surge in invention and adoption must have been explosive in the extreme. Such a sharp discontinuity could have been triggered by a sudden fall in prices or a major breakthrough in the technology of threshing machines. Available evidence suggests that no such discontinuity occurred.

Consider first the trend in prices during the 1820s.[55] When the owner of a New York agricultural implements store advertised threshing machines in 1819 (available at what he claimed to be "the only establishment of the kind in the United States"), these implements appear to have been primarily British imports and the prices were staggering. Most cost over $400, and some more than $500.[56] But as American-made machines became increasingly available in the 1820s, prices dropped dramatically. The larger and more complex American machines designed to be driven by horse power were available for $100 to $150 by 1825, while smaller machines worked by hand listed at a fraction of that cost—usually less than $50, and sometimes as low as $20.[57]

How well these worked was quite another matter. Although new machines were often advertised as signaling a major breakthrough in threshing technology, commentary by contemporary experts makes clear that throughout the period 1820–1835, all machines were prone to defects, and none towered above the rest in mechanical performance or cost effectiveness.[58] Thus, although by 1833 over seven hundred different kinds were in use, complaints about defects continued with a vigor and universality not significantly different from those of a decade earlier.

Figure 8.6 depicts one of the more popular models developed during this period. It used a relatively simple system of rollers (b) to feed the grain to a revolving beater (a). The drive mechanism (f) was designed to be combined with horse power, but a less expensive model ($75 instead of $125) was available "without the horse gear, to be worked by hand."[59] Two common ways of harnessing horse power are illustrated in Figures 8.7 and 8.8.[60] The sweep mechanism of Figure 8.7 consisted of a large horizontal master-wheel (mounted overhead so the horse would not have to step over shafts), which, by a series of belts, transmitted power to a rubbing type of thresher. The mechanisms of the tread (or "railway") device for harnessing animal power appear more clearly in the bottom diagram of Figure 8.8. The power of the sweep depended upon the strength of the horse; that of the treadmill, upon the weight of the horse. The

Figure 8.6. Pope's threshing machine, 1823

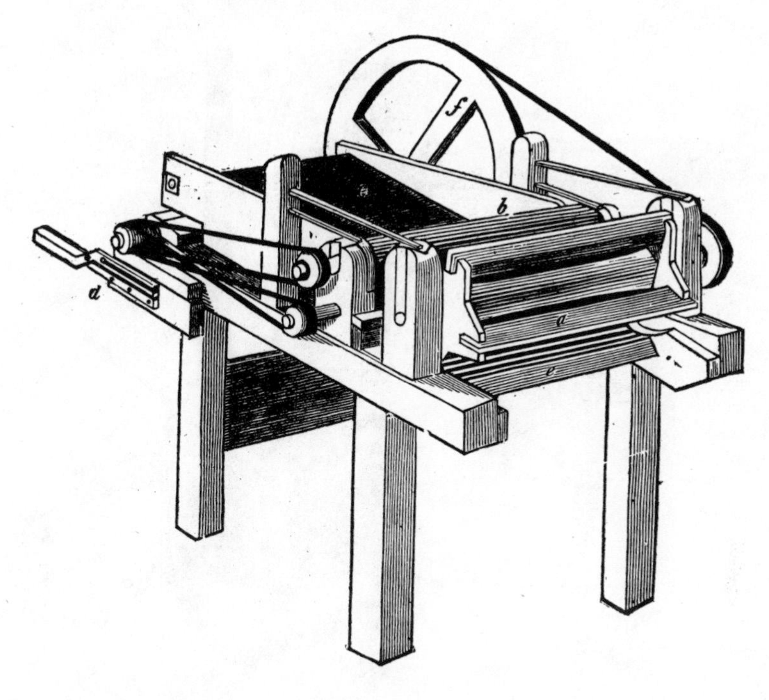

Source: American Farmer, 5 (April 11, 1823), p. 21.

former was therefore a more efficient means of generating large amounts of power, and accordingly more common on larger farms.[61]

Last, of limited importance as a threshing implement, and often overlooked in the literature, is a mechanism for separating grain in evidence in the 1820s, but with a history dating back to the Roman Empire. The "Punic cart" of Roman times broke up straw by running over it with a set of toothed rollers.[62] By the early eighteenth century, a similar principle was evident in a wooden roller used in a number of European countries, including Denmark, Germany, and the Netherlands. But now the threshing action was achieved by the slatted surface of a cone-shaped roller, one end fixed to a post, as shown in Figure 8.9. Similar threshing rollers were used in Pennsylvania and New York in the eighteenth century,[63] although the boxes tucked beneath the horses' tails in Figure 8.9 to catch their droppings evidently had limited appeal to American sensibilities. While evidently not preoccupied with cleanliness in threshing, some farmers in the New World did show an interest in modifying the crushing action of the slatted surface. Toward the end of the eighteenth century, a different conical roller was designed with protruding pegs that gave it a porcupine appearance.[64] (See Figure 8.10.) A version of the mechanism dating from

Figure 8.7. Goodsell's grain thresher, 1823

Source: American Farmer, 5 (June 23, 1823), p. 91.

the 1820s is reproduced in Figure 8.11. The threshing surface was composed of eight rows of large pins (aa) three inches square at the butt end, and eight rows of smaller pins (bb) which, being 1½ inches shorter, took none of the roller's weight as it rested on the barn floor. These rollers with a pegged surface continued to be used in the first few decades of the nineteenth century, but they never achieved widespread popularity.[65]

The Long View: Colonial Times to 1835

In the three-quarters of a century between the 1760s and the 1830s, America had been remarkably—one might even say, astonishingly—transformed. Between the outbursts of indignation provoked by Viscount Townshend's taxes and those provoked by the Tariff of Abominations, between the carnage wrought by the French and Indian War and that accompanying the forced relocation of the Cherokee Nation to territory west of the Mississippi, between Jared Eliot's pioneering but distinctly premodern *Essays on Field Husbandry* and the first American edition of Justus Liebig's *Chemistry and Its Application to Agriculture and Physiology* published

Figure 8.8. American treadmill devices, 1860

A. Horse treadmill with threshing machine

B. Dog treadmill with churn

Source: John J. Thomas, *Farm Implements and Farm Machinery* (New York: Orange Judd, 1869), pp. 188, 191.

in 1841, the fabric of American society had been monumentally reshaped. Against this backdrop of multiple upheavals and transformations, the evolution of agricultural implements might be viewed as just a case in point. But as a case, it had an unmistakable importance. In a new republic with a labor force still overwhelmingly engaged in agricultural pursuits, the economic power of the nation and the economic welfare of its people were crucially depended upon raising agricultural productivity, not just by creating more efficient implements, but by establishing a tradition of searching for improvements.

In the case of threshing implements, those improvements came in three distinctive waves between 1760 and 1835, or perhaps more accurately, as first a ripple, barely perceptible, then a wavelet, followed shortly

Figure 8.9. Dutch threshing roller, circa 1900

Source: Jan Bieleman, *Geschiedenis van de Landbouw in Nederland, 1500–1950* (Amsterdam: Boom Meppel, 1992), p. 121; by permission of the Fries Museum, Leeuwarden.

Figure 8.10. American roller thresher

Source: From the collections of Henry Ford Museum & Greenfield Village, Negative 17244.

Figure 8.11. American grain roller, 1820s

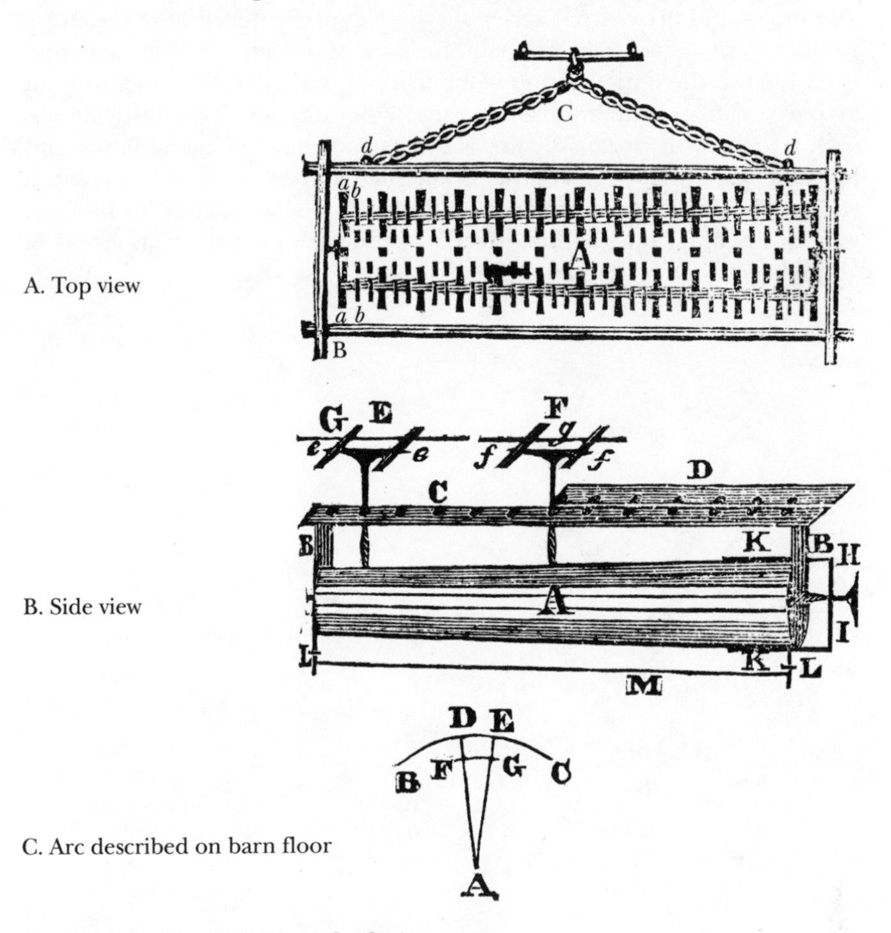

Source: American Farmer, 3 (March 8, 1822), p. 397.

by a tidal bore. Immediately prior to the Revolution, a handful of experiments were conducted by the venturesome few about which little is known beyond their negligible impact upon the literature or established practice. From the 1780s to 1815, the pace of experimentation clearly quickened. New threshing devices were designed and publicized and patented, and members of the agrarian elite evaluated their performance and sometimes employed the machines. But this period was also noteworthy for what was not accomplished. Not a single new machine appears to have had a decisive impact upon established practice or achieved commercial success or even provoked sustained commentary in the literature. All this changed quite suddenly in the years following 1815. The number of machines invented annually multiplied, their performance improved, and prices declined sharply. Some achieved commercial success and many won praise and prizes from agricultural societies. By the early 1830s, more patents were being filed annually for new threshing devices than had been filed in the entire period between 1791 and 1810. In 1800, or even in 1810, a new threshing machine in any locality was a curiosity. By the early 1830s, with dozens in use within a few miles of many farms and hundreds of models to choose from, such implements had become a commonplace. The era of complex threshers and combines, to be sure, was yet to come. But the first great surge in threshing innovations, as of 1830, lay clearly in the past.

[9]

Winnowing and Straw Cutting

Remaining Tasks

A saying familiar to British commandos of World War II was "There's always one thing left to do." The remaining problem for this book is what to do with a mixture of grain, straw, and chaff, newly threshed and lying on the barn floor. Problems yet to be addressed include:

(1) how to separate the straw from the grain and the chaff,
(2) how to separate the grain from the chaff, and
(3) how to chop up the straw that is to be fed to livestock.

In premodern America, the first of these was accomplished with comparative ease by using wooden forks or rakes to clear the bulk of the straw to one side. The other two assignments were more complicated.

The difference in weight between grain and chaff has been the key to separating these two from ancient times to the present. Given that the first is significantly heavier than the second, dropping a mixture of both through air currents causes the lighter chaff to be blown clear. The obvious air current to use was that supplied by natural wind. Indeed, the American Samuel Deane in 1790 defined winnowing as "cleaning [grain] from its chaff by wind."[1] More than thirty-one centuries before he wrote, workers depicted on the wall of the Tomb of Thebes can be seen doing exactly that, although not by procedures common in Deane's day. As illustrated in Figure 9.1, two lines of workers facing each other brought together large wooden spoons with a straight edge, lifted the mixture of chaff and grain and then, by separating the spoons, allowed the mixture to fall to the ground, presumably through blowing winds not easily depicted on a tomb wall. A different winnowing technique, dating from Roman times, would come to dominate in Europe. Wind was still the key to

Figure 9.1. Premodern winnowing techniques

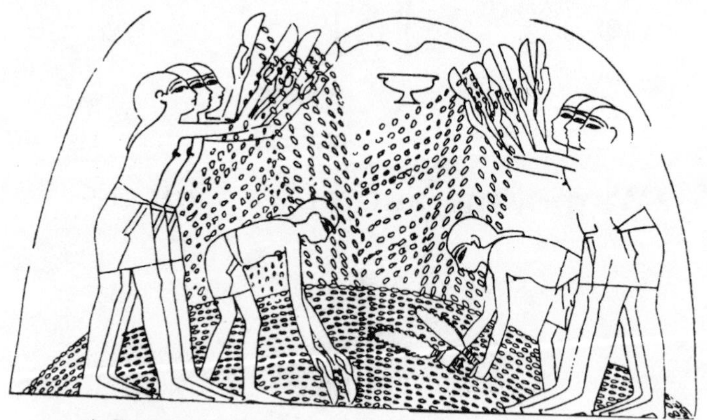

A. Egyptian winnowing with spoons, circa 1400 B.C.

B. Ancient winnowing

C. European winnowing basket, circa 1250 A.D.

D. American winnowing sieve, 1830

Sources: Norman DeGaris Davies, *The Tomb of Nakht at Thebes* (New York: Metropolitan Museum of Art, 1917), Plate XXI; Benjamin Butterworth, *The Growth of Industrial Art* (Washington, D.C.: U.S. Government Patent Office, 1892), "Winnowing and Sifting Grain"; E. M. Jope, "Agricultural Implements," in Charles Singer, E. J. Holmyard, and A. R. Hall, eds., *A History of Technology* (Oxford: Clarendon, 1956), Vol. 2, p. 68, by permission of Oxford University Press; *American Farmer,* 11 (March 5, 1830), p. 408.

separation, but now large baskets were used to toss the mixture of chaff and grain into the air. With the right toss into an air current of appropriate force, the chaff would be blown clear and the heavier grain caught in the basket as it descended. The two middle diagrams of Figure 9.1 illustrate this procedure, albeit poorly. (Well-drawn illustrations of this winnowing process are almost impossible to find, in part, no doubt, because the process is difficult to draw, involving as it does tiny particles and currents of air.) The winnowing basket resembled a giant scoop, about the width of the worker's outstretched arms, with handles on the side toward the rear. Between tosses into the wind the worker shook it with a circular or slightly forward motion to move the chaff toward the outer rim and the heavier grain toward the back. Although a slow process, it was not particularly strenuous. When most of the chaff had been removed, what remained was passed through leather or wooden sieves of the sort portrayed in the bottom diagram of Figure 9.1, a process that eliminated foreign matter, such as dirt and small pebbles acquired when the mix of grain plus chaff was swept up from the barn floor.[2]

Such a winnowing method could, of course, be used only when the wind was blowing, and whether the resulting separation was done well or badly depended crucially upon the velocity and constancy of air currents available. Too strong a wind could blow part of the grain beyond easy retrieval by the winnower. Too little wind could bring the process to a halt. And that bane of all winnowers, irregular gusts, "'tis well known disorder the whole Work, and occasion a considerable Waste of Seed."[3] Inclement weather could imperil the worker as well as the process. "When the wind blows," complained an observer of farming practices in late eighteenth-century America, "the weather is not always fair; or if fair, it is often so cold, or damp, that the health of the workman is exposed."[4]

The solution for all these difficulties was to devise ways of creating an artificial breeze. The implements initially used were crude hand-cranked fans of the sort depicted in Figure 9.2.[5] To minimize the dust stirred up and maximize the efficient use of air currents thereby generated, the next step was to enclose the fan in a box of the sort illustrated in Figure 9.2. This 1738 version had a rotary fan at one end, and a hopper (B) at the top through which the mix of grain and chaff was poured. By means of a hand crank, the fan generated a breeze that blew the chaff out the back of the box while the heavier grain descended to a sloping floor, and from there, slid out the front of the box. The optimal length for the box depended critically upon the force of the air currents produced by the fan. It had to be short enough to assure that chaff would be blown clear, but not so short that some of the grain would also be lost. Or as one contemporary characterized the desirable size, "Good Seed . . . never goes so far as the End of the Machine, it comes down within the Box, and the empty Boles and rotten Seed are carried out together straight before the Wind."[6]

Figure 9.2. Premodern winnowing fans

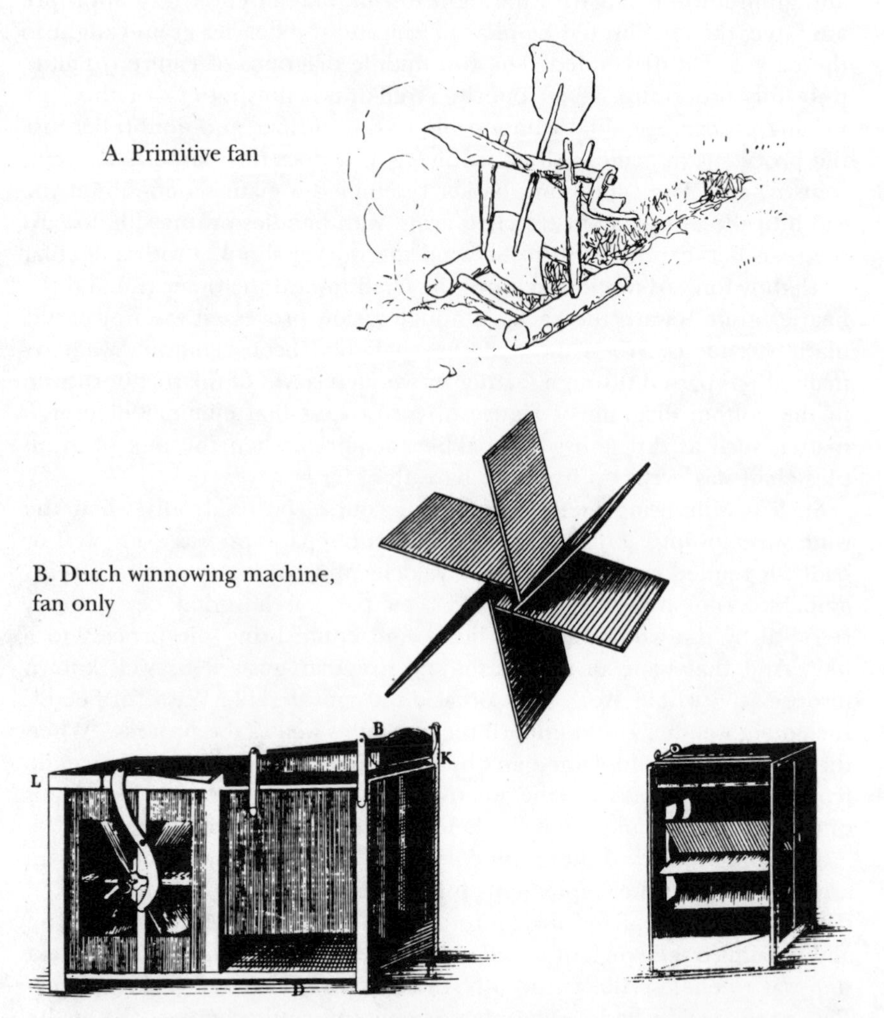

A. Primitive fan

B. Dutch winnowing machine, fan only

C. Dutch winnowing machine, 1738, front and side view

Sources: Gösta Berg, "The Introduction of the Winnowing-Machine in Europe in the 18th Century," *Tools & Tillage,* 3 (1976), p. 37; Dublin Society, *Essays and Observations* (Dublin: Dublin Society, 1740), p. 163.

The last step in this initial mechanization of the winnowing process was to add a sifting capability to the fan-in-box design. The end result is difficult to follow in diagrams of early machines, but the basic principle is readily apparent in the bottom diagram of Figure 9.3. Through the hopper at the top is poured a mixture of grain, "chaff, dirt, cockle, grass, . . .

Figure 9.3. Primitive and modern winnowing machines

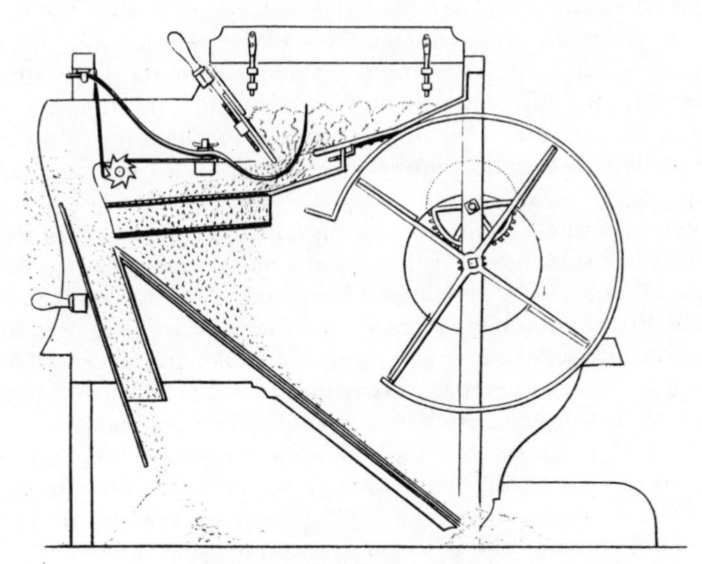

A. "Fan machine," circa 1820

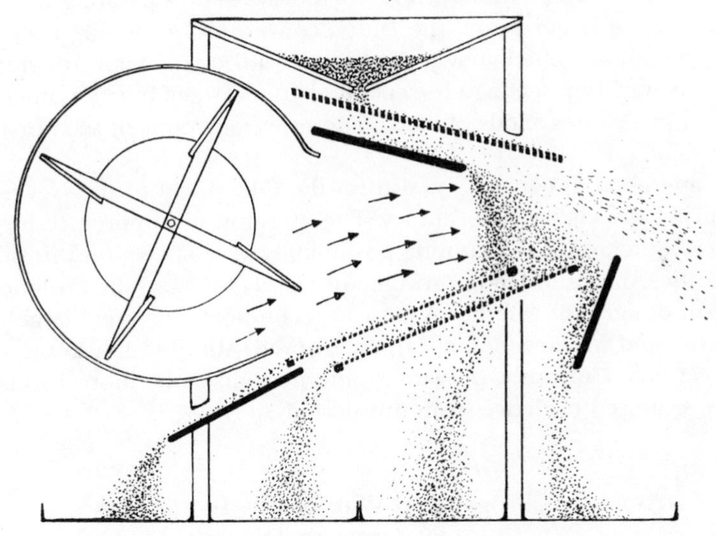

B. Sectional view of winnower, 1960

Sources: Abraham Rees, *The Cyclopaedia . . . of Arts, Sciences, and Literature* (Philadelphia: Samuel F. Bradford and Muray Fairman, 1810–1824), Vol. 1, Plate XXXVII; H. J. Hopfen, *Farm Implements for Arid and Tropical Regions* (Rome: Food and Agriculture Organization of the United Nations 1960), p. 127.

dust and straw."[7] As in the past, this conglomeration of the wanted and unwanted was subjected to a blast of air generated by the rotating paddles of a fan. (Note the arrows in Figure 9.3.) What was new were the sieves or riddles—initially one or two but over time becoming progressively more numerous, with different-sized meshes, all slanting downward to carry the mixture that remained to the next sorting mechanism. (Most machines soon included a variety of interchangeable sieves for sorting grains of different sizes.)[8]

When and where crude winnowing fans were first used in Europe, as well as their subsequent evolution into more sophisticated designs, are topics still debated by agricultural historians. Scattered evidence suggests that the first winnowing fans came from the Far East, were used in Europe as early as the sixteenth or seventeenth century, and were well known by the eighteenth century, particularly in Holland and Britain.[9] In his 1725 survey of British agriculture, for example, Bradley claimed that "every farmer" had a "fan or machine" for winnowing—no doubt an exaggeration, but one suggesting how common the mechanism had become.[10]

Sieves were evidently being added to the fan-in-box design by the early eighteenth century. The description accompanying the one portrayed in Figure 9.2, for example, noted how, for purposes of separating flax seed from "bodies of larger bulk," the Dutch converted the sloping floor to a screen. (See the hatched area at the bottom of the box below the hopper B.) In this way, flax seeds were separated from larger but unwanted objects, such as "stones, clods of earth, and especially pods of seed or roots of plants."[11]

The fan-in-box design migrated from the Old World to the New some time during the eighteenth century. The question of importance for this book is whether, once the fanning possibilities were known on this side of the Atlantic, American fan improvements closely followed the evolution of European designs, or whether American technology remained largely unmodified—and winnowing fans largely ignored—until well into the nineteenth century. Curiously, the latter pattern seems to be more consistent with the scattered evidence from this period.

American Progress in Winnowing Technology: Colonial Times to 1830

In the decade immediately preceding the American Revolution, a few references can be found in farm inventories and advertisements to winnowing "fans" or "Dutch fans," the adjective suggesting the probable geographic origins of the design.[12] Whether these early models included any sieves is not clear.[13] But by 1790 one of the few available descriptions of fans in use in post-Revolutionary America suggests a mechanism strikingly similar to that described in 1738 by the Dublin Society as typical of a "Dutch Fan."[14] In the half-century between the Revolution and the election of Andrew Jackson, the size and shape of American winnowing fans changed little. (See Figure 9.4.)[15] The main developments occurred within the box, particularly the development of a progressively more sophisticated system of oscillating sieves to achieve a better sorting of the residue left after the chaff had been blown clear. By the close of the eighteenth century, British fans were already evolving in this way.[16] American progress during the early nineteenth century might reasonably be expected to mimic British evolution. The evidence suggests a different story.

That story begins with a puzzling comment and a curious development. In his 1790 survey of New England agriculture, Samuel Deane, after noting the presence of winnowing fans in the new republic, added that "Of late, the fan is almost out of use."[17] If the new technique was clearly superior to the old, why was its application in decline? Adding to the puzzle was the growing popularity at this time of a new winnowing technique which, on close inspection, was little more than the old technology plus a large sheet. As described by one contemporary, this procedure "consisted of passing the wheat and chaff through a coarse sieve or riddle upon the barn floor, while two persons took a sheet between them, and by a particular flapping of the sheet produced a breeze that blew the chaff away."[18] Scattered references to the use of the winnowing fan admittedly persist between the end of the Revolution and the beginning of the War of 1812.[19] But if such fans represented the wave of the future, why did this nonmechanical and more laborious method of sheet shaking gain wider use[20] at exactly the time that fan technology should have been improving? The answer, albeit tentative, is that fan technology in America was not improving—or not improving in any significant way—while available fans tended to be viewed as relatively expensive and subject to mechanical defects.[21] All of this would change dramatically, but not until the 1820s, or so the evidence suggests.

Part of that evidence consists of announcements and advertisements praising winnowing fans of the 1820s as "Improved," usually with appended details on new devices for feeding or sifting.[22] But these descriptions were commonly supplied by inventors or producers of machines, a group unlikely to characterize their wares as "unimproved." More objective and, for that reason, more revealing are the opinions of agricultural organizations trying to assess the merits of inventors' claims. Taken collectively, these opinions confirm that the trend in winnowing technology in the 1820s was unambiguously for the better. A few examples:

- the Maryland Agricultural Society noted the new Watkins fan "works with three screens," and judged it to be "an exceedingly valuable improvement in this necessary implement" (1821)[23]

Figure 9.4. American winnowing machines

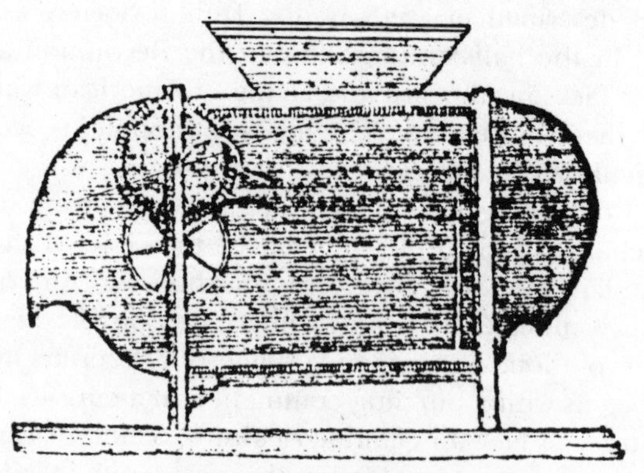

A. Dutch fan, 1774

B. Wheat fan, 1829

C. Horizontal winnowing mill, 1834

Sources: Virginia Gazette, May 26, 1774, p. 1; *American Farmer,* 11 (June 5, 1829), p. 96; *New England Farmer,* 12 (February 19, 1834), p. 249.

- the Fredricksburg Agricultural Society awarded a cup (valued at $15) to Jacob Gore for "An improved Wheat Fan . . . the superior excellence [of which] principally consists in the ease with which motion is communicated to the sifter" (1824)[24]
- the Essex Agricultural Society considered "Woodman's winnowing machine" to be "a combination of the best parts of other machines with which we have been acquainted," adding "we have never before seen the application of a third sieve in the manner used in this machine" (1824)[25]
- the Committee of the Cattle Show and Fair held at Easton considered the wheat fan of Robert Sinclair, a well-known Baltimore supplier of agricultural implements, "a good one, and well calculated for the use of most farmers" (1824)[26]
- a committee of the Maryland Agricultural Society reported favorably on several improvements in winnowing fans that included a self-feeding mechanism at the top of the hopper and a vibrating valve to smooth the flow from the hopper to the riddles (1826)[27]

As quality improved, prices fell. The trend in the general price level between 1819 and 1825 was sharply down in the first year (about 18 percent) with little decline thereafter.[28] In the same six-year period, the price of advertised winnowing machines fell by roughly half (from an average of about $40 to an average of about $20).[29]

This evidence of falling prices and quality improvements, all within the span of a few years, suggests a sharp discontinuity in the trend in winnowing fan technology within America,[30] which heretofore had been distinguished by an absence of dramatic change. By the middle of the 1820s, one of the better-known dealers in agricultural implements, while not given to understating the merits of his merchandise, could fairly characterize advances of recent years in winnowing technology as exceptional. "Great improvements have lately been made in these machines. They are now more portable, clean better and faster, and but seldom get out of order. The cockle and chaff are completely separated from the grain. By a new method of manufacturing, they are now made so as to be taken apart for foreign orders, and packed in boxes, occupying, comparatively, but a small space."[31]

The same pattern of development—little change for decades, followed by a rush of improvements in a few short years—also characterizes the evolution of straw-cutting technology in America.

Straw Cutting: The Reason Why

With grain threshed and winnowed, one of the remaining problems was what to do with the leftover stalks that had been separated out prior to winnowing. These had a variety of uses, including as feed for livestock, particularly in areas where harsh winters limited the ability of the animals to forage for themselves. In the late eighteenth and early nineteenth century, on both sides of the Atlantic one belief that gradually gained acceptance was that chopped feed was better feed. Given the primitive state of veterinary knowledge, this opinion was buttressed only by casual empiricism, occasionally masquerading as systematic experimentation. The core idea was economic: that the same animals could be raised with less food if the food was chopped up instead of "thrown in racks or on the ground, long and uncut, as gathered, and eaten that way."[32] Put differently, the same quantity, if chopped, made for "better" animals, although providing an operational definition of "better" was seldom easy. (Dairy cattle were an exception, as "better" could be defined as giving more milk for the same amount of feed.[33])

The common explanation for the superiority of chopped hay or straw was the ease with which the animal could digest it. A certain percentage of uncut straw, the argument went, would pass "through the animal, without contribution to its nourishment," whereas cut straw was "more easily and perfectly digested."[34]

For a growing body of agrarian experts, then, the benefits from cutting up animal feed were assumed to be considerable. The issue of importance to the practical farmer was whether such benefits were likely to exceed the costs incurred in chopping. The initial outlay required for a straw cutter was central to any estimate of net economic gain, as were the reliability of the machine and the manpower needed to operate it. Beginning in the latter half of the eighteenth century, all these latter calculations changed as the technology of chopping improved.

Straw-Cutting Technology: The British Challenge

By 1776 the British had developed cutting boxes for chopping straw of the kind depicted in Figure 9.5. They operated in much the same way that an office paper cutter functions today. Straw was put in the box with the ends protruding (over the edge bb), and these ends were sliced off by the downward thrust of a large blade (d) anchored at one end (at f).[35]

Subsequent technological changes were designed mainly to speed up the chopping process and refine feed mechanisms by which the straw was moved through the box. Toward the close of the eighteenth century, a series of British inventions had produced the prototype for the kind of straw cutter that would come to dominate in the nineteenth century. Figure 9.6 shows an early example.[36] What Americans preferred to label "straw cutters" the British called "chaff cutters," although what was being cut was, strictly speaking, not chaff but straw.[37] Comparing Figures 9.5 and

Figure 9.5. British cutting box, 1764

Source: Museum Rusticum, 1 (1764), Plate II, following p. 262.

Figure 9.6. British straw cutter, 1803

Source: A. F. M. Willich, *The Domestic Encyclopaedia* (Philadelphia: William Young Birch and Abraham Small, 1803), p. 49.

9.6 makes clear that the major change was the substitution of whirling blades mounted within a wheel for the old single-blade design which chopped by delivering a downward stroke. What is not well illustrated in Figure 9.6 is the means whereby the hand crank that turned the blades also kept in motion a wooden roller within the box designed to "draw the straw forward" for slicing. Although subsequently modified in many ways, the basic design of British straw cutters changed little in the next forty years.[38] (Compare, for example, the 1803 machine of Figure 9.6 with the two illustrated in Figure 9.7.)

Straw-Cutting Technology: The American Response

Early American reactions to these improved British machines were typical: the foreign implements (or most of them) were condemned as too expensive, too defective, or both, while a handful of local inventors did a bad job of trying to develop reliable low-cost alternatives.

The least expensive cutting mechanism was some variant of the British cutting box. Thus, in his edited version of the British *Domestic Encyclopaedia,* James Mease declined to reproduce the plate of Salmon's straw cutter (it was "too costly for the United States"), advocating instead the simple cutting device illustrated in Figure 9.8 (invented by a Pennsylvanian). In the American version, a treadle (not well illustrated) was used "to press the straw tight" prior to being cut by the downward stroke of a single blade (D), whose trajectory was controlled by two laths (CC) that "conducted the knife in its proper course through the straw." Scattered references can be found to the use of such boxes by agrarian pioneers like George Washington and John Taylor.[39] In the first decade of the nineteenth century, an American inventor, responding to a premium offered by the Massachusetts Agricultural Society, designed a more elaborate straw cutter for which he received a patent in 1808, but the feed mechanism was "very imperfect" and the implement was largely ignored.[40] Indeed, all straw-cutting mechanisms tended to be ignored or censured in the early years of the young republic. In his landmark surveys of New England practices published in the 1790s, Samuel Deane made no mention whatsoever of straw-cutting devices.[41] A decade later another New England writer conceded the desirability of chopping straw, but complained about the absence of simple devices to accomplish that task.[42] John Taylor of Virginia was similarly unimpressed with available options. After trying the "best" of cutting boxes "for a long time," he wrote in 1809 that he had abandoned their use "on account of the expence and inutility of the labour."[43] Another noted agrarian, Richard Peters, was not surprised. His response to Taylor's letter might fairly be characterized as "I told you so" (and some time ago). "I have always been of opinion, and so I long ago

Figure 9.7. Nineteenth-century British chaff cutters

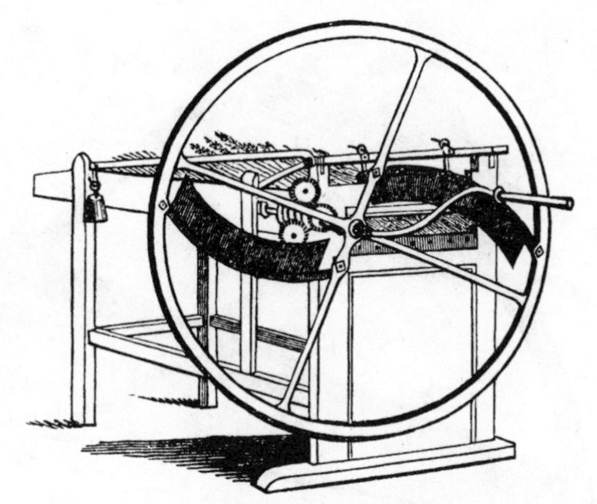

A. Weir's chaff cutter, 1839

B. McDougal's chaff cutter, circa 1820

Sources: Abraham Rees, *The Cyclopaedia . . .* (Philadelphia: Samuel F. Bradford and Murray Fairman, 1810–1824), Vol. 1, Plate VII; J. C. Loudon, *An Encyclopaedia of Agriculture* (London: Longman, Orme, Brown, Green, and Longmans, 1839), Vol. 1, p. 384.

mentioned to you as plainly *as I dared*—that your corn stalk cutter was an expensive bauble; if used on a great scale on an extensive farm. You see even the labour of slaves is thrown away in this tedious operation. It can only be useful where forage is scarce; and labour applied when there is

Figure 9.8. American straw cutter, 1803

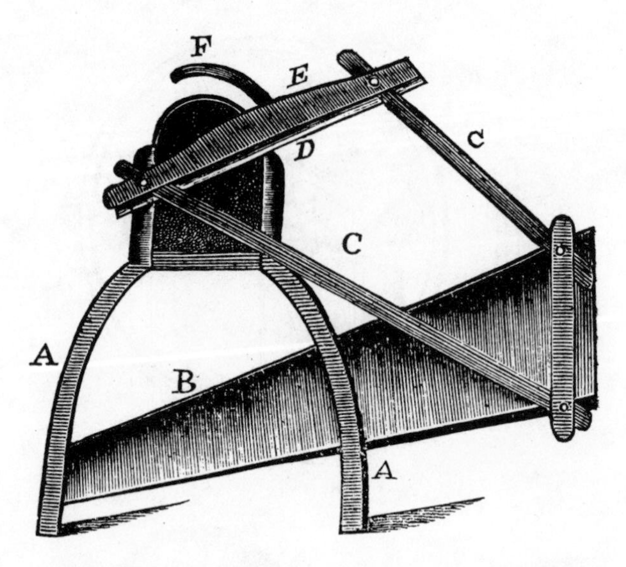

Source: A. F. M. Willich, *The Domestic Encyclopaedia* (Philadelphia: William Young Birch and Abraham Small, 1803), p. 61.

nothing else to do. And when is that interval, on even a Pennsylvania Farm?"[44]

Such fragmented evidence as has survived therefore suggests that straw cutters in America were little used and much disparaged in the era between the Revolution and the War of 1812, and with good reason. This would change, and quite suddenly, in the decade following the signing of Treaty of Ghent in December of 1814.

Broadly speaking, American inventive efforts took two directions. The first attempted to develop a high-speed, high-volume straw cutter for use by large-scale producers. The goal of the second was to design a more simple device, cheaper to make and easier to operate, suitable for use on smaller farms.

One of the first attempts to develop a straw cutter capable of handling a large volume was made by Hotchkiss. His 1817 version is illustrated in Figure 9.9. Its mechanical design is somewhat unexpected, insofar as it makes no effort to duplicate the British cutting mechanism of blades mounted in a revolving wheel. Instead, it featured a single blade (ultimately sharpened on both sides), with an up-and-down motion generated by a hand crank and a flywheel. Hotchkiss first patented a straw cutter in 1808, then subsequently tried to reduce or eliminate the more glaring defects of the original design. In 1817 he obtained a second patent for an improved model,[45] still less than perfect, with a flywheel that was too small and, more important, a cutting mechanism prone to malfunction.[46]

Figure 9.9. Hotchkiss's improved straw cutter, 1817

Source: Philadelphia Society for Promoting Agriculture, *Memoirs,* 4 (1818), p. 105.

But new and better machines were soon devised. By 1820 Jonathan Eastman had acquired the patent rights to Hotchkiss's machine for all states "west and south of New York." Through repeated experimentation, Eastman had produced by 1822 the radically different machine illustrated in Figure 9.10. The cutting mechanism might be considered a variant of British designs, but the resemblance is only superficial, as the machine's many prizes for novelty would soon attest. The cutting blades were made to rotate, but were mounted between two small wheels in a manner reminiscent of what today we would call an antiquated hand lawnmower. (See the bottom diagram of Figure 9.10, in which two blades GG are mounted between the wheels KK.)[47] By 1824 Eastman claimed to have spent five years developing a machine that, while capable of being operated by hand, was "perfectly adapted to horse or water power," and whether run at high or low speeds could cut "every kind of forage." It could also cut, albeit by accident, "pitchfork handles, flails, hatchet handles and steel pitchforks . . . without the least injury to the machine."[48] (This raises the question of its record with hands and fingers, but on that topic Eastman was silent.)

It was not cheap. By the middle of the 1820s, the prices Eastman was quoting were $45 for the smaller version, $55 for the "larger size," and $85 for the extra large.[49] By that time, however, competitive cutters were available that cost less and functioned differently.

For those with smaller farms and more limited incomes, American inventors developed a different type of straw cutter whose appearance more

Figure 9.10. Eastman's straw cutter, 1822

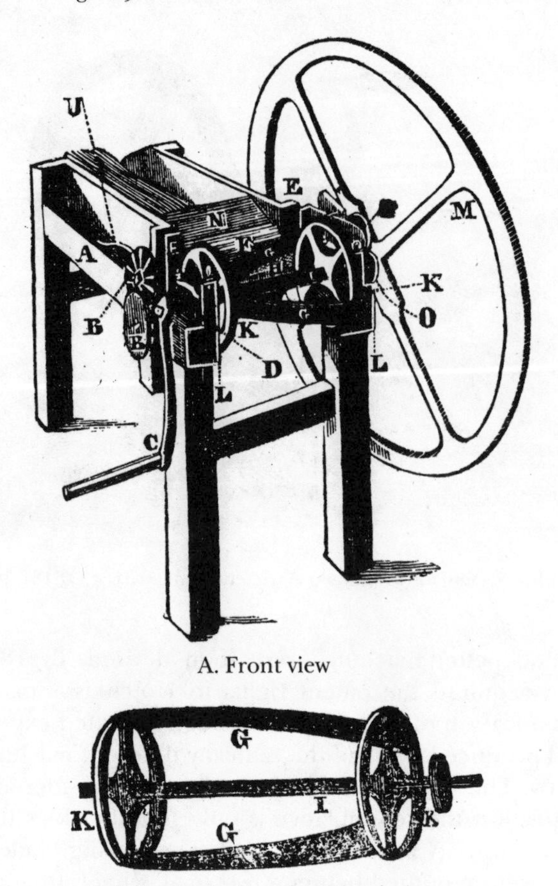

A. Front view

B. Wheels, shaft and cutting knives

Source: American Farmer, 3 (January 4, 1822), p. 326.

closely resembled the design that had come to dominate British cutting devices.[50] All of the machines illustrated in Figure 9.11 were developed in the 1820s, and in appearance were not that different from machines available in Britain several decades earlier. (Compare, for example, Figure 9.11 with Figures 9.6 and 9.7.) All were inexpensive, at least relative to the prices Eastman charged for his larger and more powerful straw cutters. Dayton's machine, for example, sold for $15 to $20, and Safford's for $14 to $25, the higher prices being charged for larger versions of the same machine.[51]

As was true for winnowing fans created during this period, inventors of straw cutters repeatedly assured prospective users that their machines

Figure 9.11. American straw cutters of the 1820s

A. Dayton's improved straw cutter, 1823

B. Willis's improved cylindrical hay and straw cutter, 1826

C. Safford's straw cutter, 1826

Sources: American Farmer, 4 (January 29, 1823), p. 350; *New England Farmer,* 4 (January 13, 1826), p. 199; ibid., 4 (February 3, 1826), p. 217.

were both "novel" and "improved." Here too a more reliable source of machine assessment is provided by the committee reports of agricultural societies specifically appointed for that task. In the case of straw cutters, such reports give the unmistakable impression of unambiguous progress throughout the 1820s.[52] Fueling the incentive to devise and commercially manufacture new machines was the gradual expansion of livestock markets, encouraging as it did increased attention to the care and feeding of animals destined to be sold.[53] All these trends, well established by the end of the 1820s, persisted through the 1830s.[54] By 1842, after surveying progress of the recent past, a Virginian noted that options of the present included a "great variety of cutting boxes," ranging in price from $5 to $70.[55]

Conclusion

In 1817 Thomas Gold, president of the Berkshire Association for the Promotion of Agriculture and Manufactures, reported on his recent inspection of three new straw-cutting machines. In his opinion, all three needed minor improvements, but despite such quibbles he expected them soon to be "perfect for work," adding "This is, indeed, the age of invention."[56] As a commentary upon evolving agricultural technology, that latter observation claimed at once too much and too little. As of 1817 the first great rush of new mechanical inventions designed for farm use lay largely in the future. From one perspective, then, the agricultural revolution in America had scarcely begun. And yet by 1817 the evidence suggests a revolution in attitude was well under way in the American farming community that, by and large, had been absent in the years between the framing of the Constitution and the War of 1812. A question was being asked, not just infrequently by the isolated few, but by progressively more farmers about every facet of agrarian production: "Is there a better way?" It was a question that, when first directed systematically and persistently to any primitive productive process, was sure to bring a revolution in its wake. The histories of evolving designs for straw cutters and winnowing fans in this country are merely instances of this larger pattern of development.

[10]

Novus Ordo Seclorum?

A New Order of the Ages: these words in Latin incorporated in 1784 into the Great Seal of the United States, seemed to most citizens at that time both a statement of accomplished fact and a promise for the future. For the next three decades, however, many wondered whether the new order would survive. Europe was repeatedly awash in wars, and beginning in 1812 Americans found themselves locked in a struggle for survival with one of Europe's mightiest powers, with the likely outcome far from clear. By 1814 within the young republic many wondered whether the *London Globe*'s promise to British readers would come to pass: that the war would end only after America had been "confoundedly well flogged," its union dismembered, its government overthrown, and its navy swept from the ocean, "not leaving a single bit of bunting or a rag or stitch behind."[1] The news of a peace treaty, when it reached America in the second week of February 1815, was therefore greeted with jubilation and relief. Church bells were rung, cannons fired, bonfires lit, and in every city, as the news spread, people rushed into the streets. In New York, where the news broke first on a Saturday evening, lower Broadway quickly filled with celebrants, many carrying candles and shouting "Huzza for the return of Peace." Next morning, when riders reached Philadelphia, church bells, initially sounded to mark the Sabbath, were furiously rung again. On Monday, Bostonians learned of the treaty, prompting stores to close, schools to be dismissed, and at noon, military units marched down State Street firing "feu de joie." (Adding perhaps a more decorous note, Harvard that evening "was splendidly illuminated.")[2] Few if any knew the actual terms of the peace treaty,[3] but no one seemed to care. What was important—indeed, what was at least "splendid," if not "glorious"—was that the republic was victorious and its future now secure, or so ran the common article of faith in this hour of national celebration. "The American Nation," as one historian characterized the moment, "with its back to Europe and its face to the West, addressed itself to the solution of the problems of the Nineteenth Century."[4]

A number of historians have portrayed the year 1815 as a turning point in the history of the United States, although the evidence they usually offer in support of that contention is less than compelling.[5] If a new age was dawning, its beginning was not particularly auspicious for those who earned their living from the land. A hurricane in September 1815 decimated the crops and livestock of New England. A "severe" winter was followed by a bizarre spring and summer featuring "the most extraordinary weather . . . ever known since the settlement of the country,"[6] with wildly fluctuating temperatures nationwide, drought in the south, and June snowfalls in the north. (Peachham, Vermont, recorded half a foot in a single day.) To many farmers, 1816 became known as "the year of two winters," or simply "Eighteen hundred and froze to death."[7]

And yet in the farming communities of the nation, something was afoot.

It could be sensed in the changing fashionability and longevity of agricultural periodicals. Between 1775 and 1810 the number of newspapers in America fairly exploded (see Table 10.1), and yet only in 1810 was the first "farm journal," *The Agricultural Museum*, established and that one lasted for only two years.[8] Now during the latter half of the second decade of the nineteenth century came two publication developments without precedent. In 1816 in Connecticut and in 1817 in Pennsylvania, with the backing of state agricultural societies, agricultural almanacs were established that contained not just directions for planting, as previous almanacs had done, but detailed agricultural information.[9] More importantly, the publication of *American Farmer* and *The Plough Boy* in 1819 heralded the beginning of "farm papers" as a permanent fixture among popular reading materials, with the number of such papers destined to expand rapidly in the 1820s and 1830s.[10]

That something was afoot in farming could also be sensed in the prominence given in agricultural literature to a topic heretofore generally neglected. Between the Revolution and the War of 1812, agricultural reformers evidenced little interest in novel farming implements. Agricultural societies offered few premiums to encourage them,[11] and queries to establish typical farming practices did not attempt to elicit any information on them.[12] When Samuel Deane reviewed "Experiments" in his pathbreaking survey of New England agriculture in 1790 and again in 1797, he made no mention of implement innovations.[13] Neither did Thomas Jefferson when, in 1808, he reviewed fashionable agricultural topics in his correspondence with Sir John Sinclair. (Jefferson singled out for special mention "the application of the new chemistry to the subject of manures,

Table 10.1. Number of newspapers published by colony or by state, 1775 and 1810

Colony or state	1775 Number publishing	1775 (Of which, still publishing in 1810)	1810
New England			
Massachusetts	7	(1)	32
Rhode Island	2	(2)	7
Connecticut	4	(3)	11
New Hampshire	1	(1)	12
Vermont	—	—	14
TOTAL	14		76
Mid-Atlantic			
New York	4	(0)	66
Pennsylvania	9	(2)	71
New Jersey	—	—	8
Delaware	—	—	2
TOTAL	13		147
South			
Maryland	2	(0)	21
Virginia	2	(0)	23
North Carolina	2	(0)	10
South Carolina	3	(0)	10
Georgia	1	(0)	13
District of Columbia	—	—	6
TOTAL	10		83
Other			
Kentucky	—	—	17
Tennesee	—	—	6
Ohio	—	—	14
Louisiana[a]	—	—	11
Mississippi	—	—	4
Total: United States	37		359[b]

Source: Isaiah Thomas, *The History of Printing in America* (Worcester, Mass.: Isaiah Thomas, Jr., 1810), Vol. 2, pp. 515–524.

[a]Includes "Louisiana" (1) and "Territory of Orleans" (10).

[b]Includes "Indiana Territory" (1), not included in state totals.

the discussion of the question of the size of farms, [and] the treatise on the potato.")[14] This silence on the subject of implement innovations appears to have been the rule, not the exception. Throughout the four decades ending in 1815, surveys of what was new or what was needed to improve American farming practices seldom included any mention of new implement designs.[15] John Taylor's highly critical survey of Southern husbandry was typical. A Virginian long recognized as a farming pioneer, Taylor published *Arator* in 1813, hailed by contemporaries as a landmark in agricultural publications.[16] And yet not one of its sixty-one essays focused upon new farming implements. Northerners usually followed suit.

About the same time that Taylor's book appeared, Robert Livingston of New York, another recognized expert on agrarian matters, wrote a review of American agriculture for *The Edinburgh Encyclopaedia* that made no reference whatsoever to new farming implements recently developed or presently in use.[17]

The main focus of attention—of agrarian pioneers such as Jefferson, Taylor, and Livingston and of agricultural societies, as evident in their premiums offered and queries circulated—were two sets of problems. One was soil exhaustion and possible remedies that ranged from fertilizer use to improved rotation schemes to pasture development using "English grasses." The other set of popular problems involved livestock care and breeding. But the difficulties scrutinized and remedies proposed, by and large, would almost surely have come to dominate agrarian practice and agrarian literature in America of the early nineteenth century *with or without modernization*, as markets became more integrated and uncleared land progressively more scarce along the eastern seaboard. (See Appendices A and B.) The same could not be said for the host of impending changes in farm implement design.

In the years immediately following the achievement of a lasting peace with Britain, the fashionable topics for discussion among agrarian experts and in agrarian literature began to change. In 1817, for the first time in its twenty-five-year history the Massachusetts Agricultural Society began to offer cash awards for novel farm implements, including "the best, simplest and least expensive machine" for threshing wheat or for sowing seeds "on an extensive scale."[18] The same society now added to its list of broadly circulated questions (also for the first time) an inquiry about "any improvement in the tools of husbandry."[19] In 1818, for the first time—or so the owner claimed—"at the only establishment of the kind in the United States" over fifty types of new and improved agricultural implements could be viewed or purchased, ranging from plows to seed drills to threshing machines to fanning mills.[20] The following year when Jesse Buel reviewed "the origins and progress" of agriculture in this country, he emphasized what Livingston and Taylor had overlooked only a few years previously: the new discoveries and recent improvements in "the instruments necessary to agriculture."[21]

Not that Buel was articulating what had become self-evident to all. Some of the brightest and best-informed agricultural experts still missed the beginnings of what was shortly to become a deluge of farming innovations. Speaking before America's oldest agricultural society in 1817, James Mease included only one reference to new farming implement possibilities (the threshing machine), and that was buried in a footnote.[22] In 1818 Noah Webster totally ignored such possibilities while reviewing American agricultural developments before the Hampshire, Franklin, and Hampden Agricultural Society—a review so comprehensive it required four successive editions of a farming periodical to be reported in full.[23] In 1819, in an address of similar length on the same topic before the Agricultural Society of Albemarle, Virginia, James Madison singled out for special mention only one novel implement and that one (the hillside plow) had been invented more than a decade earlier.[24]

But a shift in the focus of farming innovation which Buel had sensed in 1819 and Madison had missed soon became apparent to all. Little noted and commonly ignored during the first two decades of the nineteenth century, "implement improvements" became in the 1820s something of a shibboleth in contemporary assessments of agrarian progress.[25] As attention to such changes intensified, available arenas for that attention multiplied. Other "farm papers" were established—the *New England Farmer* in 1822, the *New-York Farmer and Horticultural Repository* in 1828, and the *Southern Agriculturist, Horticulturist, and Register of Rural Affairs* in 1828.[26] Perceptions of the rate of change seemed to change, with greater emphasis upon the number of improvements "within the past few years."[27] The criteria for judging which among the new were "better" also became more focused. That priority of modern capitalism, greater profits, previously muted or altogether missing in the literature of agricultural societies now resonated powerfully in recommendations to improve "outmoded" farming practices.[28] While trumpeting the cause of agrarian reform they often also urged the keeping of better farming records.[29] Nor was this surprising. As increments to profits became the norm by which to judge prospective changes, that judgment could scarcely proceed without information about related revenues and costs.[30]

The decision-making processes whereby farmers abandoned the old in favor of the new are at best poorly understood. What does seem clear is the limited importance of the written word divorced from practical demonstration. Part of the reason was the lack of clarity in what was written. The relative merits of different rotation schemes, for example, depended crucially upon climate and soil conditions difficult to specify with any great precision. (See Appendix A.) Written accounts of new implements were largely unintelligible unless accompanied by illustrations, and even then such descriptions left unresolved the crucial issue of performance in the field: performance that could be significantly affected by topography, climate, soil characteristics, and the presence or absence of obstructions such as roots and stones. Finally, the written word had limited impact upon established practice because, prior to 1820, not much was written that was considered by most practitioners to be particularly helpful. British books describing European farming practices were commonly discounted or ignored because they "contained not much that is applicable, without considerable variation, to *our soils*, climates and situations."[31] On this side of the Atlantic, the number of books, pamphlets, and periodicals published in the second decade of the nineteenth century exceeded 30,000, and by 1820 more books were being published in a single year than had been published in the entire seventeenth century. (See Table 10.2.) Yet by 1821 the list of recommended readings for farmers compiled by an acknowledged farming expert, who was himself a writer, included only three books, two newspapers, one British encyclopedia revised for an American audience by James Mease, and the published papers of two agricultural societies.[32]

Not that literature was totally irrelevant in the reshaping of farming practice. A few might read, revise their own procedures, and thereby influence their neighbors. But written advice unsupported by practical demonstrations was, at most, an insipid impetus for change, at least for those with limited incomes, an aversion to risk, and a jaundiced view of all that broke with patterns of the past.[33]

Such problems help to explain the limited impact upon American farming of the early agricultural societies. The reasons for their ineffectiveness, however, ran much deeper.

For many agrarian analysts, such societies seemed ideally suited to initiate and disseminate innovations in farming practices. To experiment was to take risks, society would gain when the results of successful experiments were widely publicized, but ease of duplication of what was publicized would usually mean that those who took the risks and paid related costs would often receive little compensation. To combat these and related difficulties—not the least of which was innovations being lost forever when the inventor passed away—the obvious solution seemed to be to establish organizations for the express purpose of gathering and disseminating information, while encouraging experimentation through the prizes these societies offered and the literature they published. Or as Washington summarized the hopes of would-be agrarian reformers in the post-Revolutionary era, agricultural societies, "composed of proper characters," would promote the growth of agriculture by "collecting and diffusing information," while encouraging and assisting "a spirit of discovery and improvement" by offering "premiums and small pecuniary aids."[34]

In theory, then, the establishment of societies was a desirable institutional development to foster change in the early phases of an agricultural revolution. In practice, they were remarkably ineffective.

In part, this reflected the intransigence of most farmers to pleas for change. On both sides of the Atlantic, the conservatism of those who worked the land had been an object of commentary and derision for generations. (Jethro Tull, for example, complained that "It were more easy to teach the beasts of the field than to drive the ploughman out of his way.")[35] But allegiance to the status quo was not necessarily irrational in a world where most incomes were low, and the prospective benefits of farming novelties were difficult to estimate. This was particularly true in

Table 10.2. Annual number of books, pamphlets, and periodicals published, 1639–1820

Year	Number	Year	Number	Year	Number
1639	2	1688	21	1737	128
1640	2	1689	71	1738	148
1641	3	1690	59	1739	169
1642	3	1691	49	1740	146
1643	3	1692	53	1741	257
1644	2	1693	61	1742	268
1645	2	1694	32	1743	250
1646	3	1695	31	1744	244
1647	3	1696	41	1745	240
1648	2	1697	44	1746	205
1649	5	1698	48	1747	224
1650	3	1699	69	1748	232
1651	2	1700	78	1749	216
1652	1	1701	77	1750	224
1653	3	1702	80	1751	221
1654	3	1703	61	1752	187
1655	4	1704	63	1753	232
1656	4	1705	59	1754	267
1657	5	1706	58	1755	349
1658	6	1707	71	1756	283
1659	4	1708	56	1757	324
1660	7	1709	84	1758	298
1661	4	1710	72	1759	324
1662	4	1711	63	1760	338
1663	12	1712	77	1761	348
1664	13	1713	94	1762	355
1665	11	1714	88	1763	344
1666	7	1715	109	1764	430
1667	7	1716	84	1765	421
1668	20	1717	99	1766	401
1669	12	1718	87	1767	381
1670	14	1719	96	1768	460
1671	6	1720	128	1769	545
1672	9	1721	140	1770	585
1673	14	1722	116	1771	453
1674	17	1723	112	1772	432
1675	19	1724	122	1773	623
1676	26	1725	146	1774	928
1677	17	1726	125	1775	1,116
1678	23	1727	172	1776	839
1679	21	1728	174	1777	703
1680	21	1729	153	1778	661
1681	15	1730	170	1779	720
1682	33	1731	140	1780	572
1683	20	1732	154	1781	567
1684	27	1733	146	1782	591
1685	22	1734	149	1783	752
1686	32	1735	149	1784	738
1687	21	1736	151	1785	719

Table 10.2.—Continued

Year	Number	Year	Number	Year	Number
1786	932	1798	2,300	1810	2,751
1787	950	1799	1,952	1811	2,384
1788	941	1800	2,663	1812	3,055
1789	1,025	1801	1,638	1813	2,876
1790	1,136	1802	1,915	1814	3,101
1791	1,206	1803	2,032	1815	2,838
1792	1,398	1804	2,150	1816	3,182
1793	1,764	1805	2,011	1817	2,978
1794	2,061	1806	2,152	1818	3,817
1795	2,227	1807	2,298	1819	3,198
1796	2,134	1808	2,505	1820	4,388
1797	1,939	1809	2,490		

Source: Charles Evans, *American Bibliography: A Chronological Dictionary of all Books, Pamphlets and Perodical Publications printed in the United States of America, from the genesis of printing in 1639 down to and including the year 1820* (Chicago: Blakeley Press, 1903–1959); and Roger P. Bristol, *Supplement to Charles Evans' American Bibliography* (Charlottesville: University Press of Virginia, 1970).

Note: These numbers exclude a small number published in a given decade for which no specific year of publication is known, as, for example, for the 1720s under "172-," during which two items were published.

the earliest phases of an agricultural revolution when examples of unambiguous success were often few and far between.

If resistance to new farming practices was formidable, the power of agricultural societies to persuade was intrinsically weak, to a significant degree because they relied primarily upon the written word unsupported by demonstrations in the field. They further undermined their own influence by structuring many of their premiums and prizes to reward suboptimal behavior. Those concerned more with profits than with appearance or prestige repeatedly pointed out that awards given for the greatest output from a given acre had "a tendency to render tillage and pasturage more showy than useful."[36] The same was true of livestock, or so they claimed, cattle awards going to "some ill-formed, over-fed, monstrous animal."[37] That characterization is undoubtedly too harsh. Most farmers would concede a grudging admiration for animals raised to exceptional dimensions with little regard to cost. But their attitude toward such prize-winning livestock was reminiscent of comedian W.C. Fields's attitude toward elephants: "I like to look at 'em, but I'd hate to own one."[38] Admiration for results did not inhibit condemnation of the prize system leading to those results, as Edmund Ruffin's scathing observations of 1823 made clear. "A reference to the list of premiums offered by societies will show that much the greater number are for practices, which, so far from being profitable, would insure a loss to every one who gained or contended for them. Such are the premiums for the most productive acre of corn, wheat, rye, potatoes, etc. the finest (i.e., the fattest) ox, cow, sheep, hog, etc. To obtain these premiums, enormous crops have been raised on many single acres, and beasts have been brought to a degree of fatness, which no consideration of market price, or *net profit* would ever have induced, or will ever cause others to imitate."[39]

The limited impact these agricultural societies had upon farming innovation and farming practice is suggested by their pattern of growth. Initial enthusiasm seemed quickly to give way in many cases to relative inactivity. The period immediately following the Revolution was one of rapid growth, with twelve societies to promote "agriculture" established in ten years. Between 1795 and 1815, however, only three more were founded. (See Tables 10.3 and 10.4.) A movement to create a comparable organization on

Table 10.3. Number of agricultural and other societies established during the eighteenth century

Societies	Prior to 1776	1776–1783	1784–1789	1790–1794	1795–1799
Agricultural	—	—	4	3	—
Agricultural, commercial, & manufacturing	2	1	2	3	—
Manufacturing & commercial	3	—	3	2	—
Medical	4	1	5	3	2
Learned & scientific	2[a]	1	—	—	—

Source: Ralph S. Bates, *Scientific Societies in the United States* (Cambridge, Mass.: M.I.T. Press, 1965); Rodney H. True, "The Early Development of Agricultural Societies in the United States," American Historical Association, *Annual Report*, 1920, pp. 295–306; J. C. Kiger, *American Learned Societies* (Washington, D.C.: Public Affairs Press, 1963); Stevenson Whitcomb Fletcher, *The Philadelphia Society for Promoting Agriculture, 1785–1955* (Philadelphia: Philadelphia Society for Promoting Agriculture, 1976); Lucius F. Ellsworth, "The Philadelphia Society for the Promotion of Agriculture and Agricultural Reform, 1785–1793," *Agricultural History*, 42 (July 1968), pp. 189–199; *Eighty Years' Progress of the United States* (New York: New National Publishing House, 1864), 25; Brooke Hindle, *The Pursuit of Science in Revolutionary America, 1735–1789* (Chapel Hill, N.C.: University of North Carolina Press, 1956); Donald B. Marti, "Early Agricultural Societies in New York: The Foundations of Improvement," *New York History*, 48 (October 1967), pp. 313–331; James H. Cassedy, "Medicine and the Learned Society in the United States, 1660–1850" in Alexandra Oleson and Sanborn C. Brown, eds., *The Pursuit of Knowledge in the Early American Republic* (Baltimore: Johns Hopkins University Press, 1976), pp. 261–278; Walter Muir Whitehill, "Early Learned Societies in Boston and Vicinity," in Oleson and Brown, eds., *Pursuit of Knowledge*, pp. 151–173.

[a]Includes the Boston Philosophical Society (1683) and the American Philosophical Society Held at Philadelphia for Promoting Useful Knowledge (1769), the latter being a combination of two societies formed earlier, the American Philosophical Society and the American Society Held at Philadelphia for Promoting Useful Knowledge.

Table 10.4. Number of agricultural and horticultural societies founded per year, 1800–1850

Year	Number	Year	Number	Year	Number
1800	1	1820	1	1840	12
1	—	1	—	1	9
2	—	2	—	2	4
3	1	3	—	3	6
4	—	4	1	4	5
1805	—	1825	—	1845	6
6	—	6	—	6	13
7	—	7	—	7	7
8	—	8	1	8	10
9	—	9	1	9	16
1810	1	1830	—	1850	19
1	—	1	1		
2	—	2	3		
3	—	3	—		
4	—	4	1		
1815	—	1835	—		
6	—	6	2		
7	2	7	—		
8	4	8	2		
9	2	9	2		

Source: U.S. Commissioner of Agriculture, *Report*, 1867, 364–403.

Note: Included are state agricultural societies plus county and township agricultural and horticultural societies and clubs still in operation as of 1867.

a national scale died in Congress in the 1790s, despite Washington's endorsement.[40] National indifference was mirrored in local inactivity. For example, the oldest agricultural society in the country, that of Philadelphia, seems to have done so little between 1793 and 1805 that its minute books record no entries, although thereafter it was revived under the direction of its new president, Richard Peters.[41]

Perhaps the best indicators of the limited appeal of these organizations were the persistent complaints from ardent supporters about a lack of members and a shortage of funds.[42] The 1807 observations of a New York association member were typical. "The patriotic wish of doing more," he acknowledged, "was cramped by a poverty of means. . . . The little that has been done by the society in this way [distributing premiums] it is believed has not been without effect; but it is to be lamented that *but a little* is all that could be done."[43]

Needed, and largely missing at least prior to 1815, were public arenas of performance that could reinforce in powerful ways the impetus to agrarian change. Following the War of 1812 these became progressively more common in response to forces that were a curious blend of the predictable and the improbable.

Private arenas of performance had existed for generations in the form of the innovator demonstrating to his neighbors the effects of new farming techniques or novel implements on his own fields. But these had obvious limitations. The transmission of change was limited to the influence of the one upon the few, and those few might ultimately reject the new because of a difference in soil conditions or topography or an incapacity to follow instructions. In the case of implements, particularly in the early stages of experimentation when successive modifications were needed to make the novel progressively more efficient, much superior to individual demonstrations were public displays and competitions where innovators could vie with one another, but also learn from one another.

The obvious public gatherings to evolve into such arenas were agricultural fairs. Long in evidence in both America and Europe, these had served primarily in the colonies (and elsewhere) as commercial and social gatherings to facilitate the sale of livestock and farm produce.[44] Following the Revolution, as Americans became progressively more interested in improving livestock (see Appendix B), it seemed inevitable that the old institution of the fair should take on the new function of exhibiting "prize" specimens of the newer breeds.

The actual development seems to have been the opposite of the obvious: the desire for exhibiting new breeds prompted the resuscitation of agricultural fairs, which, once revived, were promptly modified after 1815 in response to a rising flood of new farm implements.

As the story is commonly told, the nineteenth-century origins of the American livestock fair (or at least a livestock fair of a particular sort) can be traced to the efforts of Elkanah Watson. Interested in improved breeds of farm animals, particularly sheep, he exhibited a pair of recently acquired Merinos in the public square of Pittsfield, Massachusetts, in 1807. The evident interest of local farmers prompted Watson three years later to persuade twenty-five others to join him in a second exhibition, this time on the Pittsfield village green.[45] The success of the latter led Watson the following year to organize the Berkshire Agricultural Society. Markedly different in composition and purpose from older agricultural societies and with membership dominated by "ordinary farmers" (not the agrarian elite), this organization had as its primary objective to hold an annual exhibition and fair, with no professed interest whatsoever in learned meetings or literary publications. The first such fair was held in 1811, the year the Berkshire Society was founded. This 1811 exhibition clearly was not the "first" livestock fair or cattle show, since similar gatherings had occurred in the country at least a year or two earlier.[46] But to Watson must go much of the credit for popularizing this form of public gathering in the second decade of the nineteenth century. Combining a shrewd sense of what would interest farmers and what would draw a crowd, he quickly helped transform the initial format of a livestock exhibition into an elaborate exposition, festival and social gathering that included exhibits of farm produce and household handicrafts, presentation ceremonies, and—to assure the ladies' interest (or so Watson argued)—a pastoral ball. It was a social triumph, but a fiscal failure. The annual membership fee of one dollar did not cover all the costs, and Watson with a few Boston friends had to make up early deficits. Discouraged by this "apparent failure," the founder of the Berkshire Agricultural Society resigned as president in 1814.[47]

But Watson had misjudged the impetus for agrarian change he had unleashed. The "Berkshire" type of society was rapidly copied elsewhere, and often with the help of subsidies from state governments, such organizations by the end of the decade could be found in Pennsylvania, Maryland, Virginia, North Carolina, Ohio, Illinois, and all the counties of the New England States (except Rhode Island). By the autumn of 1819, *Niles' Weekly Register* noted with enthusiasm their rapid proliferation and gave credit where it was due. "The public newspapers teem with accounts of shows and fairs, and the proceedings of the numerous societies which have sprung from their common parent in Berkshire, Mass. to immortalize the venerable name of Watson, and disseminate useful facts into every district of our country. The best methods of managing a farm, in all its details, rearing cattle, sheep and swine, etc. and of applying the surplus labor of the people within doors, are carefully attended to, and every man's experience is thrown into the common stock of knowledge—the *power* thereby to be acquired is of incomprehensible magnitude."[48]

Notice, however, what was missing. The accounts of the original Berkshire fairs make clear that they failed to feature either the exhibition or the testing of new agricultural implements.[49] The original Pittsfield gatherings were designated in the press as "a cattle show and fair."[50] An exception to this neglect was evident at the Columbian Society's semi-annual exhibition in 1812. The plowing match it sponsored that year seems to have been the first field trial of implements in America.[51] But by the end of 1812, the society itself had folded for lack of interest. Plowing matches would shortly be revived, not by Berkshire Agricultural Societies at first, but by one of the most prestigious of the older agricultural societies. In 1816 the Massachusetts Agricultural Society (in part influenced by the Berkshire example) held its first exhibition at Brighton, which featured a plowing match as well as the more typical livestock competitions. Curiously, the reason for the match was not to test plows but to assess the relative merits of oxen as draught animals.[52]

To this point, then—the autumn of 1816—nothing in the evolution of livestock exhibitions and fairs clearly heralded, or even suggested, an impending revolution in American farming implements. This would quickly change. The very next year the Massachusetts Agricultural Society began to offer cash prizes for new farm implements, including "the best, simplest and least expensive machine" for threshing wheat and sowing seeds.[53] By the early 1820s, many agricultural societies (including variants of the Berkshire model) had followed suit, as the exhibition and testing of farm implements became a common appendage to livestock shows and fairs.[54] The growing flood of innovations, in many cases, forced a revision in the prizes offered. The Maryland Agricultural Society, for example, by 1823 abandoned any reference to specific types of implements and simply offered a cash award of $20 "For the Agricultural Machine or Implement that may be new, and be thought worthy of reward by the Society."[55] Three years later, the same society decided that market forces had obviated the need for further prizes. Or as the awards committee chose to put the matter, "no improvement was expected to result" from the offering of premiums for novel implements, because "the active competition amongst these establishments [i.e., those producing farm implements] has superseded the necessity of further stimulus."[56]

A sudden surge to prominence of agricultural implements can therefore be detected in the history of the rise of American agricultural fairs in the early nineteenth century, both in the changing priorities of their committees and in the changing composition of their prizes. The same surge is also suggested by the change in coverage of America's second and third manufacturing censuses.

When Tench Coxe surveyed the state of American manufacturing in 1794, his observations contained no hint of an agricultural revolution in progress. (His only observation about farm implements was the inclusion of "plough irons, hoes, and other farming utensils" in a lengthy list of items manufactured using iron.[57]) Coxe's review of the first serious effort at a comprehensive American census of manufacturing, made in 1810, is similarly uninformative. The census, all agreed, was woefully incomplete, but what data were assembled included nothing about the production of agricultural implements. The one reference in Coxe's written commentary is the mention of scythes, in an extensive list of products manufactured with the aid of water mills or steam engines.[58] When Rodolphus Dickinson carefully scrutinized in 1813 the 1810 census information for Massachusetts, he found the information "very extensive," but the coverage in question included neither data nor written commentary about the manufacturing of agricultural implements.[59]

This absence of information was in striking contrast to the coverage provided by the very next census in 1820. Like its predecessor, it was

Table 10.5. Reported manufacturing of agricultural implements by census of 1820

	Scythes		Plows & plow irons[c]		Agricultural implements & farming utensils[d]	
STATE	Number of firms[a]	Hands employed[b]	Number of firms[a]	Hands employed[b]	Number of firms[a]	Hands employed[b]
New Hampshire	1	6	—	—	—	—
Massachusetts	—	—	—	—	2	15
Connecticut	1	10	—	—	1	5
Vermont	2	3	1	1	—	—
New York	12	44	6	47	—	—
Pennsylvania	3	9	4	36	1	42
Maryland	—	—	1	5	—	—
Virginia	—	—	10	324	—	—
South Carolina	—	—	—	—	1	4
Georgia	—	—	2	11	—	—
Alabama	—	—	—	—	2	26
Tennessee	—	—	11	302	2	4
Kentucky	—	—	1	6	—	—
Ohio	1	3	7	45	1	2
Total of above	20	75	43	777	10	98

Source: U.S. Congress, Senate, *Articles Manufactured in the United States*, 18th Cong., 1st Sess., 1824, Report 45, 5–199.

[a]Calculated by assuming 1 firm per county reporting, which appears to be the case except in rare instances. See data in source noted.

[b]No employee data given for 2 firms, one producing plows in Adams, Ohio, and the other producing agricultural implements in Monroe, Alabama.

[c]Includes "plough plates" (1 firm), "plough-shares" (2 firms), and "bar iron plough molds" (1 firm).

[d]Includes "hoes" (1 firm) and "pitchforks" (1 firm).

known for being incomplete, but the data it assembled now included details on the production of agricultural implements in seventy-three firms employing almost one thousand hands. (See Table 10.5.) Several of these, by the standards of the day, were exceptionally large, six having more than twenty-five workers each and three having more than fifty. The production of farm implements was clearly on the rise, and the obvious way to plot its progress would be to compare information collected for the 1820 census with that collected in 1830. The latter census, however, was notorious for the paucity of its coverage. One indication of its deficiencies is the fact that of the seventy-three firms producing agricultural implements in 1820, a possible overlap with those recorded as producing in 1830 is limited to four, one each in Massachusetts and New Hampshire, and two in Vermont.[60]

But if advances in farm implement production in the 1820s cannot be tracked in the obvious way using available census data, progress can be explored for particular implements using the types of sources cited in Chapters 3 through 9. The pattern that emerged there was one of extraordinary transformation in the development and use of new implements for every single stage of grain farming in this country. Attitudes among farmers concerning the merits of the status quo were also being revolutionized, as indicated by a new willingness to experiment, and by the changing characterization of those attitudes by contemporaries.

On both sides of the Atlantic for generations would-be agrarian reformers repeatedly decried the intractable conservatism of those who tilled the land. An American denunciation of 1758 was typical. "Farmers [in the colonies] in general are neither able to bear the expense, nor frequent disappointments, of making new experiments; and what is worse, they are so bigoted to their own ways in every part of the world that as an English gentleman observes of some husbandmen in Norfolk, 'their husbandry is precisely that of their great grandfather's grandfather.' "[61] Receptiveness to novelty had not improved significantly by the early nineteenth century, or so a correspondent to the *American Farmer* complained in 1819, repeating the all-too-familiar charge that "every son treads in his father's precise footsteps," adding that the chosen path was somewhat out of step with a new age of scientific inquiry: "Our whole system of agriculture depends on the phases of the moon and the signs of the zodiac."[62] But by 1822, the derogatory tone had softened, with skepticism of the new endorsed as a rational response to an outpouring of novelties. "We would not advise farmers in middling circumstances to make expensive experiments, nor adopt any novelty in husbandry on slight grounds, without being well convinced by testimony, observation or experience, of its beneficial effects. We had better by half follow the beaten tract of our ancestors, if it be a little rugged and circuitous, than strike out at once into a wilderness of whim-whams, and theories not sanctioned by actual and repeated experiments."[63] The same commentator than hastened to add that while skepticism was sensible, unthinking allegiance to old ways was not. Or as he preferred to put the matter, his plea to personally evaluate the reputed merits of every farming innovation "should not check enterprise, but inspire caution, and teach us that every novelty may not be an improvement, although *every improvement was once a novelty.*"[64] A different perspective on the same problem published in 1823 suggested that, while allegiance to the status quo was unlikely to be affected by oral or written exhortations, it could be modified by practical demonstrations. "In vain may the *tongue* and the *pen* be employed to satisfy the practical man of the errors, which a life of labor has confirmed, and the experience of ages has consecrated. He is either deaf, and blind, and dumb to your appeals, or answers you in the language of distrust, and with the reproach that they are theories of idle speculation only. But make for him the experiment, explain to him the method, exhibit to his natural senses the successful result—he will hesitatingly yield credit to ocular demonstration, and tardily follow in the footsteps of improvement."[65]

By the end of the 1820s, an outpouring of commentary combined with the demonstrated success of many new farm implements was effecting a revolution in the judgmental procedures applied to innovations. Novelties which heretofore for most has been considered prima facie useless, now for growing numbers were considered prima facie worthy of investigation. In 1829, for example, one commentator noted with approval the "progress in agricultural and horticultural improvement" which was now "steady and flattering," and suggested an attitudinal transformation under way that was both a cause, and an effect, of the recent rapid growth in innovations. "Our farmers are acquiring the first requisites for improvement; a consciousness that their practice *is not* the best, and that they really *want* information. Our breeds of livestock have manifestly been much improved. Our implements have been multiplied; and they are better made, and better used than formerly."[66]

That judgment made in 1829 was echoed six years later by Thomas Fessenden, the editor of *The New England Farmer*, in a book reviewing American agriculture (*The Complete Farmer and Rural Economist*). As a young man, Fessenden in 1808 had compiled a list of "Some of the Most Useful Modern Discoveries and Inventions." (See Table 10.6.) Only 27 of its 145 entries concerned agricultural discoveries, and *not a single one* of that 27 involved an agricultural implement. Such ignoring of farm implements was typical of innovation surveys of this earlier period. In 1814, for example, no less an expert on American manufacturing than Tench Coxe, when listing "many curious invaluable inventions and improvements" of recent decades restricted the agricultural improvements worthy of inclusion to "the division of labor in the cultivation of cane" and "the improvement of . . . sheep breeding and farming."[67] By the 1830s, however, the composition of such compilations had radically changed. The subject of Fessenden's book was "some important improvements in modern husbandry."[68] To impress upon his readers the rapidity of recent progress in that activity he resorted in the preface to an arithmetic imagery. "The advances of agriculture of late years have not been uniform, but accelerated; its progress has been in what mathematicians would call a geometrical ratio. Every step has furnished means for quickening the pace and extending the reach of the next step, and every path has led to a longer and wider avenue of improvement."[69] In a chapter entitled "Agricultural Implements," he began by quoting an English agricultural expert about the difficulty of getting new implements adopted, given "the ignorance, the prejudice and obstinacy" of farmers.[70] Whatever relevance such disparaging characterizations had had in the past, Fessenden declared, they no longer described most of those who worked the land in the re-

Table 10.6. Fessenden's 1808 review of "Some of the Most Useful Modern Discoveries and Inventions"

Topic	Number of entries
Agriculture	
livestock, care and raising	5
orchard management	5
potatoes	4
poppies and opium	3
fertilizer[a]	2
hemp	1
other[b]	7
Total	27
Manufacturing	
steam and steam engines	6
iron and iron working	8
bridges	3
brewing[c]	7
building and operating ships	4
carriages and wagons	2
leather and leather working	2
mill machinery	2
manufacturing "sketches," by state	15
TOTAL, ALL ENTRIES	145

Source: Thomas Green Fessenden, *The Register of Arts, or A Compendious View of Some of the Most Useful Modern Discoveries and Inventions* . . . (Philadelphia: C. and A. Conrad, 1808).

[a] "On oil as manure" and "On calcoreous and gypseous earths."
[b] Includes entries on burning stubble (2), stump removal (1), seeds (2), "caterpillars on gooseberry bushes" (1), "analysis of soils" (1).
[c] Ale, beer, wines, yeast, and "ardent spirits."

gion he knew best. "The farmers of New England are too enlightened, and have too much regard for their own best interest, to be under the dominion of such profitless prejudices. Accordingly, we find not only a very increasing demand for new and improved agricultural machines, tools, etc., but that our practical farmers see that it is for their interest to procure the best."[71]

What was "best," of course, was constantly changing as cloudbursts of innovations annually descended upon the landscape. What was extraordinary was that this persistent rain of novel implements, unthinkable in 1808, had become the norm by the middle of the 1830s.

This book began by noting that, in certain academic circles, the concept of an "industrial revolution" has fallen upon hard times. It must close by acknowledging the same is true for the concept of an "agricultural revolution."[72] In the farming regions of Europe, antecedents of change—or change itself—in many cases can be tracked far back into the Middle Ages. The implication is that continuity, not discontinuity, is the appropriate expectation to bring to analysis of agrarian transformation, or so the argument runs. What counts as "revolutionary," of course, is ultimately a matter of personal preference. Judged against the backdrop of nine thousand years of human history, the changes that have occurred within the present millennium can hardly be considered anything but "revolutionary," as the graphs in Chapter 1 make clear. But if the temporal focus is narrowed to several centuries, and the changes under scrutiny seem to play out over that entire period, the willingness to characterize the process as revolutionary may be undermined.

At bottom, however, what is at issue here is not a question of preference but a question of fact: Did the agricultural changes that transformed a primitive rural economy in such a way as to place it firmly upon the path to modernization occur within a few centuries or a few decades? Here, as in so many other instances, the American experience appears to differ significantly from the European. The modernization of agriculture in the Old World in many regions was a process spanning many centuries, thus requiring a broad temporal perspective for its narration. The same is not true in the United States. When in 1815 its citizens reacted to the news of peace with Britain by ringing church bells, firing cannons and holding street parades, they evidenced no awareness of the first stages of an upheaval—monumental in scope and duration—that would totally transform the production processes on their farms and in their factories. Both an industrial *and* an agricultural revolution were afoot, but few, if any, noticed.

Such insensibility was not surprising. In 1815 such revolutions had no precedent in human history, except for processes then unfolding in a handful of European countries, and even the most perceptive of contemporary observers might easily miss the significance of that unfolding. To be sure, Americans believed the founding of their nation signaled "A New Order of the Ages," and with typical self-assertiveness they had emblazoned that slogan upon the Great Seal of their country.[73] That motto, however, would seem to understate the case. It should have been crafted in the plural, because the political, economic, social, and cultural components of their society would all soon include what was, by Western standards, much that was astonishingly new. Not the least of these was an agricultural transformation that made the farming practices of 1900 unrecognizable by the standards of 1800. As described in subsequent appendices, the first great surge in changes in fertilizer use and crop rotations and pasture development occurred in the decades immediately following 1783. But, by themselves, these did not herald the beginning of the modern era in American agriculture. That distinction belongs to the first great surge in implement innovations which was surprisingly short, most of it occurring in the fifteen years following the War of 1812. To document the facts of that initial surge in implement innovations has been the main purpose of this book. To explain why it occurred when it did, and with such speed, as O. Henry might say, is another story, or the appropriate topic of another book.

Appendix A

Remedies for Soil Exhaustion

During the colonial period, the central feature of America's economy was land abundance, resulting in the kinds of "land-extensive" farming techniques discussed throughout this work. From first settlement to the Revolution, farming in the New World featured rapid clearing, hard use, and moving on. Practices that foreign visitors decried as "wasteful" and "slipshod" were, by and large, a rational response to a set of relative factor prices significantly different from those of Europe. But if abundant land was one central feature of British North America, rapid population growth was another. This growth, after some point, was sure to produce a scarcity of uncleared land in older settled regions. That rising scarcity, in turn, combined with farming techniques that heretofore had rapidly exhausted the fertility of land already cleared, would inevitably lead to growing concerns about combating soil exhaustion. For purposes of this study, the central point is that a revision in farming practices reflecting such concerns was sure to happen in America, whether or not the economy embarked upon that extraordinary transformation loosely designated as "modernization" or "industrialization" or "a take-off into sustained growth." This inevitability of a shift in farming practices to combat soil exhaustion complicates the integration of such changes into the larger picture of America's first agricultural revolution. If their occurrence did not necessarily signal the beginning of the modern age down on the farm, what did they signal? This appendix describes the changes.

Fertilizer

The reluctance of colonists to use fertilizers, despite repeated European exhortations to the contrary, would seem to be one of many rational choices made in a setting of abundant land. The requisite economic reasoning was put concisely by Thomas Jefferson: "we can buy an acre of new land cheaper than we can manure an old one."[1] Admittedly, in areas where potential fertilizers were also abundant and cheap, they were occasionally used.[2] On farms located near the sea, for example, the colonists sometimes improved soil fertility by adding fish, seaweed, or even horseshoe crabs. "But the time will come," warned a New York analyst in 1792, "and indeed in many places now is, when the land, . . . exhausted of its richness, shall be too weak, of itself, to make plants grow with their former luxuriance."[3] This is a slight misstatement of the problem. For generations, American farming practices had virtually assured the rapid exhaustion of fertility on land newly cleared. The catalyst for change in the young republic would be a growing scarcity of uncleared acres in older settled regions, with the associated shift in relative factor prices forcing a revision in the old practice of rapid clearing, hard use, and moving on.

In the decades immediately following the Revolution, scattered evidence suggests this rethinking of old practices had begun in at least some regions of the country. A Delaware survey of the 1780s, for example, found that "manure is now more necessary, and more used than formerly."[4] The Philadelphia Society for Promoting Agriculture, in its list of premiums offered in 1790, included "For the best method of recovering worn-out fields to a more hearty state, within the power of common farmers, without dear or far-fetched manures; but, by judicious culture, and the application of materials common to the generality of farmers, founded in experience,—a gold medal; and for the second best,—a silver medal."[5] Samuel Deane, in his surveys of New England farming practices in the 1790s, enthusiastically endorsed the use of fertilizer, as did a number of other agricultural commentators at that time.[6]

For most American farmers, finding possible fertilizers to use was not a problem. Deane published a list of sixty-six available types in 1790, and John Beale Bordley included fifty-eight in his 1803 survey.[7] The problem was deciding which to use. Little was known about how plants received their nourishment, or how different fertilizers contributed to that process. Thus, for example, as late as 1821 an American observer conceded, "The theory of manures is, as yet, but little understood."[8] Equally important, for any given crop, the beneficial effects of different fertilizers depended upon a number of variables, most of which were difficult to specify with any precision. Climatic conditions might be designated as "wet" or "dry," or soil types as "stiff clay," "light sandy," or "loamy," but this is not the kind of terminology that made for precision in reporting the results of experiments with different crops and fertilizers. Trial and error was the only method of investigation, systematic records were seldom kept, and even when they were, the vagueness of the terminology used assured a lack of clarity in reported results and associated recommendations. Small wonder, then, that even in the 1820s, an agrarian commentator wrote disparagingly of "the various and contradictory opinions derived from our [fertilizer] experiments."[9]

The resulting difficulties in making rational choices from among available fertilizer options are illustrated by the histories of early American interest in gypsum and lime.

Gypsum, or plaster of Paris (or plaister of Paris), became used with increasing frequency in the years following the Revolution,[10] and by the second decade of the nineteenth century was often reported as being "generally in use" in many sections of the country.[11] A key reason for this rising popularity was the evident success this fertilizer had in restoring fertility when used in conjunction with "English grasses," particularly clover[12] (of which more in a later section). But doubts about its long-run efficacy were evident even in the early years of use, and became more pronounced with the passage of time. In the short run, whatever beneficial effects it had upon plant growth could be substantially reduced or even negated by "excessive moisture or excessive droughts," and its performance near the seacoast was highly variable for reasons that remained unclear.[13] But even in areas where its initial effects were unambiguously positive, doubts began to surface among those who pioneered its use whether "sustained use" and/or "heavy dressings" led to "impoverishment" of the soil. This charge, raised tentatively by a few in the second decade of the nineteenth century,[14] by the 1820s had become a topic of widespread concern.[15] In a world devoid of scientific knowledge about the relevant chemistry, uncertainty was the rule, at least among the best informed. In 1819 Noah Webster noted that gypsum's "real value" had "not been ascertained . . . and on some crops, its value is probably overstated."[16] Such tentativeness was echoed two years later by "one of the best practical farmers" of Virginia. "Whether it [gypsum] acts as a stimulant, or as a solvent; or whether, by its kindly influence it attracts moisture from the atmosphere, has never been determined."[17] In 1825 the *New England Farmer* summarized the prevailing wisdom about gypsum use: "while some extol it highly, others condemn it as useless."[18]

In sum, while gypsum use was on the rise in the early years of the young republic, many of its users remained uncertain about its merits, and that uncertainty, which then available scientific knowledge could not lessen or eradicate, led to a progressive questioning of its merits, presumably inhibiting more widespread use.

The history of early experiments with lime reflects similar difficulties. As with gypsum, scattered references suggest a rising interest in the possibility of using lime as a fertilizer in the final decades of the eighteenth century. Indeed, the first lime experiments may have preceded those with gypsum,[19] but certainly both fertilizers were the subject of growing commentary in the years immediately following 1783.[20] Unlike gypsum, however, lime does not seem to have become widely used in the first few decades of the nineteenth century. Its performance appears to have been much more variable, with negative effects self-evident in some cases, even in the short run. Potential users were cautioned that "too much" would produce "ruinous" effects,[21] but how much was too much remained uncertain, as was the impact of different soils and different climates upon results. Thus in 1797 Robert Livingston of New York ventured the cautious opinion that lime "promotes the growth of some plants, and is destructive to others."[22] Sixteen years later, John Taylor of Virginia warned that "heavy dressings" frequently led to "impoverishment" of the soil.[23]

Uncertainties about its benefits and examples of outright loss evidently tempered the willingness to use lime as a possible antidote for soil exhaustion. Too much could "burn" the crops, and many believed that, even without burning, repeated usage depleted soil fertility, rather than enhancing it. (John Adams, for example, reported that Germans in New Jersey believed "lime makes the father rich, but the grandson poor."[24]) The record of lime use as a fertilizer in the first few decades of the nineteenth century is therefore quite different from that of gypsum. In the years following 1815, for example, while commentators occasionally noted that lime experiments were "increasing," no one claimed that its usage had become "widespread."[25] The record, as of 1819, was summarized by *The Plough Boy* in cautious terms: "The trials of lime in this country have been quite limited, and confined mostly to the middle states, particularly Pennsylvania."[26] In the same year, Noah Webster was similarly circumspect when forecasting future use: "It must remain for future experience to determine its [lime's] efficacy, and the kind of soil to which it may be usefully applied."[27]

Here, too, then, as was true of gypsum, uncertainties inhibited adoption, but unlike gypsum, those inhibitions appear to have constrained the use of lime in major ways during the early decades of the nineteenth century.

Between the 1780s and the 1830s, many other fertilizers were tried and used successfully to combat soil exhaustion, and in many instances, usage seems to have increased over time as long as, for any given fertilizer, price remained reasonable, and the prospect of it actually cutting output seemed remote. The use of manure, for example, was stimulated by a shift in supply made possible by improved pasturing possibilities, discussed under "English Grasses" below, and the use of marl was stimulated by the experiments and publications of Edmund Ruffin, beginning in 1818.[28] But throughout this era, although purchase costs for different fertilizers were easily ascertained, prospective benefits were not, the latter being little more than educated guesses based upon personal trial and error, neighborhood examples, or pronouncements in the agricultural literature (many of which contradicted one another). The rising recourse to many different fertilizers therefore is a tribute to the low price and the evident benefits of some, and the rising demands for means to combat declining fertility on older lands. The choices made were often less than optimal, and some proved downright destructive. But mistakes were unavoidable in a world where imperfect information was the norm. As Joseph Muse reminded the Agricultural Society of Maryland in 1821: "without a set of fixed elementary principles from which to deduce a theory and system of regular management, causes being unknown, anticipated effects will be precarious, and the farmer will delve in the dark, to endless eternity, as he has done from primeval time."[29] In sum, the era of scientific agriculture had not yet arrived, and the absence of scientific knowledge assured that all efforts at expanded fertilizer use would be at first tentative and imperfect. Early attempts to combat soil exhaustion with better crop rotations met a similar fate.

Crop Rotation

The story of colonial attention to crop rotation can be quickly told. As was true for fertilizer, land abundance encouraged an absence of concern for preserving soil fertility on cleared acres, and accordingly the crop sequences normally used quickly led to soil exhaustion. Thus, for example, the Virginian John Taylor described the "murderous" effects upon fertility of a "three shift system" of Indian corn, wheat, and pasture popular in his area. "Under it, the great body of the farm receives no manure, and no rest; and the result is, that the phrase 'the land is killed and must be turned out,' has become common over a great portion of the United States."[30]

When a scarcity of uncleared land forced a reevaluation of such practices, as was true for fertilizer use, Americans turned to Europe for information about possible techniques. Farmers in the Old World had grappled with designing better crop rotations for centuries, and had discovered a variety of crop sequences that helped to maintain the fertility of the soil.[31] Just when such concerns first surfaced in America is not entirely clear, but fragmented evidence suggests a pattern of rising interest similar to that for fertilizer use. After almost complete neglect for the entire colonial period,[32] during the decades immediately following the Revolution better crop rotations became a subject of increasing interest, particularly among the agrarian elite.[33]

As with fertilizers, the main difficulty for Americans was not a scarcity of options, but a lack of knowledge concerning which of the many available options might be best for them. Advocates of "improved" crop rotation were not sure what to recommend. The growing importance of that puzzle is suggested by the fact that, when the Philadelphia Society for Promoting Agriculture sought to develop better farming practices by offering prizes for various kinds of agrarian experiments, the first prize on its list was "for the best experiment made of a course of crops, either large or small, on not less than four acres."[34] Defining what was "best," however, remained an elusive goal. When Samuel Deane included the topic of "Rotation of Crops" in his 1790 survey of New England agriculture, he went on record as an enthusiastic supporter of the general idea but included few specifics. (As one reviewer of Deane's book correctly noted, "On the important subject of a 'rotation of crops' Mr. Deane is not very full.")[35]

The search to establish what was best encountered a number of problems, of which two proved particularly troublesome.

The first was the time required for any systematic investigation. Samuel Deane spelled out the associated problems long before experiments with various crop rotations became widely fashionable. "It is not to be expected, that the best rules concerning this matter can be established, but from the experience of many years. For though it may be easy to compare the respective advantages of different courses, in a few years, so as to find which is most productive; it will take a much longer time to determine which course will be best on the whole."[36]

The second major problem (as with fertilizer use) was the sensitivity of results to a large number of variables, and the absence of precision in the terminology used to identify those variables. These difficulties, too, were well understood by the better informed of contemporary observers.

> From the great variety of soils, difference of climate, and the peculiar nature and habits of the various cultivated plants, it is manifest there can be no rotation, that will be found applicable in all its details, to any large portion of a country so much diversified as the state of Virginia.
>
> Where the circumstances of soil, situation, and crops are similar, a great obstacle to the extension of a salutary uniformity of practice, is experienced in the want of precision in the language of agriculture, especially in relation to the nature and composition of soils; so essential to be adverted to in every agricultural experiment, in order to arrive at similar results; insomuch that the best efforts may fail of giving the precise information intended to be communicated.[37]

Recognizing difficulties was one thing. Solving them was quite another. Faced with such uncertainties, agrarian experts seemed to agree upon only two points: (1) that some rotation scheme was preferable to none, and (2) which was best for a given farmer in a given region was "difficult to ascertain," with the indecisiveness lurking in the latter destined to persist for decades.[38]

Despite such problems, an American interest in designing better crop rotations seems to surface in earnest following the Revolution and gather modest momentum in the next few decades, becoming "widespread" only after 1815.[39] Although this three-stage depiction of early American attention to rotation possibilities is generally accepted (colonial neglect, post-Revolutionary experimentation by a few, post-1815 adoption by many), it is based upon limited evidence scattered over half a century. We know surprisingly little about who experimented where with which rotations, and with what impact upon accepted practice. A similar scarcity of information confounds our understanding of the third technique developed to combat soil exhaustion in the young republic.

"English Grasses"

As discussed in Appendix B, land abundance in the New World encouraged the colonists to let livestock forage on their own. The main difficulty with this strategy was the poor quality of indigenous plants as livestock feed. The solution was to import a range of "English grasses," such as timothy (Herd's grass) and clover. References to such importations and their subsequent use can therefore be found for most of the colonial period.[40] But as a growing scarcity of uncleared land ultimately forced a rethinking of soil fertility problems, so a growing scarcity of open range forced a rethinking of foraging possibilities (the latter no doubt also encouraged by expanding markets for livestock produce). The result was three modifications in established practice. The first was the development of "artificial" (as distinct from "natural") meadows, which consisted of nothing more than the systematic sowing of English grasses upon tilled land to develop better pastures. (Pasture expansion was also pursued by irrigating meadows and draining marshes.) Second, gypsum was often combined with this sowing of grasses (particularly with clover), in many cases yielding spectacular results. (William Strickland, not known for his generous attitude toward American agriculture, found the results "astonishing."[41]) Third, English grasses (often combined with gypsum applications) began to be incorporated into rotation schemes designed to combat fertility decline, particularly schemes in which English grasses were followed by grains such as wheat.

All three of these procedures seem first to surface in earnest in the decades immediately following the Revolution.[42] As with increased fertilizer use and new rotation schemes, the subsequent nature and pace of experimentation and adoption is difficult to document. What does seem clear is that the closing decades of the eighteenth century were something of a watershed for all these different practices designed to combat soil exhaustion, the typical neglect until at least the closing years of the colonial period unmistakably supplanted by a surge in interest following the Revolution. Whether the resulting changes in farming practices should be considered the opening salvos of America's first agricultural revolution is an issue discussed in Chapter 2, with the ultimate decision dependent upon each reader's preferences as to how such a revolution ought to be defined.

APPENDIX B

Breeding and Raising Livestock

The care and breeding of livestock during the early years of the young republic confounds most attempts at broad generalizations, in part because farmers owned a wide variety of animals, and in part because so little evidence has survived concerning who did what when, and with what impact upon local or regional practices.

The task of this appendix is not to survey every change in breeding practice for all farm animals, but rather, through the study of a few breeds for which the evidence is least fragmented and incomplete, to suggest the pace of changing interest in improved breeds between the late colonial period and the 1830s. Subsequent discussion is therefore limited to cattle, sheep, and swine.[1]

For these three types of domesticated animals, three points concerning their early development are relatively clear, and one is not.

First, colonial practices, by and large, appear to have been a rational response to a primitive economic setting with limited markets and abundant land. Second, by the time breed improvement became a subject of widespread interest among American farmers in the early nineteenth century, the techniques for achieving improvement were well known on both sides of the Atlantic. Third, the major impetus for actual improvements made within this country, from first to last, was the profit motive. The changing interest in improvement (as was true of soil exhaustion remedies described in Appendix A) therefore largely reflected a shift in choices made from among known alternatives in response to changing market incentives.

What is unclear is the pace of actual improvements made.

Observers with a knowledge of Old World animals tended to judge similar livestock encountered in the colonies as "stunted." This was to be expected. In a land-abundant setting with limited markets, economic incentives favored breeding practices that minimized the use of labor and capital (such as barns and fences), while exploiting the use of that factor of production which, in relative terms, was the least expensive. Accordingly, most animals were allowed to run free in the woods, or in herds loosely supervised by the community. Breeding was largely "promiscuous" (or not controlled), and the characteristics that developed over time were a tribute to Darwinian mechanisms. With little shelter (even in the winter) and little protection from predators, forced to forage for most or all of their food, colonial cattle, sheep, and swine tended to become hardy breeds, long in the legs and somewhat small in stature, with an ability to survive with little care or attention from their owners. Small wonder, then, that to European visitors cattle of the late colonial period appeared "diminutive" in size and sheep "leggy" and "very poor and ordinary," with "a coarse, strong wool."[2]

As was so often true of changing agricultural practices within the British Empire during the eighteenth century, the mother country led the way in innovations. In this case, the pioneer of special note was Robert Bakewell, who, in the decades following 1750, experimented with the selective breeding of various animals, particularly Leicester sheep and long-horned cattle.[3] His procedures became widely publicized, beginning in the 1780s, and by 1800 were common knowledge in America.[4] On this side of the Atlantic, then, almost from the founding of the new republic livestock owners with an interest in breed improvement could have pursued that goal, either by applying Bakewell's procedures to their domestic stock, or by importing better animals from abroad. The main barriers to progress were economic. Markets for livestock were only gradually expanding, and raising better breeds was expensive. Procreation had to be controlled, which meant that foraging had to be restricted. Better breeds, being less hardy, required more attention to their food and shelter. This usually necessitated additional outlays for fences and buildings to house the animals (particularly during the winter in the North). Labor costs also rose because more care had to be provided over the life of the animal. In the case of cattle developed through selective breeding, for example, according to one nineteenth-century analyst, it was "absolutely necessary . . . to pay special attention to the calves thus produced—to furnish them at all times, summer and winter, with an abundant supply of nutritious food, and to regulate it according to their growth."[5]

American reluctance to embrace new breeding possibilities was therefore not surprising. What remains unknown is exactly when a significant number of would-be pioneers first judged these possibilities worth pursuing. What does seem relatively clear is that by the close of the colonial period, an absence of interest in improving livestock breeds was the norm.[6] Immediately following the Revolution, fragmentary evidence suggests the rising fashionability of breed improvement within a small number of the agrarian elite. The Philadelphia Society for Promoting Agriculture, for example, offered premiums for the best method of "sheltering and folding cattle" and for "raising hogs . . . in pens or sties."[7] A few specimens of improved breeds were imported from abroad: Bedford hogs toward the end of the eighteenth century, some Shorthorn cattle in the 1780s and 1790s, and Merino sheep possibly as early as 1793, and certainly beginning in 1801.[8] What is most striking, however, is what failed to happen. During the quarter-century following the establishment of the republic, the growing interest in better breeds by a select few had almost no impact upon the monumental indifference to that topic on the part of almost all American farmers, including a significant percentage of the agrarian elite.

This pattern of pervasive unconcern was suddenly shattered in 1810 by what contemporaries were quick to label "the Merino craze." This speculative bubble in high-quality livestock occurred because of an improbable shift in both demand and supply, resulting from changes in government policies on opposite sides of the Atlantic. The story begins with the various acts passed by the American Congress to inhibit foreign commerce: the Embargo Act of 1807, the Nonintercourse Act of 1809, and Macon's Bill Number Two in 1810. Whatever their limitations as effective barriers to international trade, these various pieces of legislation gave a marked stimulus to United States woolen mills, particularly the few capable of producing fine cloth. As previously noted, prior to 1807 a handful of Merino sheep had been imported into America, but with no perceptible impact upon the farming community at large.[9] But as trade restrictions limiting British textile imports stimulated domestic production, the price of high-quality wool shot up, as did the price of sheep capable of providing such wool. Merinos, which sold in 1808 for $150, by 1810 were often priced above $1000.[10] Supply of the latter for years had been restricted by export prohibitions of the Spanish government. But a Napoleonic onslaught reduced Spain to chaos, and local revolutionary juntas began to sell off Merino flocks in large numbers. Here was an opportunity for windfall profits which a handful of well-informed Americans were quick to exploit, including the American Consul in Lisbon. Between September 1, 1810, and August 31, 1811, almost 20,000 Merinos were imported into America.[11] As supply increased, prices fell, but interest in profitable opportunities through Merino ownership continued to spread, buttressed by sustained demand from American woollen mills. In the autumn of 1811, the "Merino Society of the Middle States" was established, and by the summer of 1814 sheep breeding had "become a highly interesting concern of the farmer, in most parts of the United States."[12] The following year, peace brought a flood of British textiles, a collapse in domestic woollen prices, and the end of any American delusions about making quick profits by purchasing Merinos. Toward the close of 1815, Philadelphia newspapers were advertising a product which, twelve months earlier, would have been unthinkable: "Merino mutton."[13]

This boom-and-bust sequence in Merino prices illustrates how shifts in American attitudes toward improving livestock were crucially dependent upon the profit motive. To be sure, most of the pioneers in Merino importations early in the century had patriotic reasons for wanting to improve American stock.[14] But the public at large remained unmoved by their efforts. Only when breed improvement became linked to prospective economic gain did that indifference change, and then—for Merinos—it changed with a rush. To Thomas Jefferson, the resulting frenzy to import and own high-quality sheep was hardly admirable, reflecting as it did "that kind of patriotism the strongest feature of which is to enrich the patriot himself."[15]

But the pursuit of personal enrichment appears to have been central to most attempts at livestock improvement in the early years of experimentation in the young republic. The histories of cattle and swine provide further cases in point. As was true for sheep, fragmented evidence indicates sporadic importations of a few breeds in the closing decades of the eighteenth century: a handful of Shorthorns in the 1780s and 1790s, and by 1800, a few Woburn or Bedford swine.[16] The extent to which the acquisition of these animals reflected patriotic as distinct from profit motives is impossible to say. But as was true for sheep, these new breeds of cattle and swine had no discernible effect upon the farming community's indifference to livestock improvement in the years immediately following their arrival.

That indifference apparently waned in the decade following 1815, as the pace of foreign importations quickened. The increased interest in better cattle breeds appears to have been largely the result of three factors: (1) the gradual expansion of markets for livestock products, (2) the growing interest stimulated by the rise of cattle fairs and exhibitions, as discussed in Chapter 10, and (3) the improved pasturing possibilities outlined in Appendix A. The much publicized cattle importations of Colonel Lewis Saunders in 1817 illustrates that (as with sheep) concern for personal enrichment was central to the pursuit of better breeds. By his own account, Saunders followed developments in overseas livestock markets, and saw in British price trends an American opportunity. The market information prodding him to action was an extraordinary price paid by a consortium of four English farmers for a prize Shorthorn. "It seemed to me that if four farmers were willing to pay $5,000 for a bull, there was a value in that breed that we were unappraised of, and that I should endeavor to procure it [the breed, not the bull]."[17] Accordingly, Saunders instructed the Liverpool firm of Buchanan, Smith & Co. to acquire a dozen of "the best young animals for breeding." The resulting shipment—eight Shorthorns and four longhorns,[18] evenly split between bulls and heifers—appears to be one of the first efforts to improve American stock by foreign importations in the years immediately following the conclusion of the War of 1812.

As the history of American livestock development is commonly told, in the 1815–1825 period other "initial" or "major" importations included, for cattle, Alderney, Ayrshire, Devon, Guernsey, Hereford, and Jersey; for sheep, Saxony, Southdown, and Tunis; for swine, Berkshire and Essex.[19] Other evidence also suggests a rising interest in livestock improvement at this time. Increasingly the criteria for judging cattle at local fairs were challenged as inconsistent with the profit motive. By the middle of the 1820s, a subject of spirited debate was whether livestock improvement in America was best pursued by foreign importations or by applying selective breeding practices to domestic stock.[20] Also by the mid-1820s, contemporary observers were noting "the spirit excited within the last five or six years" concerning better livestock.[21] By 1829, according to "a firm in Boston which slaughters and packs more beef than any other in New England," the quality of neat cattle had "improved more than ten percent under the influence of Cattle Shows, for a few years past."[22] To this judgment the editor of *American Farmer* added his own, that "equal improvement has also been made in sheep and wool, in swine, and numerous other growths of the farm."[23]

Early American interest in breed development, therefore, can be divided into three broad stages: (1) the colonial period, when interest was almost nonexistent, (2) the decades between the Revolution and the War of 1812, when some interest was evidenced by an elitist few (particularly in importing better breeds from abroad), and (3) the period 1815–1830, when, for the first time, breed improvement appears to have become a subject of general concern.[24]

Judged in the context of this general pattern of a gradual rise in interest, experimentation, and importation spanning half a century, the Merino mania that began about 1810 is clearly a momentary aberration, although it may well have supplied a not inconsequential stimulus to the wider interest in livestock breeding that seems to have developed in the second decade of the nineteenth century. What remains unknown is how important that stimulus was, primarily because we know so little about the quickening interest after 1815. The sporadic records of importations shed little light upon how local attitudes and practices were modified by the presence of foreign stock. As for efforts to apply selective breeding techniques to domestic stock, almost nothing is known about what was done, by whom, and with what success.[25] In sum, in the histories that trace American livestock development during the first half-century of the young republic, the fashionable generalizations about "neglect" prior to 1800, and "rising interest" after 1815[26] appear to be defensible, although based upon very limited evidence. These generalizations, in turn, suggests a coincidence in timing between the first surge in farm implement improvement that has been the subject of this book, and the first major surge in efforts to develop better livestock breeds. But the latter efforts—insofar as they were an integral part of the first American agricultural revolution—do not seem to be a promising avenue for the further exploration of the larger revolution, primarily because the limited nature of surviving evidence is sure to confound such an exploration.

Notes

1. The Problem

1. The population ranges from 7000 B.C. to 1900 are from United Nations 1973: 10. Data for 1900 to 1965 are given as point estimates in the same source, although these numbers, too, are appropriately viewed as a best guess within a range of likely totals somewhat narrower than the range for earlier totals. Data beyond 1965 are from United Nations 1992: 22 (which are also point estimates). For a discussion of estimates of total world population prior to 1900, see W. F. Willcox 1931: 33–82 and appendix I.
2. Simon Kuznets 1966: 46. The linkage between the population surge and changing birth and death rates are summarized in ibid., chapter 2. For a recent effort to link the fall in death rates to improved economic conditions in the early stages of industrialization, see Robert Fogel 1994. In the twentieth century, some countries have experienced major declines in death rates without the modernization of their economies. This happened, for example, when the spraying of DDT radically reduced malaria in Sri Lanka (then Ceylon). In eighteenth-century Britain, the question of which variable changed first and why is the subject of a vigorous and unresolved debate. See, for example, Peter Razzell 1993. Some have argued that population growth triggers technical change in agriculture, rather than viewing the primary causation as running from technical change to better living standards to lower death rates. Whatever the merits of the former view—and it has been repeatedly criticized by historians—the population trend of Figure 1.3 makes abundantly clear that the surge of the past few hundred years is without precedent, and therefore the causal forces at work were clearly different from those of earlier centuries that linked population growth and agrarian change (whichever way the causation ran between these two). For examples of the argument making the primary causal link from population growth to agrarian change, plus criticisms of that view, see David Grigg 1982: 17, 37–43, 99; Ad van der Woude 1987; Ester Boserup 1987; Jan de Vries 1972.
3. Students of technical change, particularly economists, have often tried to distinguish between "invention" and "innovation." The first, in studies of economic growth, tends to be identified as any new product or technique with a possible economic use. The second (often a variant of the first after many revisions) concerns "the process by which new products and techniques are introduced into the economic system"

(R. Nelson 1968: 339). In principle the key distinction seems to be between any novelty with a potential economic application and those novelties (possibly modified) that have a practical economic use. In practice, of course, the differences between the two can become blurred. For purposes of this book, the distinction is not crucial. As will become clearer below, the focus of this work is the change in attitude on the part of producers with respect to their willingness to search for new ways to produce. The mix of novelties so generated between useful and useless is less important than that the search for novelties become an established tradition.

4. For recent summaries of this debate, as well as position statements by some of the main participants, see Joel Mokyr, ed. 1993, P. O'Brien and R. Quinault, eds. 1993, D. C. Coleman 1992: chapter 1; D. Greasley and L. Oxley 1994; N. F. R. Crafts and T. C. Mills 1994.
5. This point is by no means lost on all those engaged in the current debate about the transformation of the British economy. See, for example, G. N. von Tunzelmann's review of Joel Mokyr, ed. 1993 in *Journal of Economic History*, 54 (June 1994): 457–459.
6. For a more detailed discussion of these and other technical changes in Western Europe prior to 1750, see Joel Mokyr 1990: chapters 3 and 4.
7. Estimates suggest that at the beginning of the nineteenth century, roughly four out of every five members of the American labor force were engaged in agriculture (Thomas Weiss 1992: 22).

2. The Approach

1. Arnold Toynbee 1964: 1.
2. See, for example, E. Ross and R. Tontz 1948. In the British case, some emphasize the revolutionary aspects of changes made before 1750, while others, focusing upon productivity gains, consider the British agricultural revolution to have taken place after 1750. Thus, Eric Kerridge claims that most of the initial revolutionary changes occurred between 1560 and the early eighteenth century, whereas J. D. Chambers and G. E. Mingay append to a book entitled *The Agricultural Revolution* the years 1750–1880. See Kerridge 1967 and 1969, Chambers and Mingay 1975; also Mingay 1963 and 1969. While scholars now generally agree that some productivity improvement occurred before 1750 in England, both the timing and the causes of this improvement are subjects of debate. See, for example, the essays in B. Campbell and M. Overton 1991a, particularly Campbell and Overton 1991b, Robert C. Allen 1991, and Gregory Clark 1991.
3. One possible source that yields surprisingly little information is the work of serious landscape artists of the time. None of America's leading artists had any interest in conveying the kind of sweaty detail that can be found in some European renderings of agrarian activity, such as those of Pieter Breughel. Farm implements, when they appear at all, are usually pictured from a distance, as minor items in a larger panorama. The following observation, while part of a study of New England art, is applicable to almost all of American landscape painting of the early nineteenth century. "Even agricultural scenes concentrate, usually, on the farmhouse and outbuildings, as tokens of prosperity or models of emulation . . ., but the arduous realities of work are barely suggested." (J. E. Cantor 1976: 5.)
4. One of the issues no longer in dispute is that the transformation of West European agriculture was a long, drawn-out process, with the antecedents of what would ultimately become revolutionary changes reaching far back into the Middle Ages. See, for example, E. Kerridge 1967 and 1969, G. E. Mingay 1963 and 1969, B. H. Slicher van Bath 1960 and 1969, F. Dovring 1966.

5. Slicher van Bath, for example, dates the development of many features of intensive husbandry in the Low Countries from "the thirteenth, fourteenth or fifteenth centuries" (1960: 137). Should these developments, by themselves, be confused with an agricultural revolution of the modern sort, Slicher van Bath adds the necessary corrective: "It is not a picture of wealth, but of scarcely controlled poverty" (153). More recent scholarship finds even earlier beginnings for important innovations in Flemish agriculture. See E. Thoen 1988; A. Verhulst 1990.
6. T. S. Ashton 1948: 58.
7. D. McCloskey 1981: 106.
8. Historians have repeatedly focused upon later changes in American agriculture, giving scant attention to the earliest stages of a transformation in attitudes and implements that set in motion forces that would lead to successive waves of innovation. This slighting is typical even in such classics of American agricultural history as P. Bidwell and J. Falconer 1925, L. C. Gray 1958, C. H. Danhof 1969, or more recently R. D. Hurt 1982, all of which give few details on how and why this first American agricultural revolution began. Also symptomatic of this slighting is the fact that in detailed studies of farming, such as Peter Coclanis's 1989 study of the South Carolina Low Country between 1670 and 1920, the reader gets almost no sense of when an agricultural revolution first occurred, or why. With such limited attention to details, estimates of timing are predictably diverse. Paul Bairoch, for example, dates the "outset" from 1760–1770 (1973: 460), while Leo Rogin detects a major change in implement use "about 1820" (1931: 229).

3. Plowing

1. James Small 1784: v.
2. Julian P. Boyd, ed. 1956: Vol. 13, 26–27. See also Edward Dumbauld 1976: 114–124, for other details of this tour.
3. For sketches made by Jefferson at the time, see Julian P. Boyd, ed. 1956: Vol. 13, 27 and the plate following 16. Later in America, Jefferson would consult with David Rittenhouse to clarify the mathematics (New York State Agricultural Society 1868: 34). A more complete set of diagrams and a detailed description of the moldboard shape that he proposed may be found in Thomas Jefferson 1799. For a discussion of the mathematics in question, see E. A. White 1918: 150, 173–174.
4. New York Agricultural Society 1868: 66.
5. Thus, for example, "the geographical distribution of the plow about the year 1500 was almost co-extensive with the civilizations of the Old World." (E. C. Curwen and G. Hatt 1953: 287.)
6. This necessity of plowing a particular field twice because of what a single passage had left undone was a common theme in the ancient world. Virgil, for example, praised the husbandman who, "by a frequent toil, / To obedience brings a disobedient soil," noted that this toil included "Crosswise again, with plow reversed" (quoted in translation in L. H. Bailey, ed. 1907: 373). Crosswise plowing was not always necessary, particularly on light soils, in which case the field might be plowed lengthwise one year, and across the next.
7. For discussions of related puzzles, see E. C. Curwen and G. Hatt 1953, F. L. Kramer 1966, and A. Gailey and A. Fenton, eds. 1970.
8. See, for example, N. E. Lee 1960: 43; E. C. Curwen and G. Hatt 1953: 72; M. Partridge 1973: 36; R. Ellison 1982: 173–184; Paul Leser 1931, chapter III.
9. New York State Agricultural Society 1868: 8.
10. The literature detailing early plow designs and development in the West is voluminous, in most instances including pictures or depictions of various designs used to

construct the ancient ard. See, for example, N. E. Lee 1960: 43; E. C. Curwen and G. Hatt 1953: 70–71; Sian E. Rees 1979: 125–128, 177; F. L. Kramer 1966: xiv; Gaetano Forni 1980: 60; E. M. Jope 1956a: 82–87; K. D. White 1967: 127–131; P. V. Glob 1951. For illustrations of plow development in China and Japan, see F. Bray 1979, Wang Xing-Guang 1989, and J. Iinuma 1982. The ard is still in use in third world countries, and depictions of modern versions may be found in Inge Schjellerup 1986; I. Mukhiddinov 1979; A. Sordinas 1978; and G. Lerche and A. Steensberg 1983. A more systematic treatment of different body types may be found in František Šach 1968; H. C. Dosedla 1984; and H. J. Hopfen 1960: 44–54.

11. See, for example, E. M. Jope 1956a: 85.

12. The basic types of ard share—socket, tang, and chisel—are illustrated in H. J. Hopfen 1960: 52. Other illustrations of ancient shares may be found in E. M. Jope 1956a: 85; and K. D. White 1967: 133–136.

13. Little or no attention will be given to other plow parts, not because their design is unimportant, but because understanding their form and function is less crucial to understanding the evolution of American plows prior to 1830.

14. E. J. Stowe 1948: 68.

15. For other illustrations of nineteenth-century American share designs, see J. J. Thomas 1869: 122. More modern varieties are depicted in H. P. Smith 1955: 97. For a discussion of design possibilities, see W. R. Gill and G. E. V. Berg 1967: 181. For an indication of the complex interaction between share design and correct hitching to draft animals, see J. J. Thomas 1869: 128–130.

16. "The angle of the point of the share with its proper clearance beneath the point is known as the 'suck.' The sharper the angle and the greater the clearance, the more suck. There is a working limit both ways to the amount of suck. If there is too little the bottom will tend to run out of the ground, while if there is too much the bottom will jump and jerk [or dive]." (Theo Brown 1925: 125.)

17. The dimensions of the knife coulter depicted in Figure 3.8 are a neck *(ab)* "about 10 inches long," and a blade *(bd)* varying in length from 18 to 22 inches, about three inches at its widest point, but tapered to a point *(d)*, flat on the landside, and on the furrow side "bevelled off towards the cutting edge." (New York State Agricultural Society 1868: 43–44.)

18. Other more modern American coulter designs may be found in A. A. Stone 1945: 48 and H. P. Smith 1955: 117. The revolving coulter has a long history, an early illustration appearing in Walter Blith 1652: plate facing 203.

19. W. T. Emmet 1983: 27.

20. See, for example, R. D. Doner and M. L. Nichols 1934.

21. New York State Agricultural Society 1868: 33.

22. For a review of these efforts plus an exposition of the mathematics involved, see E. A. White 1918: 149–182. For a review of more recent American efforts, see W. T. Emmet 1983: 2–30.

23. E. A. White 1918: 149. See also W. T. Emmet 1983: 2–30; H. P. Smith 1955: 95.

24. W. R. Gill and G. E. V. Berg 1967: 117. About the prevalence of art (as opposed to science) in modern design efforts, these authors add: "Even in more advanced societies, today, the moldboard plow is designed by empirical methods. Generally, these empirical methods are trial-and-error attempts; the tool is varied in some manner and acceptable designs are identified when the resulting soil condition is adjudged to be satisfactory. Quantitative descriptions or representations of the final soil condition are seldom used and, in addition, the forces required to move the tool are frequently not quantitatively assessed. Generally, no effort is made to describe the reaction of the soil. Consequently, design today merely accepts what occurs; it does not control what occurs. Thus, even though the need for design is great, design in the true sense of the word is not accomplished and probably will not be accom-

plished until quantitative information is available" (ibid., 211). See also Theo Brown 1925: 124–125.

25. See E. M. Jope 1956a: 86–87; K. D. White 1967: 133; N. E. Lee 1960: 84–87; E. C. Curwen and G. Hatt 1953: 83; K. D. White 1984: 59–60, 138–142.
26. K. D. White 1967: 140.
27. E. M. Jope 1956a: 86–87; N. E. Lee 1960: 87.
28. See K. D. White 1967: 134, 195.
29. To claim, as Lynn White, Jr. does, that "the Romans made little progress in solving the distinctive problems of the north" (1962: 49) is to dismiss a number of innovations of considerable importance in the development of heavy plows for heavy soils. The ard could also be used in such conditions to create a wide furrow by tilting it slightly to one side, but the resulting drag was enormous. For a sketch of this latter process, see E. M. Jope 1956a: 87.
30. For further discussion, see K. D. White 1967: 125.
31. See for example E. M. Jope 1956a: 93; N. E. Lee 1960: 86; G. E. Fussell 1966: 180; E. C. Curwen and G. Hatt 1953: 84; Lynn White, Jr. 1962: chapter 2; J. B. Passmore 1930: 1–6, 71; A. J. Spencer and J. B. Passmore 1930: 9.
32. J. M. G. van der Poel pointed out that this moldboard design was called by the Romans "auris" (or "ear"). (Private communication, December 17, 1994.) Some argue that one type of plow dating from this period had virtually no moldboard, its design not unlike Kent turn wrest plows of a later era. For illustrations of the latter, see A. Rees 1810–1824, Vol. 1, Plate XXI; and C. W. Johnson 1848, Vol. 2, 4. For a discussion of this type of plow design, see N. E. Lee 1960: 91; G. E. Fussell 1966: 178; and P. W. Blandford 1976: 50–53, 180.
33. For a discussion of how both plows with wheels and without (the "swing" plow) continued to be used during the Middle Ages, see G. E. Fussell 1966: 178–183; Lynn White, Jr. 1962: 46; E. M. Jope 1956a: 88–89. A third design was the "foot" plow, the name derived from the "foot" that supported the beam at the front (instead of a forecarriage with wheels). For a discussion and illustrations of this design, see J. M. G. van der Poel 1964. In the Netherlands, the wheeled plow was used on heavy soils and the foot plow on lighter soils, with the swing plow unknown before importations from America in the 1850s. (J. M. G. van der Poel, private communication, December 17, 1994.)
34. One curious exception to this curvature from top to bottom is the front-to-back curvature suggested by the shading in Mortimer's 1708 illustrations, one of which is reproduced in Figure 3.14. (See John Mortimer 1708: 46–47.) Among the earliest European plow illustrations is the one included on the border of the Bayeux Tapestry (Hilaire Belloc 1914: 12), but it is poorly drawn and pulled by a single animal, suggesting the implement is not a heavy plow working in heavy soil. (One also wonders about the artist's agrarian knowledge, for the plowman is incorrectly portrayed following the sower who is following a horse pulling a harrow.) For other illustrations of European plows dating from the eleventh to the early eighteenth century, see J. C. Webster 1938: Plate XVII; E. M. Jope 1956a: 89–90, 94–95; F. M. Kelly 1936: Plate III; Jean Longnon 1969: Plate F. 3v; Franco Cardini 1989: 73; Thomas Kren 1983: 81; M. Partridge 1973: 38; P. Lindemans 1952, Vol. 1, 172, 175, Plates VI–XII; Edward Maxey 1601: frontispiece; G. E. Fussell 1952: 31; A. G. Haudricourt and M. J.-B. Delamarre 1955: 369–397, 428. The plow diagrams in R. Bradley 1725 appear to have been taken from J. Mortimer 1708. (One of the diagrams from the latter is reproduced in Figure 3.14.) Bradley mainly assembled information from earlier writers, and his works accordingly include little original information.
35. This latter idea was suggested to me by J. M. G. van der Poel (private communication, December 17, 1994).
36. Walter Blith 1652: 199.

37. John Mortimer 1708: 44.
38. John Fitzherbert's *Boke of Husbandry* (often incorrectly attributed to his younger brother Anthony). For a comprehensive bibliography of early English books on husbandry, see Donald McDonald 1908. Less detailed bibliographies may be found in G. E. Fussell 1952 and J. A. Symon 1959.
39. Some English plows had only one handle, as illustrated in Figure 3.14B, as did most Dutch plows. (J. M. G. van der Poel, private communication, December 17, 1994.)
40. G. Markham 1613, chapter III. Markham also provides a second set of illustrations of a share and coulter appropriate for "gray clay," the share in question not unlike the "fin share" of Figure 3.9.
41. W. Blith 1652: 203, 208–209. Blith had published an earlier version in 1649, entitled *The English Improver*. The 1652 "improved" version included "an additional discovery of the several tools, and instruments in their forms and figures promised," plus "a second part; containing six newer pieces of improvement."
42. W. Blith 1652: 191. Passmore argues that Blith was "alive . . . to the necessity of the winding mould-board, which followed the furrow slice." (J. B. Passmore 1930: 10; see also 36–39, 63.) As noted above, Blith did complain about moldboards being "too straight," and at several points, he advocates a "twist" to the moldboard. But at no point does he offer specific instructions for designing such a moldboard. More importantly, subsequent English builders of plows evidenced no inclination to follow his advice during the next half-century. See, for example, M. Partridge 1973: 39.
43. John Mortimer 1708: 44–45. The full list of recommendations reads as follows: "First, The plough reckoned the most proper for stiff black clays, is one that is long, large and broad, with a deep head, and a square earth-board, so as to turn up a great furrow; the coulter long and very little bending, with a very large wing; and the foot long and broad, so as to make a deep furrow.

"Secondly, The plough for the white, blue, or grey clay need not be so large as the former, only somewhat broader at the breech, the coulter to be long and bending, and the shear narrow, with a wing coming up to arm and defend the earth-board from wearing.

"Thirdly, The plough for the red, white, sands or gravel or any light moulds, may be lighter and nimbler than the former, the coulter more circular and thinner, and the wing not so large."
44. Walter Blith 1652: plate facing 203. The share in question Blith labeled a "pure Dutch share," distinguishing it from an English socket share, also illustrated. The share design in question was only one of a number of Dutch designs in use at this time. See J. M. G. van der Poel 1967: passim.
45. Lynn White, Jr. claims an agricultural revolution took place in the early Middle Ages, the main elements of which he summarizes as follows: "By the early ninth century all the major interlocking elements of this revolution had been developed: the heavy plough, the open fields, the modern harness, the triennial rotation—everything except the nailed horse-shoe, which appears a hundred years later. . . . The agricultural revolution of the early Middle Ages was limited to the northern plains where the heavy plough was appropriate to the rich soils, where the summer rains permitted a large spring planting, and where the oats of the summer crop supported the horses to pull the heavy plough" (Lynn White, Jr. 1962: 78). The pictorial record suggests this account is greatly exaggerated. Surviving illustrations are consistent with the hypothesis that: (a) to pull plows, throughout Europe oxen continued to be the dominant choice by a wide margin until the fifteenth or sixteenth century, and (b) thereafter, the use of horses for this purpose gradually became more common. If primitive plows produced such high friction that they were almost impossible to pull at any speed, then the main advantage of the horse—increased speed of operation—would have been minimal. Consequently, the conversion to horse-drawn plows begin-

ning in the sixteenth century in parts of England and much earlier in the Low Countries may be indicative of improving plow designs that reduced friction and encouraged this recourse to a higher-cost draught animal. (J. M. G. van der Poel indicates that the shift to horses in the Netherlands and Belgium was evident as early as the thirteenth century; private communication, December 17, 1994.) See, for example, Figures 3.11 and 3.12, also E. M. Jope 1956a: 91–92; G. E. Fussell 1966: 181; M. Partridge 1973: 38. (White cites, in favor of his hypothesis, the horse or donkey appearing on the border of the Bayeux Tapestry (63). But this plow is drawn by a single animal and consequently cannot possibly be a "heavy" plow.)

46. The 1652 and 1733 Hertfordshire designs are reproduced in Figure 3.14. The 1708 design appears in J. Mortimer 1708: 46. While some English writers stress the growing popularity of this design (for example, R. Bradley 1725: 340), John Mortimer in 1708 considered the Dray plow illustrated in Figure 3.14B to be "the most common plow" as well as "the best plow in winter for miry-clays." (J. Mortimer 1708: 47.)
47. Sketchy evidence also suggests the best of these premodern plows were repeatedly defective in performance. The Dray plow of Figure 3.14B, for example, although a "common" plow, when used on hard soil in the summer was "always flying out of the ground." (J. Mortimer 1708: 47.)
48. J. B. Passmore 1930: 7.
49. A. Young 1804: 36–37. Young conceded that the plow was useful for "breaking up strong flinty fallows in a dry season," but "here ends [its] merit." (Ibid.)
50. P. W. Blandford 1976: 50.
51. Again, compare illustrations in Figure 3.14. The multiple coulters of Tull's plow were to cut the turf into ribbons. (The design quickly proved to be a failure.) In designing his "horse-hoe," Tull did little more than add an elaborate hitching device to a Hertfordshire-like plow. See J. Tull 1733: plate vi, and the discussion of cultivators in Chapter 6; also G. E. Fussell 1973.
52. Compare Figure 3.14D with Figures 3.12 and 3.5. Much has been made by some English writers of the moldboard modification mentioned by John Mortimer in 1708; namely, "about Colchester" a plow was in use "very peculiar for its earth-board [moldboard] being made of iron, by which means they make it rounding, which helps to turn the earth or turf much better than any sort of plough that I have seen." (J. Mortimer 1708: 46.) Many experts consider this to be the first reference to a twisted iron moldboard. (See, for example, G. E. Fussell 1952: 44; M. Partridge 1973: 39.) Passmore argues that this light two-horse plow was introduced into East Anglia about 1600, probably from Holland, remained popular in that region thereafter, and was "an important step forward" in plow design, primarily because its moldboard used boards plated with iron "which attempted to follow the twist of the slice." (J. B. Passmore 1930: 6–10; boards plated with iron were in use in Holland by "the beginning of the fourteenth century or even earlier"; J. M. G. van der Poel, private communication, December 17, 1994.) But if the plow was truly revolutionary, Blith's failure to emphasize its revolutionary nature half a century after its supposed introduction is peculiar, and its absence more than a century later from the books of agrarian pioneers, such as John Mortimer's *Whole Art of Husbandry* (1708), R. Bradley's *Survey of the Ancient Husbandry and Gardening* (1725), and Jethro Tull's *Horse-Hoeing Husbandry* (1733) is equally remarkable. A more likely hypothesis, consistent with Mortimer's 1708 observation (noted above) about it being "very peculiar," is that, whatever the shape of its moldboard, this plow was suitable for only limited kinds of soils and failed to influence design developments in a major way because it did not closely duplicate the double curvature of the modern moldboard. For other opinions on the absence of moldboards with a double curvature in Europe prior to the eighteenth century, see K. D. White 1967: 140, and H. J. Hopfen 1960: 58.

53. One of the earliest pleas for "mechanicks" to find and specify "exact Rules" for the making of plows was made by Samuel Hartlib in 1652. (Hartlib 1652: 4–5.) But Hartlib contributed nothing to this search himself. In the writings of one of Hartlib's contemporaries, Walter Blith, one can find general "maxims" combined with design recommendations. But the maxims are nothing more than a handful of unsurprising generalizations (e.g., sharper and narrower cutting parts perform better than thick and dull ones) and their associated implications for design remain vaguely specified. ("Plough irons should be well steeled, sharp, and well pointed.") For a summary of Blith's statements, see J. B. Passmore 1930: 36–39. For similar skepticism about the development of rules for making plows prior to 1700, see M. Partridge 1973: 38.
54. 1910, Vol. 1, 395.
55. Foljambe is often credited with building the original model, but he sold the patent to Stanyforth, who subsequently manufactured and sold the plow. The terms of a written agreement between the two men suggest that Foljambe was the inventor and Stanyforth the financier. (See G. Marshall 1982: 131–136.) Some authors claim the English Rotherham plow was "built according to principles brought from Holland." (O. Beaumont and J. W. Y. Higgs 1958: 2; see also G. E. Fussell 1966: 184–185.) Others admit that possibility but are less sure (for example, J. B. Passmore 1930: 15; M. Partridge 1973: 40). Geoffrey Marshall, on the other hand, argues that Dutch influence on the original Rotherham design was, at best, trivial (G. Marshall 1978: 162–165) and the Dutch expert, J. M. G. van der Poel, agrees with Marshall (private communication, December 17, 1994).
56. F. G. Payne 1957: 83. For another endorsement of importance in more muted language, see O. Beaumont and J. W. Y. Higgs 1958: 2.
57. The moldboards of early Rotherham plows appear to have consisted of an iron plate fitted to a wooden base. (See G. Marshall 1978: 152; M. Partridge 1973: 40.) The process of using molds to gain better uniformity is described by Marshall (ibid.). "Several parts of the plough are claimed to have been made 'by a mould': the beam, breast and mould-board. The common timbers used in plough construction were ash and oak. Both of these can be soaked or steamed to take a pre-set curve and it seems likely that the curve of the beam, the forward sweep of the breast and the convex shape of the mould-board were all produced by this method. Curved mould-boards used to be made in the Netherlands by soaking a prepared piece of timber in linseed oil and then putting it into a press."
58. From an advertisement for the implement (incorrectly specified as "The Rotheran-plough") in R. Maxwell 1743: 457–458.
59. As explained in the original patent application, the Rotherham was "a new sort of plough, by which farmers may plough with greater ease, and with less charge than with ploughs hitherto used; that the number of horses used at present to work the ploughs will be sufficient to work three of their ploughs, by which means three acres of ground instead of two acres may be ploughed at the same charge, and within the same time." Quoted in G. Marshall 1978: 151.
60. The Rotherham evidently also required fewer repairs, mainly because of "the absence of a share beam which appears to have been a part of the former ploughs requiring frequent repair or replacement." (G. Marshall 1982: 133.) Dispensing with the assistant to manage the team was considered significant enough to be celebrated in contemporary verse.

> No wheels support the diving pointed share;
> No groaning ox is doom'd to labour there;
> No helpmates teach the docile steed his road;
> (Alike unknown the plow-boy and the goad);
> But, unassisted through each toilsome day,

With smiling brow the plowman cleaves his way.
(Robert Bloomfield 1798: 8.)

61. The Rotherham was made somewhat more manageable by a novel bridle design with multiple adjustments allowing the plow to be turned to and from the land; or as phrased in one of the earlier advertisements, "by which the Ploughman can give the Plough more or less Land." (R. Maxwell 1743: 458.) This was particularly crucial for a plow that—without the proper bridle setting—tended to be unstable. See G. Marshall 1982: 134; G. E. Fussell 1966: 185. "It was a tribute to the soundness of the Rotherham design that the early iron ploughs were closely modelled on it, or on Small's adaptation of it." (F. G. Payne 1957: 83.)

62. P. W. Blandford 1976: 55. For a discussion of the influence of the Rotherham design on subsequent plow inventors and inventions in the British Isles, see G. Marshall 1978: 152–162.

63. See P. W. Blandford 1976: 56; G. E. Marshall 1978: 153–154. Arbuthnot also made several improvements in the clevis of the original Rotherham. For a description of his procedures and improvements, see J. B. Passmore 1930: 14–18, 40–42.

64. J. Small 1784: 104. Whether Small used mathematics to derive this principle, and if so, what mathematics he used, is not known. He did give precise instructions for plow construction based upon a "protractor and scale" and a "bevel" that reflected his training as a carpenter. (Ibid., 108 ff.) The final shape of his moldboard can be expressed in mathematical terms (see E. A. White 1918: 175–177), but there is no evidence that Small used the equation in question, as opposed to the general principle articulated above. As late as 1810, one British analyst of plow designs complained that the mathematical principles were still unknown. (See G. E. Fussell 1952: 39–40), and a century after Small's initial efforts to produce a better plow, the New York State Agricultural Society concluded that Small "perfected his plow gradually and experimentally" and "his improvements were empirical, and were not at first guided by any mathematical principles whatever" (1868: 34–35). For further discussion, see G. E. Fussell 1952: 48–52.

65. For further discussion of Small's plow innovations and procedures, see G. E. Marshall 1978: 153; P. W. Blandford 1976: 56; M. Partridge 1973: 40; J. A. Symon 1959: 383–385; G. E. Fussell 1952: 46–52.

66. Marshall notes that from Small's drawings "it is obvious that he used the basic frame of a swing plough that was inspired by the 'Rotherham' plow." (G. E. Marshall 1978: 153.)

67. See, for example, C. W. Johnson 1848: Vol. II, 6–10; J. C. Morton 1854, Vol. 2, 631–633.

68. In 1803 Ransome received a patent for "case-hardening" or "chilling" shares. For a description of the accidental process leading to this discovery, see P. A. Wright 1961: 36.

69. C. W. Johnson 1848, Vol. 2, 10.

70. See J. A. Symon 1959: 357; M. Partridge 1973: 41. For Ransome's criticisms of Small's moldboard design, see J. B. Passmore 1930: 19–20.

71. In its survey of British plow developments, the New York State Agricultural Society concluded that after the first decade of the nineteenth century, "The changes made in the plow for the next quarter century were very slight" (1868: 62).

72. "During this period there developed a differentiation between autumn and spring ploughing, with the production of a distinct type of plough for each process. A long-breasted plough . . . , with a long gently twisting mould-board, was designed to lay the slices on edge and unbroken in the autumn, whilst a plough with a bluff elliptical breast . . . , a short mould-board twisting sharply and sometimes armed with knives, was used to burst up the soil in the spring and to leave it as well pulverized as possible." (A. J. Spencer and J. B. Passmore 1930: 10.)

73. The double-furrow plow was hardly novel, one being included in Walter Blith's illustrations in 1652. (W. Blith 1652: plate facing 203.) But the double-furrow plows of this later period were, according to their inventors, "improved."
74. The first recorded British plowing matches using the dynamometer date from the 1780s. (See G. E. Fussell 1952: 55.) For illustrations of plows from this period, see A. M. Bailey 1782, Vol. 1, Plates I, V, VI; *The Farmer's Magazine and Useful Family Companion*, 1 (1776), plate opposite 97; *Encyclopaedia Britannica* 1771 (1st ed.): Plates VIII, IX; ibid. 1778–83 (2nd ed.): Plates V, VI; ibid. 1803 (3rd ed.): Plate VI; A. F. M. Willich 1803: 288–298. For details on the evolution of British plows in the eighteenth and early nineteenth century, see J. B. Passmore 1930: 14–22; G. E. Fussell 1952, chapter 2; M. Partridge 1973: 40–48; P. W. Blandford 1976: 54–58.
75. Plow innovations in the Low Countries before 1800 are described, with illustrations, in Paul Lindemans 1952: 164–188, and J. M. G. van der Poel 1967. The influence of the Low Countries upon British developments is discussed in B. H. Slicher van Bath 1960: 130–153; G. E. Fussell 1959: 611–622. (See also Walter Blith 1652: 194.) While Dutch plows improved marginally in the eighteenth century, those developments have been ignored in the above discussion, because—as will be demonstrated shortly—neither British nor Dutch innovations in the eighteenth century prompted American farmers to undertake a wholesale reexamination of American plow designs.
76. *American Farmer*, 1 (August 27, 1819): 170.
77. The wood ashes from the undergrowth were a temporary source of potassium for the soil, as were the dead trees when subsequently burned. (See Conway Zirkle 1969: 88.)
78. J. P. Greene, ed. 1965, Vol. 2, 1038–1039. This preference for hoes over plows persisted for many years on some southern plantations. See, for example, F. A. Michaux 1805: 347–348.
79. See, for example, L. C. Gray 1933, Vol. 1, 173, 217, 281; H. W. Quaintance 1904: 3; William Strickland 1801: 37; Charles F. Grece 1819: 36; A. B. Benson, ed. 1937: 89; Israel Acrelius 1759: 147; J. P. Greene, ed. 1989: 63–64. Wayne Randolph suggests that the continued use of the hoe was not uncommon in tobacco culture in the Chesapeake area, although this was clearly not the case for wheat. (Private communication, October 7, 1994.)
80. For various estimates of the "typical" depth of American plowing in the eighteenth century, see W. G. Bek 1922: 297; Charles Campbell, ed. 1840, Vol. 2, 18; *New American Magazine*, 4 (March 1760), 95; *Columbian Magazine*, 3 (March 1789), 157; *American Museum*, 9 (January 1791), 42; *Universal Asylum*, 6 (January 1791), 28; Thomas Cooper 1794: 113; *American Farmer*, 1 (October 8, 1819), 219; New York State Board of Agriculture, *Memoirs*, 2 (1823), 69–72; L. C. Gray 1933, Vol. 2, 795. For a discussion of how contemporary beliefs discouraged deeper plowing by emphasizing the poor fertility of "dead earth" lying further below the surface, see S. Deane 1790: 219–220; *Agricultural Museum*, 2 (February 1812), 221–222; Massachusetts Board of Agriculture 1854: 11n.
81. See David Brewster, ed. 1832: Vol. I, 336; W. G. Bek 1922: 297; L. Rogin 1931: 14; Jesse Buel 1840: 113.
82. See, for example, L. Rogin 1931: 14, 57n. 251; P. W. Bidwell and J. I. Falconer 1925: 124; *American Farmer*, 2 (March 23, 1821), 413–414. When, in 1772, exceptionally wet weather forced Landon Carter to use four horses to plow "to go deep enough and do but a tolerable day's work," he noted the "very great expense in feeding" and doubted that corn cultivation, on a regular basis, could "bear the expense." (J. P. Greene, ed. 1965, Vol. 2, 671.) The central economic point was that both the price of capital and the price of labor were relatively high compared to the price of land. See also *Southern Agriculturist*, 4 (May 1831), 276.

83. A. B. Allen 1846: 229–230.
84. Contemporary depictions of American plows can be found in *The Pennsylvania Town and Country-Man's Almanack* 1756: cover (reproduced in Sinclair Hamilton 1958: plate 16); crest of Philadelphia Society for Promoting Agriculture (1785) (reproduced in S. W. Fletcher 1959: title page); watercolor of Salem, North Carolina (1787) (reproduced in D. W. Meinig 1986: 292); *Der Neue, Gemeinnützige Landwirthschafts Calender* (1796) (reproduced in P. H. Cousins 1973: frontispiece); *Columbian Magazine*, 1 (October 1786), opposite page 77; ibid., 1 (June 1787), opposite page 491; painting of the Battle of Lexington (1798) (reproduced in G. G. Deák 1988: plate 141); *The Farmer's Almanack* (Boston) 1794: cover.
85. Photographs of early American plows held in museums can be found in J. T. Schlebecker 1972: 8; and by the same author 1975: 29; P. H. Cousins 1973: 2, 8–9; R. D. Hurt 1982: 8, 10; W. Randolph 1991: 228, 229; R. H. Gabriel 1926: 73, 207; J. T. Adams 1961, Vol. 2, 54; J. Van Wagenen, Jr. 1927: frontispiece. Photographs of early French Canadian plows can be found in R. L. Séguin 1989: Plates VI, VII.
86. S. Deane 1790: 218. Compared with European models, these American strong plows often had short beams and more erect handles, "enabling the plowman to more easily plow around obstructions, and to till the land to the very roots of the stumps." (I. P. Roberts 1897: 45.)
87. See, for example, S. Deane 1790: 216; L. Rogin 1931: 6.
88. The moldboard is sheathed with iron strips, and strips of iron are nailed to the left side of the beam for reinforcement. The landside and share form a single piece of iron.
89. See I. P. Roberts 1897: 48.
90. For early evidence of the use of the lock-coulter, see Wayne Randolph 1991: 227. The metal share was attached either to a wooden landside or the share was one piece with "an iron-bar landside (*bar share*)," the latter sometimes giving rise to the name "bar share plow." (See P. H. Cousins 1973: 4, 6, 8.) The colonial swing plow depicted in Figure 3.1, for example, has a forged iron landside "which with the share and point is one piece . . . [and] a coulter set upright in the landside base with its upper end wedged into a slot in the beam." (F. L. Lewton 1943: 64.)
91. Both of these plows struck the Dutch expert, J. M. G. van der Poel, as resembling Dutch plows in many ways, suggesting that Dutch influences on American designs may have been more pervasive than commonly believed.
92. A later observer of American farming practices explained the New World bias for swing plows as follows. "One of the leading questions in England, respecting the most useful forms of the plough, lies between the wheel plough, and . . . the swing plough . . . and it is obvious that each kind has the advantage, for some purposes. For instance, if the object be to throw up a thin uniform furrow slice . . . the wheel plough has greatly the advantage. But if the land be stony and uneven, the wheel plough cannot be used with good effect. Hence, the roughness of surface, and the stony lands, of most parts of the older settlements of this country, are among the reasons why the wheel plough is so little used by our cultivators, rather than want of attention to the instrument itself." (*New York Farmer and Horticultural Repository*, 2 [September 1829], 213.) Other reasons for the American preference for swing plows included the greater weight and cost of wheel plows, and, in many cases, the need for greater animal power to drag them. (See W. Gaylord and L. Tucker 1840, Vol. 2, 65.) For evidence that some wheeled plows were used, see F. W. Halsey, ed. 1906: 21; H. Glassie 1968: 150–151.
93. On the absence of coulters, see P. H. Cousins 1973: 4 and L. Rogin 1931: 8. The peculiarly shaped share was joined to the rest of the plow as follows: "While in some (presumably late) examples the pyramidal share does appear with an iron-bar landside, it is typically found with a wooden landside that is uniquely tapered and fitted

into the share in such a manner as to be riding on and supported by it." (P. H. Cousins 1973: 4.)

94. See P. H. Cousins 1973: 4–5; J. M. G. van der Poel 1967: Figures 18, 19, 21, 23a, 25, 28, 29 39a, 44, and 45. Cousins also links the share design of American heavy plows more to the designs existing in the Low Countries than to those in use in Britain. (Ibid.) How and when these European influences affected colonial plow development remains unknown, although such influences may date from the seventeenth century. For example, a reference can be found in 1649 to "Dutch plowing" on "light" ground, but the nature of the implement is not specified. (William Bullock 1649: 38.)

95. For commentary on the tendency of this type of plow design to foster shallow plowing, see *American Farmer*, 2 (August 25, 1820), 169; ibid., 2 (February 2, 1821), 358.

96. Richard Peters in *Universal Asylum*, 6 (January 1791), 28.

97. E. Arber 1910: Vol. 2, 541; L. C. Gray, 1933: Vol. 2, 454.

98. *The Plough Boy*, 3 (June 2, 1821), 5–6.

99. Wayne Randolph of Colonial Williamsburg and Bruce Craven and Jochen Welsch of Old Sturbridge Village, all of whom have field experience with early American plows, have emphasized to me how smoothly some of the oldest models perform, compared with the poor performance of others.

100. In addition to the moldboard shapes of Figures 3.1 and 3.18, see the commentary by Richard Parkinson 1805: 492; J. T. Schlebecker 1972: 10.

101. Richard Peters in *Universal Asylum*, 6 (January 1791), 28. See also Jack P. Greene, ed. 1965, Vol. 2, 611, 845; *New American Magazine*, 4 (March 1760), 95; G. K. Holmes 1899: 315.

102. *Eighty Years' Progress* 1864: 30; see also Wayne Randolph 1991: 226; L. Rogin 1931: 19; C. W. Burkett 1900: 152; U.S. Senate, Committee on Patents, *Arguments before the Committee*, August 3, 1878, 45th Cong., 2nd Sess., Misc. Doc. 50, 272; C. L. Flint 1874: 120.

103. See U.S. Bureau of the Census, *Eighth Census of the United States*, 1860, Agriculture, xviii; *The Plough Boy*, 3 (July 21, 1821), 61.

104. R. D. Hurt 1982: 8; see also A. B. Allen 1847: 229–230; W. Rasmussen 1960: 67; *Eighty Years Progress* 1864: 27, 30. Also making difficult the task of pulling these plows through the soil was the tendency of many to clog. See G. Cope 1881: 217.

105. H. J. Carman, ed. 1939: 60.

106. J. Eliot 1934: 64.

107. Donald Jackson, ed. 1976: xxxiii; B. H. Latrobe 1905: 60–61; William Willcox, ed. 1972: Vol. 16, 127–128; W. E. Brooke 1919: 22–23. Both Washington and Franklin designated the plow they ordered as a "Rotheran or patent plough," although Washington corrected the spelling to "Rotherham" in his later correspondence with Arthur Young. For evidence of early efforts by Washington to design plows himself, with limited success, see P. L. Haworth 1925: 94.

108. Richard Peters, quoted in *Universal Asylum*, 6 (January 1791), 31. For other contemporary articles advocating deeper plowing, see *Columbian Magazine*, 1 (May 1787), 369–371, 431–434; ibid., 2 (July 1788), 371–373; ibid., 3 (March 1789), 158; *New-York Farm Magazine*, 2 (June 1797), 291–292; *Agricultural Museum*, 2 (September 1811), 73–77; ibid., 2 (February 1812), 217–249.

109. W. E. Brooke 1919: 21–23; P. L. Ford, ed. 1892–1899, Vol. 9, 268–269. For other evidence of British imports during this time period, see *Agricultural Museum*, 2 (October 11, 1811), 99; *The Plough Boy*, 2 (September 16, 1820), 122.

110. Jefferson gave a detailed description of this moldboard in a letter to Sir John Sinclair in 1798, published in American Philosophical Society, *Transactions*, 4 (1799), 313–322, and reproduced by James Mease in A. F. M. Willich 1803: 288–293. For

other comments on the designing and performance of this moldboard, see E. M. Betts 1953: 49–56. Some confusion exists about the extent to which Jefferson pioneered the use of mathematics in plow designs. L. C. Gray argues that James Small's work probably influenced Jefferson's design, while the New York State Agricultural Society claims the key ideas were original with Jefferson. (L. C. Gray 1933, Vol. 2, 794; New York State Agricultural Society 1868: 34–35.)

111. In his letter to Sinclair, Jefferson noted the difficulty with wooden implements, "no two [of which] will be alike" and indicated a desire to have the moldboard made of cast iron. (American Philosophical Society, *Transactions*, 4[1799]: 315, 320.) In 1804 James Mease wrote to Jefferson about the possibility of designing a moldboard pattern suitable for casting, but there is no record of whether Jefferson obliged or not. (E. M. Betts 1953: 48.) In 1814 he was following up this idea, having supplied John Staples with a "model of the Moldboard of a plough of a form of my own," and ordered two dozen to be "cast . . . in iron." (Ibid., 62.)

112. New York Society for the Promotion of Agriculture, Arts, and Manufactures, *Transactions*, 1 (1794), Part 2, vi, 168; F. L. Lewton 1943: 63; A. F. M. Willich 1803: 292–293.

113. New York State Agricultural Society 1868: 66–67, 98–99; H. W. Quaintance 1904: 6.

114. While historians are generally agreed that money changed hands, the sum paid by Peacock to Newbold remains a subject of dispute. See New York State Agricultural Society 1868: 69; U.S. Bureau of the Census, *Eighth Census of the United States. 1860. Agriculture* 1864: xviii; I. P. Roberts 1897: 46–47; Solon Robinson 1869, Vol. 2, 918; *American Agriculturist*, 7 (April 1848), 122.

115. Writing in 1817, Jefferson indicated that his son-in-law "first introduced [this hillside plow] about a dozen or fifteen years since," which suggests the invention was made between 1802 and 1805. (Jefferson's letter was reproduced in *American Farmer*, 2 [June 16, 1820], 93.)

116. Letter to Charles W. Peale, April 17, 1813, reproduced in Charles A. Browne 1944: 409.

117. How this plow functioned is roughly described in Philadelphia Society for Promoting Agriculture, *Memoirs*, 4 (1818), 18, and much more clearly explained by Thomas Jefferson in ibid., 16–17.

118. For European antecedents, see *American Farmer*, 15 (May 24, 1833), 82–83; B. P. Poore 1866: 513; Richard Peters in Philadelphia Society for Promoting Agriculture, *Memoirs*, 4 (1818), 13; *Southern Agriculturist*, 8 (May 1835), 272; J. T. Schlebecker 1975: 99. For a suggestion of an early American antecedent, see F. L. Lewton 1943: 64. For Dutch examples, see J. M. G. van der Poel 1967: Figures 18, 19, 47, 48, and 49.

119. Upon viewing this illustration, the Dutch expert J. M. G. van der Poel noted that, as drawn, the moldboard appears to be immobile, and the illustration is probably more a product of the artist's imagination than a replication of an actual plow design. (Private communication, December 17, 1994.)

120. While Jefferson scrupulously avoided any suggestion that his ideas influenced Randolph's experiments in design, the former's travels in Europe had made him familiar with such design possibilities and an advocate of contour plowing. See B. L. C. Wailes 1854: 153; also E. M. Betts, ed. 1944: 48, who speculates that this invention was made "probably with the aid of Jefferson."

121. For examples of other plow designs explored between 1783 and 1812, see Philadelphia Society for Promoting Agriculture, *Memoirs*, 1 (1808), 90–91 (coulter plow); ibid., 257–259 (three-furrow plough). A list of plows patented during these three decades is given in New York Agricultural Society 1868: 68–72.

122. See New York State Agricultural Society 1868: 65.

123. *New England Farmer*, 8 (March 26, 1830), 281. For other criticisms of the performance of Jefferson's moldboard, see New York State Agricultural Society 1868: 32, 93. E. M. Jope is therefore incorrect when he argues that Jefferson worked out "the optimal shape of a mould-board," as is Leo Rogin when he argues that Jefferson "demonstrated the value of basing plow construction upon mathematical principles." (E. M. Jope 1956a: 90; L. Rogin 1931: 28.) For a commentary upon how Jefferson's attempt to use mathematics to define an optimal design influenced later efforts of a similar sort, see *The Plough Boy*, 2 (September 16, 1822), 122.

124. New York State Agricultural Society 1868: 68. Many analysts of Newbold's failure have focused upon prevailing beliefs among some American farmers about poisoning and weeds, and missed completely the central economic problem. See, for example, Solon Robinson 1869, Vol. 2, 918; E. H. Fowler 1895: 353; C. S. Phelps 1917: 44; H. W. Quaintance 1904: 6; Wayne Rasmussen 1960: 67. That these beliefs were not a crucial barrier to rapid adoption of cast-iron plows once the economic incentives were reversed is demonstrated by the experience following 1815, of which more below. See also J. T. Schlebecker 1975: 99; P. W. Bidwell and J. I. Falconer 1925: 209.

125. See, for example, Philadelphia Society for Promoting Agriculture, *Memoirs*, 4 (1818), 81–82; *American Farmer*, 2 (December 29, 1820), 320; *Farmer's Register*, 1 (December, 1833), 487.

126. Jared Sparks, ed. 1837: Vol. 12, 330.

127. Letter dated March 27, 1798, reproduced in Frank Gilbert 1882: 16.

128. E. M. Betts 1953: 53. For an earlier example of similar indifference to plow designs by members of this society, see S. W. Fletcher 1950: 92–93.

129. "Experiments" in S. Deane 1790: 85. Jefferson's letter is reproduced in American Philosophical Society, *Transactions*, 4 (1799), 313–322. Moore's original 1802 article is reproduced in *American Museum*, 2 (February 1812), 217–249. Samuel Deane does note the absence of American use of trench plows available in England, but in his later article on "ploughing" he advocates more frequent plowing and more cross-plowing, deeper plowing and contour plowing but makes no reference whatsoever to the desirability of using a trench plow.

130. Jefferson, 313.

131. Philadelphia: James Humphreys, 1803.

132. A. F. M. Willich 1803: 288–293. In his review of American possibilities, Mease takes most of the allotted pages to reproduce Jefferson's detailed description of his moldboard of least resistance, adding in the last two columns of the survey (1) a quick aside about the existence of Robert Smith's cast-iron moldboard, and (2) a few general thoughts by Smith on plow construction.

133. See, for example, S. Deane 1790: 219; Richard Peters in *Universal Asylum*, 6 (January 1791), 27–31.

134. For one of the few passing references to a plow received (with no accompanying details), see Philadelphia Society for Promoting Agriculture, *Minutes*, March 9, 1790, 60.

135. See, for example, the list in *Universal Asylum*, 6 (March 1791), 162–168.

136. *Columbian Magazine*, 3 (March 1789), 157–158; S. Deane 1790: 216–222; James Mease in A. F. M. Willich 1803: 283–293.

137. *American Farmer*, 1 (August 20, 1819), 161–163; (August 27, 1819), 169–171; (September 3, 1819), 179–180.

138. Ibid., (August 27, 1819), 170.

139. Ibid.

140. Ibid.

141. *The Plough Boy*, 1 (September 18, 1819), 122–123.

142. Few details are available on the difference between the first and second patent. For a description of the 1819 patent, see Frank Gilbert 1882: 20–36.

143. The contrast between Newbold's landside and Wood's is illustrated in Figure 3.21.
144. J. T. Schlebecker 1975: 99. See also *American Farmer*, 1 (September 24, 1819), 202; New York State Agricultural Society 1868: 78.
145. The details of the various joining procedures used are outlined in the 1819 patent application, and summarized in Frank Gilbert 1882: 67 and J. T. Schlebecker 1975: 99.
146. See *Massachusetts Agricultural Repository and Journal*, 5 (1819), 363; also *American Farmer*, 1 (July 9, 1819), 120. In his patent Wood described his moldboard as "a plano-curvilinear figure." (Frank Gilbert 1882: 22.)
147. Frank Gilbert 1882: 41. Compare Figures 3.19 and 3.21B. The geometry of Wood's moldboard is described in his 1819 patent application. (See Frank Gilbert 1882: 26–29.) For a discussion of what is unique in Wood's moldboard design, see New York Agricultural Society 1868: 89. For criticism of this design by a contemporary of Wood's, see ibid., 94–98. The critic, J. Dutcher of Durham, New York, favored the geometric shape portrayed in Figure 3.10C.
148. Compare, for example, Frank Gilbert 1882: 39–40, and New York State Agricultural Society 1868: 85.
149. Prospective buyers were urged to "provide himself with a suitable number" of shares, "in order that when one becomes too dull for breaking up sward-land another may be substituted." (*The Plough Boy*, 2 [September 16, 1820], 123.)
150. The reported sales figures were 1,550 in 1817, 1,600 in 1818, 3,600 in 1819 and an even larger (although unspecified) number in 1820. "These sales very largely exceeded the sale of any other plow then in existence." (New York State Agricultural Society 1868: 89.) One of the earliest manufacturers of Wood's plows was Thomas Freeborn of New York City, the plows variously known as "Wood's Cast-Iron Plough," "Wood's Freeborn Plough," or "Freeborn's Plough." For details on the leasing of rights and manufacturing of Wood's plows, as well as assessments of their performance, see *Niles' Register*, 15 (November 7, 1818), 178; *Massachusetts Agricultural Repository and Journal*, 5 (1819), 361–364, 372; ibid., 6 (1821), 170; *American Farmer*, 1 (July 9, 1819), 120; ibid., 2 (August 25, 1820), 169; ibid., 2 (September 15, 1820), 194; ibid., 2 (January 26, 1821), 352; *The Plough Boy*, 1 (February 12, 1820), 290; ibid., 1 (May 13, 1820), 398.
151. As its shape suggests, this plow was suitable for tilling "mellow" or sandy soils relatively free of obstacles. (When it did encounter roots, it could easily be drawn back because it had no coulter and a share with very little slant and a short nose.) This plow had perhaps more names than any other implement then in use, known variously as the Carey (Cary) plow, Dagon (Deagan, Dagen, Degan) plow, and Connecticut or Enfield plow. For a discussion of its history and performance characteristics, see *American Farmer*, 2 (August 25, 1820), 169; ibid., 2 (December 8, 1820), 290; ibid., 2 (December 29, 1820), 319; ibid., 2 (March 9, 1821), 400; ibid., 3 (June 8, 1821), 85; L. C. Gray 1933, Vol. 2, 795; L. Rogin 1931: 8–11, 27, 56; J. T. Schlebecker 1975: 99–100; C. L. Flint 1874: 111–112.
152. After inspecting the original Smithsonian photographs of these two implements, Jochen Welsch, a supervisor of agriculture at Old Sturbridge Village, concluded that the only differences of note were that the earlier of the two plows had inferior handle placement (probably requiring a less erect position in plowing) and possibly inferior vertical suction.
153. Address by William A. Dangerfield to the Prince George's Agricultural Society, reported in *American Farmer*, 3 (June 8, 1821), 85. The previous year a correspondent to the same newspaper praised the Dagon plow, but still expected it to be "superseded by the Freeborn or Wood plough." (Ibid., 2 [December 29, 1820], 319.)
154. Among the better-known experimenters in plow design at this time were Henry Burden, Gideon Davis, Josiah Dutcher, David Hitchcock, John Gibson, and David

Peacock. For discussions of designs by these and other plow innovators of the 1820s, see *The Plough Boy*, 1 (April 29, 1820), 382; ibid., 1 (May 6, 1820), 390; ibid., 1 (May 13, 1820), 398; ibid., 1 (May 20, 1820), 402; ibid., 2 (September 30, 1820), 138; ibid., 2 (February 3, 1821), 283; ibid., 2 (March 10, 1821), 327; ibid., 3 (June 2, 1821), 4; ibid., 3 (July 21, 1821), 61; *American Farmer*, 2 (June 2, 1820), 80; ibid., 2 (August 18, 1820), 160; ibid., 2 (August 25, 1820), 169; ibid., 2 (November 10, 1820), 262; ibid., 2 (December 8, 1820), 290; ibid., 3 (October 12, 1821), 229–230; ibid., 4 (April 12, 1822), 31; ibid., 4 (July 5, 1822), 119; ibid., 6 (March 4, 1825), 393–394; ibid., 10 (November 7, 1828), 267; *Massachusetts Agricultural Repository and Journal*, 6 (1821), 231; *New England Farmer*, 4 (August 26, 1825), 37; ibid., 8 (May 7, 1830), 329; *New-York Farmer and Horticultural Repository*, 2 (September, 1829), 212–214; New York Agricultural Society 1868: 72–81, 89–92. One of the more crucial improvements involved the development of self-sharpening. Cold-chilling, as noted previously, was developed in Britain in the first decade of the nineteenth century. The Americans appear to have invented the process independently in the second decade. (Edwin Stevens of New Jersey received a patent in 1817; see J. T. Schlebecker 1975: 99.) For examples of self-sharpening methods used at this time, see *American Farmer*, 5 (December 26, 1823), 313; ibid., 6 (October 8, 1824), 232; ibid., 6 (December 31, 1824), 323; ibid., 9 (February 15, 1828), 383; ibid., 10 (November 7, 1828), 267; *New England Farmer*, 4 (November 4, 1825), 114. One of the better descriptions of one of the self-sharpening techniques that began to be used in the 1820s may be found in *The Agriculturist*, 6 (February, 1847), 43.

155. See Figure 3.14D.

156. See, for example, H. G. Ashmead 1884: 207.

157. Less sophisticated versions of plow cleaners had been in use in many parts of Europe for centuries. For illustrations of Dutch versions, see J. M. G. van der Poel 1967: Figures 28, 32, 41.

158. See, for example, the "mole-plough" illustrated in A. F. M. Willich 1803: 296.

159. For illustrations of later subsoil plows, see *New England Farmer*, 16 (June 5, 1839), 379 (a British design); S. E. Todd 1866: 245, and J. J. Thomas 1869: 135–136 (American designs).

160. Variations of this general design (some with more than one "coulter") are discussed in *American Farmer*, 1 (April 16, 1819), 18–19; ibid., 2 (May 26, 1820), 66; ibid., 3 (September 21, 1821), 205; ibid., 4 (March 29, 1822), 13; ibid., 4 (April 12, 1822), 22–23; ibid., 4 (May 24, 1822), 71; ibid., 5 (March 28, 1823), 1–3, 9–10; *New England Farmer*, 1 (June 14, 1823), 366.

161. *American Farmer*, 4 (December 27, 1822), 325; ibid., 5 (November 21, 1823), 275; *Massachusetts Agricultural Repository and Journal*, 6 (1821), 235.

162. *Massachusetts Agricultural Repository and Journal*, 4 (1817), 313–315; *American Farmer*, 1 (May 21, 1819), 64; ibid., 1 (February 4, 1820), 358–359; ibid., 2 (November 10, 1820), 262; ibid., 4 (May 17, 1822), 60; ibid., 9 (February 15, 1828), 384; *New England Farmer*, 1 (August 10, 1822), 14; *New-York Farmer and Horticultural Repository*, 1 (October, 1828), 238.

163. Two-horse plows, for example, continued to sell for $15–$17.

164. This drop far exceeded the decline in the general price level, wholesale prices falling by roughly 20 percent between 1819 and 1825, with most of that decline occurring in 1819. (U.S. Bureau of the Census, *Historical Statistics* . . . 1975, Vol. 1, 201, Series E52.) A comparison of plow prices across time is complicated (a) by the many variables that can affect price (with or without coulter, wrought iron or cast iron, with or without extra shares), and (b) by the tendency of many advertisements not to include information about items listed under (a). The above summary judgment nevertheless seems reasonable, based upon price quotations in the following sources: *The Plough Boy*, 1 (July 10, 1819), 45; ibid., 1 (September 18, 1819), 123;

American Farmer, 1 (November 12, 1819), 265; ibid., 1 (February 25, 1820), 384; ibid., 2 (June 2, 1820), 80; ibid., 2 (June 23, 1820), 104; ibid., 2 (August 18, 1820), 160; ibid., 3 (December 14, 1821), 301; ibid., 4 (July 5, 1822), 116; ibid., 4 (November 22, 1822), 280; ibid., 4 (December 27, 1822), 325; ibid., 6 (October 8, 1824), 232; ibid., 8 (May 12, 1826), 62; ibid., 9 (March 23, 1827), 8; J. Bronson, "Freeborn's Patent Plough," *Massachusetts Agricultural Repository and Journal,* 5 (July, 1819), 362–364.

165. As one farmer characterized the problem, "The share came back in a different shape. It no longer ran like the same plough; it often had too much, or too little *pitch,* or the share had warped in hardening." (*The Plough Boy,* 3 [June 2, 1821], 5–6.)

166. See *The Plough Boy,* 2 (September 23, 1820), 133; W. Drown 1824: 39–40; *American Farmer,* 7 (May 13, 1825), 59.

167. See *Niles' Register,* 15 (Novemer 7, 1818), 178; *The Plough Boy,* 2 (September 23, 1820), 133; *American Farmer,* 2 (September 15, 1820), 194; J. T. Schlebecker 1975: 100; L. Rogin 1931: 27; *Genesee Farmer,* 2 (January 7, 1832), 8. As noted previously, the price of heavy plows did not fall significantly in the early 1820s. But with the cost of a new two-horse cast-iron plow ranging between $15 and $17, while the expense of maintaining an additional horse for one year was estimated to be more than twice as great ($35), the economic advantages of converting from a wooden to an iron implement were overwhelming, even if these figures are off by 20 percent. For the horse maintenance estimate, see *The Plough Boy,* 2 (March 17, 1821), 331. For sources of price data on cast-iron plows, see footnote 164. For a more sophisticated estimate of additional expenses incurred by owning an additional horse, see *American Farmer,* 7 (October 14, 1825), 234. These include lower maintenance costs ($15–$20 per year), but also include an allowance for "losses" plus interest on the purchase price of the horse (for a $100 mare, estimated at 6 percent per year).

168. U.S. Commissioner of Patents 1860: 24; B. P. Poore 1866: 522.

169. See, for example, *Massachusetts Agricultural Repository and Journal,* 4 (1817), 300–301; ibid., 5 (1819), 13–15, 249–251; ibid., 6 (1821); *American Farmer,* 3 (November 16, 1821), 272; New York Agricultural Society 1868: 89.

170. See L. Rogin 1931: 24–26; A. Baker and F. W. White 1994. Although the belief that "iron poisons the soil" persisted into the 1820s, when the cost advantage in favor of iron became decisive, whatever were the reservations associated with that belief, they were not sufficient to inhibit widespread adoption prompted by the cost considerations noted. For persistence of this belief, see U.S. Senate, Committee on Patents, *Arguments before the Committee,* August 3, 1878, 45th Cong., 2nd Sess., Misc. Doc. No. 50, 273.

171. See *The Plough Boy,* 1 (September 18, 1819), 123; *Massachusetts Agricultural Repository and Journal,* 5 (1819), 13–15; Jesse Buel 1819: 55–59; L. Rogin 1931: 22,n. 76.

172. For an indication that the benefits of deep plowing over shallow was still a subject of uncertainty and debate at this time, see *American Farmer,* 1 (May 28, 1819), 66; ibid., 2 (May 26, 1820), 65; ibid., 2 (January 26, 1821), 346; ibid., 3 (May 4, 1821), 44; ibid., 5 (March 28, 1823), 2; ibid., 9 (August 17, 1827), 169; *New England Farmer,* 2 (November 8, 1823), 113; *New-York Farmer and Horticultural Repository,* 1 (March, 1828), 66–67; T. G. Fessenden 1835: 282–288.

173. New York State Agricultural Society 1868: 66–72; U.S. Bureau of the Census, *Eighth Census of the United States,* 1860, Agriculture 1864: xviii.

174. Address before the Williamsburgh Agricultural Society of South Carolina, reported in *American Farmer,* 5 (May 9, 1823), 49.

175. William Drown 1824: 42.

176. T. G. Fessenden 1835: 335.

177. Pictures of these implements may be found in S. E. Todd 1866: 242–243; J. J.

Thomas 1868: 126, 137–138. Illustrations of Daniel Webster's oversized plow (of negligible significance in the evolution of plow design) may be found in New York State Agricultural Society 1868: 104–105. For a review of plow designs and developments dating from the immediate postbellum era, see J. J. Thomas 1868: 118–142. More recent reviews may be found in J. B. Davidson 1907: 387–398; A. A. Stone 1945, chapter 1, and H. P. Smith 1955, chapter 9. For an 1846 description of a factory specializing in agricultural implements, see *The Southern Cultivator*, 4 (March, 1846), 39. For descriptions and illustrations of John Deere's early efforts to create and produce a steel plow, see Wayne G. Broehl, 1984, Part I.

178. See New York State Agricultural Society 1868: 48–49, 93, 114–117, 128–129, 149; P. Bidwell and J. I. Falconer 1925: 209; Jesse Buel 1840: 113; I. P. Roberts 1897: 47.

179. Samuel Witherow and David Pierce, quoted in I. P. Roberts 1897: 49–50. (Italics in original.) Three decades later, the New York Agricultural Society tried to articulate the design principles for achieving this priority. "We believe that we have been the first to announce that the great object in all plows is to form the curve in such a way as to make all the parts of the furrow slice to travel with different velocities in order to produce pulverization, but that these different velocities should be no greater than is required for the disintegration of the soil, in order to avoid an unnecessary expenditure of power." (1868: 128–129.)

180. New York State Agricultural Society 1868: 93. First developed by Joel Norse, the Eagle plow was significantly improved by Governor Holbrook of Vermont in the mid-1840s, and by the 1850s "became the standard plow of New England." (J. T. Schlebecker 1972: 13; see also New York State Agricultural Society 1868: 92–93; L. Rogin 1931: 29.) Schlebecker's statement that the Eagle's "moldboard was based on a design worked out by Thomas Jefferson" is misleading. (Ibid.) Norse did begin by following Jefferson's rule for moldboard construction, but found the results totally unsatisfactory. His subsequent modifications were the result of trial and error, calculated to improve the plow's ability to pulverize the soil.

181. See, for example, the extended article on plowing with illustrations in the *American Agriculturist*, 10 (May, 1851), 153–157.

4. Sowing

1. Mark 4: 3–8.

2. S. Deane 1790: 258. In terms of "birds of the air" being a continuing problem, British farmers expressed their frustration in a popular couplet:

> One for the rook, one for the crow,
> One to rot, and one to grow.

(P. A. Wright 1961: 7.) A loss rate of 50 percent to birds is undoubtedly an exaggeration, but the problem was serious enough that depictions of sowing from premodern times often include such devices for repelling birds as dogs, workers with slingshots, or even a scarecrow armed with a bow. See for example E. G. Millar 1932: F. 170b, and J. Longnon and R. Cazelles 1969: F. 10v. The latter includes a scarecrow armed with a bow, which served as a dubious deterrent, to judge by the number of birds the artist included at the heels of the sower. For other early depictions of broadcast sowing in Western art, see J. C. Webster 1938: plates 33a and 34a (pre-1200 A.D.) and H. Belloc 1914: 12 (early twelfth century).

3. C. W. Johnson 1848, Vol. 2, 73n, quoting a government report on Scottish practices. The importance of skill in the broadcast sowing of all seeds is difficult to exaggerate.

The first edition of the *Encyclopaedia Britannica* explains why. "The most common method of sowing is by the hand. This method requires great skill and address in the sower: For, at the same time that he gives his arm a circular motion, to cast the seed with strength, he must open his hand gradually, that it may not fall in a heap, but be properly scattered and spread. It is remarkable, that good sowers, by the force of habit, take their handful out of the sheet so very exactly, that they will sow any quantity of seed on an acre, according as it is designed to be thinner or thicker. But this dexterity in a few sowers, is itself an objection to the method of sowing by the hand; because long practice and observation are necessary to make a good sower: This remark is too well justified by experience; for good sowers are extremely rare, and, in some places of the country, hardly to be got." (1771: 59–60.)

4. See Norman E. Lee 1960: 45, and R. L. Anderson 1936: 159–162. None of the surviving depictions of this Babylonian implement is particularly clear. For depictions other than the one reproduced in Figure 4.2, see Norman E. Lee 1960: 45, and M. S. Drower 1954: 557. (The latter depicts an "Assyrian seed-plough," circa 700 B.C., which was a variant of the Babylonian implement.)
5. M. S. Drower 1954: 550.
6. Some experts have argued that variants of the Babylonian seed drill were used in ancient India, China, and Japan (for example, H. J. Hopfen 1960: 78–79), but one of the leading authorities on the history of grain drills "has not seen sufficient evidence" to justify that conclusion (R. L. Anderson 1936: 161, n. 6).
7. The Italian pioneers included Camillo Tarelo (1566) and Tadeo Cavalini (1580). For a detailed review of these early experiments, see R. L. Anderson 1936: 162–169.
8. Others who experimented with seed drill designs after Locatelli and before Jethro Tull include Walter Blith, Alexander Hamilton, Gabriel Plattes, Daniel Ramsey, and John Worlidge. (See G. E. Fussell 1952: 42–43, 94–101; R. H. Anderson 1936: 167–169.) There is no evidence that any of these precursors to Tull devised a working machine that could perform satisfactorily in the field.
9. Tull began experimenting in 1701, published his results in 1731 (*The New Horse-Houghing Husbandry* . . .), and then added plates to the revised edition of 1733 (and changed the spelling to *The Horse-Hoeing Husbandry: An Essay on the Principles of Tillage and Vegetation, Wherein is shewn A Method of introducing a Sort of Vineyard-Culture into the Corn-Fields, In order to Increase their Product, and diminish the common Expense; By the Use of Instruments described in CUTS*). See E. Cathcart 1891: 14–32; L. H. Bailey 1907: 374–377; G. E. Fussell 1973.
10. Quoted in T. H. Marshall 1929: 46.
11. Quoted in E. Cathcart 1891: 15.
12. The book, first published in 1731, was immediately "pirated without acknowledgment" and reprinted in Ireland, a copyright violation for which prevailing laws provided no effective remedy. See E. Cathcart 1891: 27–28.
13. One of the few exceptions is M. and C. H. B Quennell 1961: 29–31.
14. A worker had to follow Tull's drill while it was in operation "whose chief Business is, with a Paddle to keep all the Shares and Tines from being clogged up by the Dirt sticking to them, and also to observe whether the Seed be delivered equally and justly to all the Channels" (Jethro Tull 1733: 380).
15. The other end of the screw that holds this spring in place ("B" in Figure 4.5B) appears as a small circle on the opposite side of the flap, as that flap is pictured in Figure 4.5A.
16. The amount of seed fed into the funnel "is altered by the number of notches, and by their depth, or breadth, or by both." (J. Tull 1733: 332.)
17. E. Cathcart 1891: 15–16; J. Tull 1733: 319. According to Tull, his diversion "in youth was music"; according to Cathcart, the result of such diversions was an organist and musician "of no inconsiderable attainments." (E. Cathcart 1891: 3, 14.)

18. J. Tull 1733: 326.
19. Quoted in T. H. Marshall 1929: 54. Tull's proposed implements were similarly spurned by the Dutch (although many of his ideas proved influential). See J. M. G. van der Poel 1983: 78.
20. Ibid., 56.
21. J. Tull 1733: 381.
22. J. J. Thomas 1869: 154.
23. S. E. Todd 1867: 262.
24. J. Eliot 1934: 116–117.
25. For assistance with mathematical computations, Eliot consulted the president of Yale College. For help in the construction of the drill, he hired "a very ingenious man of this town [Killingworth], a wheel-wright by trade." (Ibid., 117.)
26. J. Eliot 1934: 213.
27. Ibid., 231.
28. Philadelphia Society for Promoting Agriculture, *Memoirs*, 6 (1939), 112–114.
29. The drawings in Figure 4.11 are reproduced from an 1811 report "On a simple Wheat Drill, by Mr. John Lorain; with a Plate." (Philadelphia Society for Promoting Agriculture, *Memoirs*, 3 [1814], 32–36.) But Lorain was reporting on a drill "originally invented in Sussex County, New Jersey, and first described . . . in the Transactions of the Society of Arts of New York." (Ibid., 36.) That description, by Walter Rutherford of New York, is dated 1792, and reports on a drill first introduced "some years ago" (New-York Society for the Promotion of Agriculture, Arts and Manufactures, *Transactions* 1, 2d rev. ed. [1801], 86–87.) This particular drill is probably the one referred to by several American writers in the late eighteenth century as "the New Jersey seed drill." (See for example *The Columbian Magazine*, 1 [April 1787], 369–371.) It may or may not have been the "grain drill" with "cups cut into iron rollers" supplied by John Taylor of Virginia to Thomas Jefferson in 1797. (E. M. Betts 1953: 74.)
30. Philadelphia Society for Promoting Agriculture, *Memoirs*, 3 (1814), 33.
31. On Lorain's drill, the notches or "mortices" were one inch long, three-quarters of an inch wide, and three-eighths deep "at the further or outer end, and gradually lessening in depth, inward, to nothing." The bottom of the hopper was "soal leather, and is punched with holes directly over the mortises in the axle placed under it. The leather is made to fit close on the axle. The coulters are twenty-one inches long; the upper part is of wood, the lower end is sheathed with iron, in front, and is seven inches high, and five inches broad. The back of the lower part of the coulters is grooved to receive the end of the funnel, n, o, through which the seed drops into the drill, made by the coulter." (Philadelphia Society for Promoting Agriculture, *Memoirs*, 3 [1814], 35.)
32. D. Matteson 1931: 4, 6. This "barrel drill . . . drops the grains thicker or thinner in proportion to the quantity of seed in the barrel." (Ibid., 6.) For a further description of Washington's barrel efforts, including some of the defects of the delivery mechanism, see Bureau of Agricultural Economics 1937: 11–13; P. L. Haworth 1915: 107–109.
33. J. B. Bordley 1799: 105–106. A clearer picture of this box than the one provided by Bordley can be found in *American Farmer*, 2 (May 19, 1820), 60.
34. S. Deane 1797: 87.
35. The peculiar shape of the funnels reflected the intent to lengthen the slide of descending seeds, and thus (it was hoped) increase the chances of clumps of seeds separating out into a more uniform flow. One of the land wheels provided the power for rotating the cylinder in the conventional manner. In this case, to the axle of the land wheel was fastened a "cast-iron wheel [ff in Figure 4.12A] to turn the axis of the seed-box, which has a similar wheel but only one-fourth its diameter [h in Figure

4.12B], so that the axis of the seed-box revolves four times to one revolution of the [larger] wheel." (A. F. M. Willich 1803: 374.)

36. A. F. M. Willich 1803: 381. Although Willich labels his drawing of this machine "Dr. Darwin's Improvements of the Drill-plough," the improvements appear to have originated with a "Mr. Swanwick, of Derby." (Ibid., 380.)

37. See for example J. P. Greene 1965: 706 (on Landon Carter's efforts in 1772); H. J. Carman 1939: 114 (on a drill plow in "the back parts" of Pennsylvania in the 1770s); Philadelphia Society for Promoting Agriculture, *Memoirs*, 6 (1939), 147 (on George Morgan's rice drill); J. C. Fitzpatrick 1931–1944: Vol. 30, 51 (on a "hand-machine" for sowing clover seed recommended to Washington in 1788 by a friend); E. M. Betts 1953: 74 (on a grain drill sent to Thomas Jefferson by John Taylor); *Literary Magazine and American Register*, 1 (February 1804), 330 (on a cotton seed drill).

38. *Universal Asylum*, 4 (March 1790), 164.

39. S. Deane 1797: 86.

40. Philadelphia Society for Promoting Agriculture, *Minutes* (November 7, 1786), 27. This may have been Cooke's grain drill, which Mease reports as having been imported by the Philadelphia Society, and himself characterizes as a "ponderous machine" that was "too complicated" and "little used, even in England, where it was invented." (One of Mease's many editorial asides in A. F. M. Willich 1803: 373.) A sketch and description of Cooke's machine from Arthur Young's *Annals of Agriculture* is reproduced in *The Columbian Magazine*, 3 (May 1789), 352–359. The complexity of British machines at this time is evident in the "drill-plough" depicted in the first and second editions of the *Encyclopaedia Britannica*, and in the various renderings of "The Universal Sowing Machine" in the third edition, which also includes a sketch of "Cooke's Drill Machine." (*Encyclopaedia Britannica* 1771: Plate IX; ibid., 1778–1783: Plate VI; ibid., 1803: 2d Plate VII.) Cooke's machine featured "the spoon type of feed, seed-pipes, drill-shoes, and spike-like shovels to cover the seed." (R. L. Anderson 1936: 172.) For a recent description of Cooke's machine with accompanying diagrams, see J. Vince 1983: 120–121. For an illustration of another early British seed drill (Willey's Drill Plough), see A. M. Bailey 1782, Vol. 1, Plate VII.

41. From Walter Rutherford's 1792 report on a "New-Jersey" seed drill. (New-York Society for the Promotion of Agriculture, Arts and Manufactures, *Transactions*, 1 [1801], 86.) This is not to deny that some Americans continued to hope that the "drill husbandry" of Britain could be successfully applied in America, including Benjamin Franklin's son. (See W. B. Willcox 1972, Vol. 16, 60–61.) But such hopes were not only not realized; they were thoroughly dashed by the total absence of any British machine successfully applied in this country.

42. G. E. Fussell 1952: 102.

43. R. L. Anderson 1936: 174–175; T. H. Marshall 1929: 56. For a brief survey of the developments in British seed drill design, see R. L. Anderson 1936: 171–175; M. Partridge 1973: 110–118; G. E. Fussell 1952: 101–114. As late as 1848, however, a British expert would note that for satisfactory results, drilling machines still required special weather and soil conditions, and thus "the system of broad-casting still, therefore, very generally prevails." (C. W. Johnson 1848, Vol. 2, 77.) For other evidence of continuing use of broadcast techniques in Britain, see A. Rees 1810–1824, Vol. 34, "Sowing"; J. C. Loudon 1839: 387.

44. J. J. Thomas 1869: 154.

45. One novel technique for sowing developed in Britain and ignored in America was "dibbling": a labor-intensive process whereby a man, walking backwards, used two long pointed sticks ("dibbling irons") to make two rows of holes into which a precise number of seeds were placed, usually by children trailing the adult "dibbler." This practice appears to have evolved from the use of setting-boards (begun around the beginning of the seventeenth century), and became relatively widespread during the

latter part of the eighteenth century, although the pace and nature of this evolution remains obscure. Most historians note in passing the total absence of this practice in America, but few offer any explanation. The most likely answer would seem to be the labor demands of the technique and the relatively high cost of labor in America compared with Britain. Even in England the profitability of this technique, which obviously saved seed, was crucially dependent upon whether the price of seed was high or low. (See for example A. F. M. Willich 1803: 230.) For descriptions of the technique and its development, see S. Deane 1790: 380; J. B. Bordley 1803: 100–101; A. F. M. Willich 1803: 329–330; J. C. Loudon 1839: 372, 410, 816–817; C. W. Johnson 1848: 82–84; T. H. Marshall 1929: 51; R. L. Anderson 1936: 158, 176–177.

46. *American Farmer*, 2 (February 23, 1821), 384.

47. Philadelphia Society for Promoting Agriculture, *Memoirs*, 4 (1818), 44.

48. The origins of Bennett's design remain obscure. In his 1803 survey, the American Bordley reported on two British "machines to scatter seed in the broadcast manner" (J. B. Bordley 1803: 106–107), but his description was perfunctory, no illustrations were included, and no similar machines of any importance appeared in America for over a decade. In the two great illustrated works of this era readily available to Americans (A. F. M. Willich 1803, and A. Rees 1810–1824), no broadcast machines are depicted. Under "Sowing", however, Rees does describe a broadcast sowing device by "Mr. Bennet" (one "t") that is similar to, but definitely not identical with, the "Bennett" machine of Figure 4.13. (See A. Reese 1810–1824: "Sowing.") The English appear to have had a machine similar to the latter prior to 1817 (M. Partridge 1973: 113), but how similar it was, or whether it affected the design of (or was identical to) the American Bennett machine is not clear. What is clear is that by the 1830s, an almost identical sowing machine was widely popular in the British Isles. (See, for example, J. C. Loudon 1839, Vol. 1, 387; C. W. Johnson 1848, Vol. 2, 74; J. C. Morton 1856, Vol. 2, 894–895.)

49. Reports, endorsements, and advertisements may be found in Philadelphia Society for Promoting Agriculture, *Memoirs*, 4 (1818), 44–47; *The Plough Boy*, 1 (September 18, 1819), 123; ibid., 3 (November 24, 1821), 208; *Massachusetts Agricultural Repository and Journal*, 6 (1821), 324; *American Farmer*, 2 (June 2, 1820), 80; ibid. 2 (June 23, 1820), 104; ibid., (February 23, 1821), 384; *New England Farmer*, 1 (January 25, 1823), 207; ibid., 1 (May 3, 1823), 319; ibid., 2 (April 3, 1824), 287; ibid., 3 (March 4, 1825), 255.

51. See, for example, *American Farmer*, 4 (July 5, 1822), 116, 120.

51. For other implements subsequently developed in America for the mechanical throwing of seeds (as opposed to mechanical placement of seeds onto the soil), see L. Rogin 1931: 199–204, 209–212 and J. T. Schlebecker 1972: 31.

52. Once again the role of British designs in influencing American developments is unclear. As early as 1801 the English had a turnip drill with many of the features Sinclair later included in his machine. (See A. F. M. Willich 1803: 205.) A poorly drawn variant of this English "drill barrow" appeared in an 1819 American newspaper, which emphasized in the accompanying article that the implement was "extensively used in England" but "hardly known with us." (*The Plough Boy*, 1 [July 10, 1819], 45.) But no evidence exists to link Sinclair's inventive efforts to a knowledge of British farm implements. For seed drills, as for other implements, the preference of Americans for simplicity of design is striking when compared with available British options. Compare, for example, all of the American one-furrow drills of the 1820s depicted in Figure 4.17 and subsequent figures with British options available two decades earlier, as depicted in the *Encyclopaedia Britannica* 1803: 2d Plate VII.

53. "In the bottom of the hopper there is a solid cylinder, whose surface is indented. The grain, or seeds are placed in the hopper, and gently pressing upon the cylinder, they fill the indentations or cavities on its surface. And as it revolves, the seed fall out

and drop into the tin conductor, which conveys them into the hollow backed coulter, that descends below the bottom line of the wheel, and cuts open the mellow soil to receive the seeds as they come from the conductor. . . . The cavities may be placed on the surface of the cylinder, so as to drop the seeds at any distance. This is regulated by the circumference of the wheel, and the number of the cavities on the cylinder. Suppose the first to be four feet, and the last two in number placed at opposite points, then as to the revolution of the wheel marks a line of four feet, the seed would be dropped two feet distant from each other. One cylinder can be taken off and others put on, with cavities suited to various seeds, such as corn, beans, peas, turnips, radishes, beets, carrots and parsnips. Flat seeds should be mixed with ashes or compost to separate them." (*American Farmer*, 2 [June 26, 1821], 349.)

54. *American Farmer*, 2 (June 26, 1821), 349.
55. For favorable comments, see *American Farmer*, 2 (December 21, 1820), 312; ibid., 2, (June 26, 1821), 349.
56. Some of Smith's efforts to improve his machine can be found in *American Farmer*, 8 (December 26, 1826), 313–314. Other references are largely confined to the years 1825 to 1828. See *American Farmer*, 7 (November 25, 1825), 286; ibid., 7 (February 3, 1826), 364; ibid., 7 (February 24, 1826), 387, 392; ibid., 9 (April 3, 1827), 27; ibid., 9 (July 20, 1827), 139; ibid., 9 (September 28, 1827), 224; *Southern Agriculturist*, 1 (April, 1828), 177. Smith's machine was marketed by Robert Sinclair of Baltimore.
57. *New England Farmer*, 17 (April 3, 1839), 311.
58. J. J. Thomas 1869: 154. The history of the development of force-feed mechanisms in America is briefly sketched in R. L. Ardrey 1894: 27–29, 230–232.
59. A British expert of the 1820s summarized the associated difficulties as follows. "In the internal parts of machines of this sort [mechanical sowing drills] there should be great simplicity and incapability of being put out of order, the moving cylinders, or other parts supplying their places, being made so as to admit of great variation in the degrees of their motion or velocity. The utility of tin cylinders for this purpose is somewhat doubtful, and the shaking or sifting principle has not been found to succeed well, except in the case of round seeds, such for instance as those of the turnip, cabbage, rape, clover, and some other similar kinds, and even in these cases sometimes, with some of these seeds, a seed will stick in a hole which will permit another sort to pass in a free manner, and in those sorts which most readily pass, there is rarely a deficiency of some small seeds of which two are capable of entering the same hole at the same time, which, by sticking fast together, impede and prevent the delivery in such machines. On this account, some sowing machines sow too much of one sort of such seeds and scarcely half enough of another. Pierced cylinders of the tin kind might probably be made to sow peas, which are nearly round in their form; but they would not be found to answer the same purpose for wheat or beans, and still less for barley or oats. And the same may perhaps be the case in sowing some other sorts. It is therefore probable, that wooden or metal cylinders, which have grooves on their surfaces, and brushes to promote and regulate the delivery, are the best modes which have yet been discovered for effecting the purpose of sowing in a proper manner in these machines, though they are far from approaching any sort of perfection in the business." (A. Rees 1810–1824: "Sowing.") For a supporting American view, see *Massachusetts Agricultural Repository and Journal*, 5 (July 1819), 347–348.
60. For evidence of British and American uncertainties that persisted in the years immediately after 1815, see *The Plough Boy*, 1 (December 4, 1819), 213, 222, 229; ibid., 3 (July 28, 1821), 69; *American Farmer*, 1 (June 11, 1819), 83; *Massachusetts Agricultural Repository and Journal*, 4 (July 1817), 343–345; ibid., 5 (July 1819), 347–348; S. Deane 1822: 114–115; A. Rees 1810–1824: "Broad-cast husbandry" and "Sowing"; *New England Farmer*, 3 (May 20, 1825), 341. For evidence of continued American

reliance upon broadcast sowing in the 1820s for clover, wheat, barley, hemp and turnips, see *American Farmer,* 2 (March 31, 1820), 5–6; ibid., 2 (May 9, 1820), 60; ibid., 2 (July 7, 1820), 117; ibid., 4 (April 26, 1822), 38; ibid., 5 (April 4, 1823), 10; ibid., 6 (June 4, 1824), 84; ibid., 10 (May 16, 1828), 65; *New England Farmer,* 2 (April 17, 1824), 300; ibid., 3 (September 4, 1824), 44; ibid., 5 (May 25, 1827), 346; ibid., 6 (January 18, 1828), 204. For later commentary on uncertainties regarding the superiority of drilling versus broadcast sowing, see J. C. Loudon 1839: 808–809; C. W. Johnson 1848: 76–79; S. E. Todd 1867: 263; L. H. Bailey 1907: 207; L. Rogin 1931: 205.

61. *The Plough Boy,* 1 (July 10, 1819), 45; ibid., 1 (September 18, 1819), 123; ibid., 1 (October 30, 1819), 171; *American Farmer,* 1 (April 2, 1819), 8; ibid., 1 (December 10, 1819), 293; ibid., 2 (June 2, 1820), 80; ibid., 2 (July 7, 1820), 114; ibid., 3 (June 22, 1821), 99; ibid., 4 (May 3, 1822), 42; *New England Farmer,* 4 (March 24, 1826), 279; ibid., 6 (August 17, 1827), 25; *Massachusetts Agricultural Repository and Journal,* 4 (1816), 213–214; ibid., 5 (1817), 21–25; *Southern Agriculturist,* 2 (April 1829), 153; *New York Farmer and Horticultural Repository,* 2 (May 1828), 123; ibid., 7 (July 10, 1829), 404. For descriptions and diagrams of drill machines developed after 1830, see *The Cultivator,* 3 (October 1836), 125; ibid., 7 (September 1840), 141; ibid., 8 (March 1841), 43–44; *American Farmer,* 1 (February 1846), 245; ibid., 3 (September 1847), 77; ibid., 5 (July 1849), 60; ibid., 6 (June 1851), 477; *The Cultivator,* 5 (May 1848), 158–159; ibid., 8 (June 1851), 209; ibid. (3d series), 1 (October 1853), 302; *Southern Cultivator,* 9 (September 1851), 38; *American Agriculturist,* 21 (March 1862), 76.

5. Harrowing

1. R. Bradley 1725: 342; E. M. Betts 1953: 65; J. P. Greene 1965, Vol. 2, 846; J. J. Thomas 1869: 142; A. Rees 1810–1824: 18, "harrow;" J. C. Loudon 1839: 416, 816; L. Rogin 1931: 59–60.
2. R. L. Ardrey 1894: 21; M. S. Drower 1954: 540; N. E. Lee 1960: 41.
3. K. D. White 1967: 146–149. A bronze model of a rectangular harrow recovered from a Roman grave suggests that the Romans employed both shapes. See W. Haberey 1949: 98.
4. The earliest picture of the harrow appears to be on the border of the Bayeux Tapestry, which Hilaire Belloc dates as "certainly prior to 1200" and "probably as late as 1160." (H. Belloc 1914: xvii–xviii). The harrow, badly drawn, is roughly rectangular in shape. Curiously, the depiction reverses the normal order, with the harrow preceding the sower who in turn precedes the plow. (See ibid., p. 12.)
5. J. Buel 1823: 248.
6. For contemporary descriptions of North American harrows, see A. B. Benson 1937: 383, 477; S. Deane 1790: 118–120; *The Columbian Magazine,* 1 (January 1786), 225n; ibid, 3 (April 1787), 246–247. For a comment on European historical developments, see M. Partridge 1973: 78–83; C. W. Johnston 1848: 12; O. Beaumont and J. W. Y. Higgs 1958: 4–5; E. M. Jope 1956: 94. For other pictures of European harrows see G. E. Fussell 1952: 30; P. Lindemans 1952, Vol. 1, plate XXII; J. Bell 1983: 203; F. Cardini 1989: 45; F. M. Kelly 1936: plate X; W. O. Hassall 1970: plate 40. J. M. G. van der Poel has pointed out that in the southern sandy regions of the Netherlands, triangular shapes were often used "for light ground." (Private communication, December 17, 1994.)
7. H. J. Carman 1939: 60; S. Deane 1790: 118. American farmers of this era could not even agree on whether grain seeds should be harrowed in shallow or plowed in deep.

See *Boston Magazine,* 1 (March 1784), 186–188; *The Columbian Magazine,* 1 (January 1787), 122–124.

8. S. Deane 1797: 141–145; A. F. M. Willich 1803: 252. For comments by contemporaries on harrows with no suggestion of experimentation or innovation in usage or design, see R. Parkinson 1805: 493; *The Plough Boy,* 4 (August 6, 1822), 75 (an 1805 account); *American Farmer,* 9 (August 10, 1827), 161–162 (an 1807 account); *Agricultural Museum,* 1 (October 10, 1810), 125; *Massachusetts Agricultural Repository and Journal,* 1 (1807), 202 (results of an 1806 survey); ibid., 3 (1815), 116 (results of an 1813 survey).
9. *American Farmer,* 2 (May 19, 1820), 61. Jeffrey's response is dated March 1, 1819.
10. Better practice among better farmers, such as George Washington, was to have a wider setting of teeth on harrows used for preparing the ground for sowing, and a narrower setting on those used for harrowing in grain. (See P. L. Haworth 1915: 94.)
11. *New England Farmer,* 3 (February 25, 1825), 245. (Italics in the original.)
12. R. K. Meade 1818: 188. This solution had been anticipated by the British, as evident in the illustrations of A. F. M. Willich 1803: 253. A different (and less efficient) variant was (1) to retain the external form of a rectangle, but (2) slant all of the teeth-carrying boards within that rectangle. This, too, had been devised within the British Isles well before the War of 1812. See for example the illustration in R. Forsyth 1803: plate x, reproduced in L. Carrier 1923: 266.
13. A similar but less elegant harrow seems to have been developed by Jesse Buel in the early 1820s. (See J. Buel 1823: 248.) The Abbot harrow was a solution not much in evidence in British designs. (See for example A. F. M. Willich 1803: 253, and A. Rees 1810–1824, Vol. 1, Plates IX and XXVII.) This harrow was evidently designed as a kind of all-purpose implement, from breaking rough ground to cultivating between rows of standing corn. (*American Farmer,* 6 [April 9, 1824], 21–22.) The later Geddes harrow was primarily used for rough or uneven ground, in part because "the center or draught-rod forms a set of hinges [missing in the Abbot harrow], by which it becomes adapted to uneven ground, or by which it may be easily lifted to discharge weeds, roots, or other obstructions." (J. J. Thomas 1869: 143.)
14. See S. Deane 1790: 118; also *The Columbian Magazine,* 3 (March 1789), 158.
15. *The Plough Boy,* 1 (July 31, 1819), 69. "As the Harrow is intended to run on each side of the corn, you may make the teeth next [to] the corn cut as near as you please by screwing up the coupling bolts which are long enough to admit also of a considerable extension; the drawing then as close as practicable diminishes much the labour of the hand hoes, which after these Harrows, have little else to do, than merely to weed the narrow space left between the teeth running next [to] the corn." (Ibid.) Although no mention is made of British influence in the American literature, the basic geometry of this harrow design appeared in the *Encyclopaedia Britannica* fifteen years earlier (1803: Plate V).
16. *American Farmer,* 7 (August 12, 1825), 161–163. Wayne Randolph of Colonial Williamsburg has underscored the importance of adding handles to harrows, of the sort depicted in Figure 5.8. These improve performance "by allowing tracking adjustment and control and trash dumping without stopping." (Private communication, October 16, 1995.)
17. *The Plough Boy,* 3 (March 23, 1822), 340; *American Farmer,* 1 (July 6, 1819), 123; ibid., 1 (December 17, 1819), 304; ibid., 1 (December 31, 1819), 316; ibid., 2 (March 31, 1820), 5; ibid., 2 (April 28, 1820), 35–37; ibid., 2 (March 16, 1821), 401; ibid., 3 (April 6, 1821), 15; ibid., 3 (April 20, 1821), 30; ibid., 3 (October 5, 1821), 219; ibid., 6 (April 23, 1824), 40; ibid., 7 (December 16, 1825), 305.
18. *The Plough Boy,* 2 (July 22, 1820), 59; *New England Farmer,* 4 (October 7, 1825), 86; *American Farmer,* 2 (August 4, 1820), 149–150; ibid., 2 (November 10, 1820), 261–262; ibid., 2 (December 15, 1820), 307; ibid., 3 (June 22, 1821), 99; ibid., 3 (Au-

gust 10, 1821), 156; ibid., 7 (March 25, 1825), 1–3; ibid., 8 (June 9, 1826), 92–93; ibid., 10 (July 27, 1828), 113–114.

19. *The Plough Boy*, 1 (September 18, 1819), 122–123; *American Farmer*, 1 (April 2, 1819), 8; ibid., 2 (June 23, 1820), 104; ibid., 2 (August 18, 1820), 160; ibid., 4 (November 22, 1822), 280; ibid., 8 (May 5, 1826), 55–56, 62; ibid., 9 (August 17, 1827), 169.

20. Correspondence from Thomas Gold, reported in *American Farmer*, 9 (August 17, 1827), 169.

6. Cultivating

1. Samuel Johnson, *A Dictionary of the English Language* . . . (London: W. Strahan, 1755).

2. As the above wording implies, these definitions were common, but by no means universal. For examples of British usage, see A. F. M. Willich 1803: 294; A. Beatson 1820: 53–54, 58, Plate I; J. C. Loudon 1839, Vol. 1, 403–407; C. W. Johnson 1848, Vol. 2, 30–32, 226–230; H. Stephens 1855: Vol. 1, 567–568, Vol. 2, 62; J. C. Morton 1856: Vol. 1, Plates VIII and IX, Vol. 2, Plates XXVII and XXVIII. Jethro Tull used "hoe-plough" and "horse-hoe" to designate his horse-drawn inter-row weeding implement, and subsequent British writers tended to follow his lead. See, for example, *Museum Rusticum et Commerciale*, 6 (1766), 397–403. For one of the earliest British descriptions of a "cultivator", meaning useful for "pulverizing tenacious soils," see Society for the Encouragement of Arts, Manufactures, and Commerce, *Transactions*, 19 (1801), 142–144. This meaning of "cultivator" is adopted by the American Bordley in his 1803 survey of British implements (J. B. Bordley 1803: 97–98). But as Americans developed specialized inter-row weeding implements after 1815, the preferred name for these implements, as indicated, was "cultivator" (with no clarifying adjectives appended). For examples of American usage, see Philadelphia Society for the Promotion of Agriculture, *Memoirs*, 5 (1824), 232n; *Southern Agriculturist*, 8 (August 1835), 438; R. L. Allen 1849: 95–96; J. J. Thomas 1855: 144; L. Rogin 1931: 66–68; R. H. Hurt 1982: 36–37. Successive editions of Noah Webster's dictionaries indicate both the popular confusion and the slow development of precise terminology. His first edition (1828) had no reference to an implement under "cultivator," and his reliance upon British sources is suggested by the entry "Horsehoe: To hoe or clean a field by means of horses." (Notice that the latter refers to a process, with no indication of the implement used.) Webster's first reference to "cultivator" as an implement appears in the 1852 edition of the dictionary, and the entry is not a paragon of clarity or accuracy ("Cultivator: A kind of harrow").

3. Tull's three great principles were soil pulverizing, sowing in drills, and mechanical (and constant) weeding. Although he had definite ideas about the underlying principles justifying his approach, most of his ideas were incorrect, and unsurprisingly so given the imperfect state of chemistry in Tull's day. See E. Cathcart 1891: 19, n. 4, 33–35.

4. J. Tull 1733: 403.

5. Quoted in T. H. Marshall 1929: 48.

6. T. H. Marshall 1929: 52.

7. W. Hewitt 1766: 397–398.

8. For further details of this implement, see ibid., 397–403.

9. A. F. M. Willich 1803: 294. The American James Mease, who edited this edition of Willich, added a personal observation about this implement: "The editor has used an earthing hoe fixed in a beam, exactly like [this 'improved hand-hoe'] with great advantage." (Ibid.) A similar principle applied to a horse-drawn implement is evident in the Paring plough depicted in the 1803 edition of the *Encyclopaedia Britannica*

(Plate VII). For other illustrations of early British horse-hoes, see A. M. Bailey 1782, Vol. 1, Plates IX, X, XIV; *The Farmer's Magazine and Useful Companion* (London), 2 (1777), plate opposite 148.

10. J. Eliot 1934: 119.

11. The evidence suggests that many British farmers retreated to a similar solution. In its discussion of the "Advantages of Horse-hoeing," the first edition of the *Encyclopaedia Britannica* depicted a "hoe-plough of a very simple construction" which was little more than a common plow (1771: Plate VI, Figure 7).

12. J. P. Greene 1965: Vol. 2, 440–445, 453, 626.

13. H. J. Carman, ed. 1939: 38, 99, 165, 166, 222, 316, 326. For other commentary on the colonial practice of sowing one crop (usually rye) between the standing rows of another (notably corn), see ibid., 99, 104; also A. B. Benson 1937: 89. Sowing wheat between rows was more common in the Chesapeake region. (Wayne Randolph, Colonial Williamsburg, private communication, October 16, 1995.)

14. See S. Deane 1790: 128–129, 143, 184–188. Deane defines a "horse-hoe" as nothing more than a common plow ("horse-plough") with a different hitching system to facilitate running the implement closer to, or further from, the standing plant. Like Jared Eliot, he was not a fan of Tull's design. "The advantage of this instrument [the horse-hoe] above a horse-plough is said to be principally the steadiness of its going, by which a furrow may be drawn very near to a row of plants, without danger of injuring them. This was the opinion of Mr. Tull, the inventor. But as it cannot be so well governed by the handles as the common horse-plough, the safety of the plants must chiefly depend upon the steadiness of a horse's going. I therefore prefer the horse-plough, on the whole, for loosening the ground betwixt rows. It will answer, at least, every purpose of the horse-hoe." (Ibid., 135.)

15. J. B. Bordley 1799: 39, 535.

16. See *Universal Asylum*, 6 (February 1791), 85; ibid., 9 (October 1792), 258; *New-York Magazine*, 6 (February 1795), 99. The last of these is an address to the Agricultural Society of the State of New York by its president, Robert Livingston. See also the replies to queries of the Massachusetts Agricultural Society from Worcester and the Western Society of Middlesex Husbandmen in *Massachusetts Agricultural Repository and Journal*, 2 (1807), 18, 19, 23; also William Strickland 1801: 39.

17. S. Deane 1790: 128, 187; 1797: 76.

18. The diagrams in Figure 6.4 come from a British source first published in America in 1803. Identical diagrams (minus the details on share points) appear in J. B. Bordley 1803: Figure 13. The bottom diagram, minus the struts to vary width, was more typical of what in American was meant by a "double-mouldboard plow." (Wayne Randolph, Colonial Williamsburg, private communication, October 16, 1995.)

19. In 1775 Landon Carter refers to "running my plows without mold boards" between tobacco rows "to cut those dreadful Potatoe Vines." (J. P. Greene 1965, Volume 2, 920.) The implements that Carter used may or may not have been shovel plows. One of the earliest references to that implement by that name occurred in an estate inventory of 1785 (estate of Francis Shackleford of Caswell County, N.C., dated November 12, 1785, Caswell County Will Book, Vol. B, p. 102; citation supplied by Wayne Randolph, Colonial Williamsburg, private communication, October 16, 1995). While touring America in the closing years of the century, Richard Parkinson noted the presence of the shovel plow, but considered the implement so unexceptional that he dismissed it as "not worth describing." (R. Parkinson 1805: 492–493.)

20. The British literature on early horse-hoe development often stressed the damage to crops that could result from using the wrong horse with an unskilled driver. For farms relying upon low-skilled workers, particularly slaves, the shovel plow's ease of operation was clearly an advantage.

21. A three-pronged version may have been used as early as 1757 by Landon Carter.

J. P. Greene, ed. 1965, Vol. 1, 184.) Carter also refers to "a machine plow of five hoes" for tobacco cultivation, but not even Carter makes later references to the implement. (Ibid., 1, 172.) For a detailed discussion plus primitive illustrations, see *American Farmer*, 7 (March 25, 1825), 1–2. The author of the latter assumes that readers are familiar with the basic design and purpose of the "trowel hoe."

22. David Cohen suggests that the "small triangular harrow" was a typical implement for the inter-row weeding of corn "in eighteenth-century New York" (1992: 114), but in support of that contention he cites only one source (the Verplank Family Papers). Other isolated references to harrows being used for this purpose are not impossible to find. For example, in response to queries about corn cultivation by the Philadelphia Society for Promoting Agriculture, James Tilton of Delaware suggested that the "flake harrow" was sometimes used to remove weeds between the rows. (*The Columbian Magazine*, 3 [March 1798], 158.) In Virginia, blacksmith Jas. Anderson produced a number of "harrow hoes" in the 1780s. (Wayne Randolph, Colonial Williamsburg, private communication, October 16, 1995.)
23. Caleb Kirk's 1808 "ripper" described above seems to disappear from the literature almost immediately. The author of *American Husbandry* (1775) made a single reference to "a machine . . . between a plough and a harrow" found in North Carolina which he expected "to become more general," but it did not. (H. J. Carman, ed. 1939: 249.) John Beale Bordley in 1784 advocated "a shim" for potatoes and peas, and fifteen years later explained that the implement in question consisted of a single blade, twelve inches wide, which, because of its fragile structure, was suitable only "against young weeds." (J. B. Bordley 1784: 20, and by the same author, 1799: 51.) See also Richard Peters's passing reference to a "scarifier" in Peters 1811: 71.
24. Philadelphia Society for Promoting Agriculture, *Memoirs*, 3 (1814), 34.
25. 2 (September 1811), 75–76. The "skimmer" appears to be similar to Bordley's "shim" mentioned in the above footnote.
26. *Niles' Register*, 11 (August 22, 1812), 408.
27. For a discussion of the use of interchangeable teeth in harrows, see *American Farmer*, 6 (May 9, 1824), 21–22.
28. Examples of shovel plow use are legion. For examples of continued reliance upon double-moldboard plows, see *The Plough Boy*, 1 (December 18, 1819), 230; *Massachusetts Agricultural Repository and Journal*, 5 (July 1819), 344: *American Farmer*, 4 (November 1, 1822), 253; *New England Farmer*, 1 (January 4, 1823), 179.
29. For details on patents, see *American Farmer*, 2 (October 13, 1820), 231–232. For early discussions, endorsements and advertisements, see *American Farmer*, 1 (June 18, 1819), 91; ibid., 2 (April 28, 1820), 10; ibid., 2 (August 18, 1820), 160; ibid., 3 (September 14, 1821), 196.
30. "This form (the jumping shovel) is invaluable for breaking up new cane land, and indeed in new land of any kind that is full of roots. It cuts a great majority of them, and when it comes to one that is too large, it skips over and instantly enters the soil again." (*The Agriculturist*, 3 [October 1844], 305.) For a discussion plus picture of the same device (a coulter) added to a regular shovel plow to achieve a similar "jumping" effect, see *American Farmer*, 8 (July 14, 1826), 129–130. For early references to a shovel plow with a narrow blade of six or seven inches, see *American Farmer*, 1 (September 17, 1819), 195; ibid., 3 (January 11, 1822), 332; ibid., 7 (August 12, 1825), 161. The blade in Figure 6.9 (b) is seven inches by four inches. For later references to a "bull tongue plow," see *Southern Agriculturist*, 4 (January 1831), 5 (three inches wide); ibid., 4 (November 1831), 263–264 ("bull-tongue or gofer plough" for opening furrows five inches wide); *The Cultivator*, 9 (June 1842), 93 (one of the few illustrations of the blade from the front and side); *American Agriculturist*, 2 (July 1843), 117 ("bull-tongue or scooter plow" about four inches wide).

31. Address to the South Carolina Farmers' Society, *American Farmer,* 3 (January 11, 1822), 329–332.
32. *American Farmer,* 5 (March 28, 1823), 2.
33. *American Farmer,* 3 (January 18, 1822), 341. For later references to a similar implement called a "scraper," see *Southern Agriculturist,* 4 (November 1831), 564; *American Agriculturist,* 2 (July 1843), 117–118.
34. *American Farmer,* 2 (September 15, 1820), 195–196. For a later description of a skimmer of similar dimensions, see *Southern Agriculturist,* 4 (October 1831), 517. For a description and picture of a similar implement called a "sweep," see *American Agriculturist,* 1 (March 1843), 363; also ibid. (July 1843), 117; J. A. Taylor 1857: 54–56.
35. J. B. Bordley 1799: 51n.
36. As noted above (footnote 21), these modern "high-speed sweeps," in design, are very similar to some forms of a "hoe plow" or "trowel plow" that, in the South, may have been in use in the late eighteenth century, and certainly were in use by the 1820s.
37. *American Farmer,* 2 (November 10, 1820), 262.
38. *The Cultivator,* 3 (January 1837), 180.
39. *The Cultivator,* 2 (June 1835), 50.
40. Whether a wheel was desirable or not depended upon the condition of the field. "This [a wheel on the end of the cultivator] is useless in very uneven or rocky ground; but when the surface is tolerably smooth it is very desirable, as it makes the cultivator move easier and steadier, and with it the teeth can be exactly gauged, to work the ground any required depth." (*American Agriculturist,* 5 [June 1846], 173.)
41. For advertisements of Beatson's "scarifier" (or "improved scarifier" or "scarifier and cultivator") by American farm implement dealers in the 1820s, see *New England Farmer,* 1 (January 25, 1823), 207; ibid., 1 (May 3, 1823), 319; ibid., 2 (April 3, 1824), 287; ibid., 3 (March 4, 1825), 255; *American Farmer,* 8 (May 5, 1826), 55. For American endorsements and public exhibitions of the same implement, see *American Farmer,* 5 (December 26, 1823), 313; ibid., 5 (March 19, 1824), 409; Philadelphia Society for Promoting Agriculture, *Memoirs,* 5 (1824), 230–231. Partridge argues that British "cultivators" were imported into America in the first half of the nineteenth century, but offers no supporting evidence. (M. Partridge 1973: 77.) The advertisement that announced a $20 price for Beatson's scarifier offered inter-row cultivators at $8 to $12. (*American Farmer,* 8 [May 5, 1826], 55.) How original Beatson's scarifier was, by British standards, is far from clear. The idea of pulverizing soil with an implement combining a heavy harrow frame with long, powerful teeth, slanted slightly forward, was well known by the late eighteenth century. See, for example, *Encyclopaedia Britannica* 1771: 58–59; ibid., 1803: 276 and Plate V.
42. The description was part of a general endorsement for "cultivators" give by the editor of the *American Farmer,* which he considered to be "one of the most efficient instruments in the hands of the skilful farmer" (1 [September 17, 1819], 194n). The option of using a harrow frame with "scarifying teeth" was also explored. For example, into Abbott's harrow (depicted in Chapter 5) could be inserted "teeth made flat and sharp like the coulter of a plough" for "scarifying ground not ploughed." (*New England Farmer,* 2 [March 27, 1824], 276.) The timing of the development of these "field cultivators," as they would sometimes be called, predates by two decades the examples cited by Leo Rogin (1931: 66–68).
43. For evidence of availability of expandable cultivators, see *The Plough Boy* 1 (Septemter 18, 1819), 122–123; *American Farmer,* 2 (June 23, 1820), 104; *New England Farmer,* 1 (July 12, 1823), 397; ibid., 2 (April 3, 1824), 387; *American Farmer,* 8 (May 5, 1826), 55–56.
44. Among the options available by the mid-1820s were "reversible" teeth; that is, when the cutting side became dulled, the tooth could be reversed to bring into play a

second cutting surface identical to the first. See *New England Farmer*, 4 (November 4, 1825), 116; *American Farmer*, 8 (June 9, 1826), 92.

45. *The Plough Boy*, 2 (June 3, 1820), 3; ibid., 2 (September 30, 1820), 138.
46. Prices are from *American Farmer*, 4 (November 22, 1822), 280. For evidence of availability of triangular cultivators, see *American Farmer*, 2 (November 10, 1820), 262; ibid., 3 (April 6, 1821), 15; ibid., 3 (April 27, 1821), 40; ibid., 3 (June 29, 1821), 106; ibid., 3 (December 21, 1821), 312; ibid., 4 (May 3, 1822), 44; ibid., 4 (November 22, 1822), 280; ibid., 6 (May 7, 1824), 56; *New England Farmer*, 4 (November 4, 1825), 116.
47. *American Farmer*, 9 (December 28, 1827), 328. The cities were Philadelphia, New York, Richmond, Raleigh, Charleston, Savannah, and Statesburg, South Carolina.
48. *American Farmer*, 8 (April 21, 1826), 33.
49. By 1840 Buel was evidently well satisfied with the range of options, and advised farmers to keep "different kinds, for instance, such as are fitted to skim the surface, and destroy weeds—others to break up and pulverize the surface; and others, again, to gather the roots of quack and other perennial pests." (J. Buel 1840: 150.)
50. For illustrations of these later developments, including walking, riding, straddle-row, self-sharpening, and disk cultivators, see *New England Farmer*, 18 (July 17, 1839), 20; *The Agriculturist*, 3 (April 1844), 112; *Southern Planter*, 5 (1845), 205; J. J. Thomas 1855: 144–145; S. E. Todd 1866: 247–248, 252; New York State Agricultural Society 1868: 262; E. H. Knight 1880: 93; B. Butterworth 1892: "Cultivators" and "Cultivators—Walking"; R. L. Ardrey 1894: 167; H. W. Campbell 1907: 400; J. B. Davidson 1907: 393–399. More recent reviews of twentieth-century equipment, with illustrations, can be found in A. A. Stone 1945: 172–217; H. P. Smith 1955: 224–249.

7. Reaping

1. Quoted in P. T. Dondlinger 1919: 74.
2. For a detailed example of the difference weather could make to the harvesting of wheat, see the farm journal entries in E. Ruffin 1851: 109–113.
3. *American Farmer*, 2 (May 12, 1820), 54.
4. N. E. Lee 1960: 30; S. M. Cole 1954: 503; G. R. Quick and W. F. Buchele 1978: 2–4.
5. For depictions of ancient reaping knives and sickles, see E. C. Curwen and G. Hatt 1953: 24; M. S. Drower 1954: 542; S. M. Cole 1954: 503; G. R. Quick and W. F. Buchele 1978: 6; S. E. Rees 1979: passim; A. Steensberg 1943: passim. For illustrations of curved sickles in use from Egyptian times to the Middle Ages, see N. D. Davies 1917: Plate XVIII; G. R. Quick and W. F. Buchele 1978: 3–5; P. Lindemans 1952, Vol. 2, Plates V, VI; J. C. Webster 1938: Plates 31, 34a.
6. An alternative and early attempt to reduce the strain on the wrist is depicted in F. Hobi 1926: 4.
7. According to V. G. Childe, balanced sickles were known in the Mediterranean area by the second millennium B.C. (1951: 39–48). See also Sian E. Rees 1979: part 2, 439 (on Egyptian evidence); K. D. White 1967: 77, 80 (on Roman evidence); and E. M. Jope 1956a: 95; E. C. Curwen and G. Hatt 1953: 113; J. M. G. van der Poel 1983: 31; A. Steensberg 1943: 115 (on diffusion through Europe).
8. Many of Anderson's depictions of farming activities, including this one, first appeared in farmer's almanacs of the 1820s, suggesting that a shortage of woodcuts of farming scenes made editors willing to compromise on accuracy. See, for example, *State of New-York Agricultural Almanack*, 1821 (Albany: Packard and Van Benthuysen, [1820?]); ibid., 1823 (Albany: E. and E. Hosford, [1822?]); ibid., 1824 (Albany: Packard and Van Benthuysen [1823?]).
9. This description applies to the use of a "toothed" or "serrated" sickle. Smooth-edged

sickles required a slightly different procedure, omitted here because the serrated blade appears to have been the preferred choice in American agriculture by a wide margin. For a discussion contrasting the cutting techniques for the two different blades, see H. Stephens 1855: Vol. 2, 329–333. The teeth of the serrated sickle are inclined toward the handle, so cutting occurs only when the implement is drawn toward the reaper (and not when it is introduced into the standing grain). Compared with the serrated sickle, the smooth-edged sickle, although faster, had several drawbacks: (a) the curvature of the blade had to be very exact to achieve the desired cutting effect, (b) the implement had to be sharpened in the field when dulled from use, and (c) the slashing motion required invariably resulted in more grain loss due to shattering. See H. Stephens 1855, Vol. 2, 330–331; A. Rees 1810–1824: "Sickle." One of the best set of diagrams illustrating the difference between the smooth and the serrated blade may be found in R. L. Séguin 1989: 657–659. An additional drawback of the smooth-edged sickle, seldom emphasized in the literature but demonstrated to me with devastating clarity by Bruce Craven of Old Sturbridge Village, is the danger of slashing at a collection of stalks held by one hand with an implement that can "cut to the bone" (Craven's phrase).

10. The late Bronze and early Iron Age clearly coincided with a drop in average temperature throughout Europe. (See, for example, C. Flon 1985: 53.) The simple hypothesis linking climate change to scythe development in Northwest Europe is advanced by (among others) A. Steensberg 1943: 100–101; S. E. Rees 1979: 741; J. M. G. van der Poel 1983: 31. Many of these authors emphasize the special livestock needs of Roman occupation forces. A more complex hypothesis about scythe development elsewhere in Europe that includes rising fodder demands from (a) improvements in animal husbandry, (b) the rise of large estates, and (c) a growing labor shortage, may be found in K. D. White 1967: 102–103; A. Steensberg 1943: 179–180; and S. E. Rees 1979: 477.
11. S. E. Rees 1979: 478.
12. See K. D. White 1967: 98–99; S. E. Rees 1979: 473–479; E. M. Jope 1956a: 95.
13. Many experts argue that the long-handled scythe was not widely used in Europe until the Middle Ages. See, for example, A. Steensberg 1943: 115; E. C. Curwen and G. Hatt 1953: 116. For other illustrations of long-handled scythes in use in premodern Europe, see P. Lindemans 1952, Vol. 1, Plates III, XXVII, XXVIII, Vol. 2, Plates IV, IX; J. C. Webster 1938: Plates 34b, 38, 50, 93, 94; J. Longnon 1969: Plate 6v.
14. A picture of various handles with various grips can be found in H. J. Hopfen 1960: 103.
15. One of the first American discussions of these properties can be found in S. Deane 1790: 254. Another variable was the angle between blade and handle, the more obtuse the angle, the more material taken with each stroke, but the harder the work. See H. J. Hopfen 1960: 104. For descriptions of parts of the scythe and how the implement is used, see J. Vince 1983: 126–127; E. J. Stowe 1948: 78.
16. The photograph reproduced in Figure 7.10, because it was taken from behind the reapers, does not show the correct position for the hands. For an early illustration of reaping with the scythe that does show both hands, see W. H. Pyne 1806: Plate I, 38; for a later and more elegant illustration, G. E. Fussell 1952: illustration 66.
17. This description was often quoted in American literature, the process referred to as mowing "in the English manner." See, for example, S. Deane 1790: 179; *Boston Weekly Magazine*, 36 (July 2, 1803), 147. Wayne Randolph of Colonial Williamsburg has pointed out that some later versions of this implement had much longer blades, possibly for use in very light stands of grain. (Private communication, November 7, 1995.)
18. Ibid.

19. Pictures of the implement can evidently be found in some thirteenth-century illuminations. See G. R. Quick and W. F. Buchele 1978: 8. "There is some evidence of scythes-with-bow in Belgium in the beginning of the fourteenth century." (J. M. G. van der Poel, private correspondence, February 25, 1994.)
20. See, for example, S. Deane 1790: 179; *New England Farmer,* 7 (August 22, 1828), 33; *Maine Farmer and Journal of Useful Arts,* 5 (August, 1837), 178. For evidence of the popularity of this implement in nineteenth-century Britain, see J. C. Morton 1856, Vol. 2, 10–11; G. E. Fussell 1952: illustration 66; P. W. Blandford 1976: 116; *New England Farmer,* 6 (July 11, 1828), 402.
21. An editorial aside by the American James Mease in A. F. M. Willich 1803: 459. The British also preferred ash for cradle teeth. See, for example, C. W. Johnson 1848, Vol. 2, 189–190.
22. Initially the term "cradle" was commonly applied to the appendage only, the implement thus designated as a "scythe with cradle." By the second decade of the nineteenth century, the term "cradle" tended to refer to the entire implement. For examples of early usage, see J. F. D. Smyth 1784, Vol. 2, 113; S. Deane 1790: 64; the editorial aside of James Mease in A. F. M. Willich 1803: 270; and A. Rees 1810–1824: "Cradling."
23. Relative to British cradles, the American version tended to have longer teeth (almost the length of the blade) and more of them (five or six, compared to three or four in the typical British case). See, for example, A. F. M. Willich 1803: 459; Alexander Coventry 1978: 116; J. T. Schlebecker 1972: 17; C. W. Johnson 1848: Vol. 2, 189–190; M. Partridge 1973: 135. For other illustrations of grain cradles, see R. D. Hurt 1982: 41; J. C. Loudon 1839: Vol. 1, 69, 373; J. C. Morton 1856: Vol. 2, 10; P. A. Wright 1961: 40; G. R. Quick and W. F. Buchele 1978: 10; E. Sloane 1964: 103; D. Jackson 1976: 392. For other illustrations of grain cradles in use, see H. J. Hopfen 1960: 6; E. Tunis 1961: 78; E. Tunis 1957: 71; R. L. Séguin 1989: 668–669.
24. Observations of an 1818 visitor to America, reproduced in C. W. Burkett 1900: 168–169.
25. A. Coventry 1978: 116, describing harvesting witnessed in Connecticut in 1785.
26. E. Sloane 1958: 68. In both Britain and America, cradling was commonly done away from the standing grain. See S. Deane 1790: 64; A. F. M. Willich 1803: 459; J. C. Morton 1856: Vol. 2, 10–11.
27. The similarity in purpose of the two implements suggests that the cradle's design may have evolved as a logical extension of the scythe-plus-bow, the cradle being better suited for gathering grain in light stands. Clearly the cradle's claim to novelty is constrained by the previous existence of the scythe-plus-bow, and Isaac Phelps Roberts would seem to overstate the case when he suggests that implement was "a greater stroke of absolute genius than the self-rake reaper or the binder." (Reported in J. van Wagenen 1953: 226.)
28. See A. Steensberg 1943: 226; A. F. M. Willich 1803: 270–271, 459; E. J. T. Collins 1969: 456; J. A. Perkins 1976: 56; D. Jackson 1976: 278.
29. The use of the cradle for harvesting small grains, particularly wheat, is variously dated as occurring throughout the colonies in the later colonial period (W. D. Rasmussen 1960: 79; L. Rogin 1931: 69; J. A. Henretta 1973: 18), in the South and Middle Colonies prior to the Revolution (L. C. Gray 1958: Vol. 2, 798; P. Bidwell and J. I. Falconer 1925: 125) but in New England only after the Revolution (P. Bidwell and J. I. Falconer 1925: 125; K. M. Sweeney 1988: 62; H. S. Russell 1976: 309), or alternatively, as being "first perfected" during the last quarter of the eighteenth century (P. T. Dondlinger 1919: 80). In the South, cradles actually begin to appear in Virginia inventories by the late 1720s. (Wayne Randolph, private communication, November 7, 1995.)
30. See J. T. Lemon 1972: 178; K. W. Sweeney 1988: 62–63; J. P. Greene 1965, Vol. 1, 180; D. M. Mattheson 1931: 4–5; P. L. Haworth 1925: 95; N. F. Cabell 1800: 24–25.

31. See, for example, A. Coventry 1978: 116 (which gives an example of Connecticut use in 1785); D. S. Cohen 1992: 128 (quoting an example of usage in 1786); *The Colombian Magazine*, 1 (October 1786), 82; ibid., 1 (May 1787), 432; ibid., 3 (April 1789), 217; *American Museum*, 5 (1789), 389; F. Knight 1847: 61, 84 (examples from Washington's 1792 correspondence from Richard Peters and to Arthur Young); J. F. D. Smyth 1784, Vol. 2, 113.
32. S. Deane 1790: 64. Wheat appears to have been harvested primarily with sickles at this time, and almost never with scythes. For an exception to the rule (occasioned by a labor shortage), see La Rochefoucauld 1800, Vol. 3, 210–211.
33. J. B. Bordley 1799: 102.
34. An editorial aside in A. F. M. Willich 1803: 459. (Italics added.) Elsewhere in the same volume the cradle is described as "useful in harvesting" (270–271). See also an 1807 document reproduced in *American Farmer*, 9 (August 10, 1827), 161. As one might expect, the number of teeth was not uniformly five throughout America. In his report on agricultural practices in Virginia, for example, Cabell argued the common number was three. (N. F. Cabell 1800: 25.) E. Franklin and S. Evans, in their study of Pennsylvania practices, suggest the typical number was four. (E. Franklin and S. Evans 1883: 349.)
35. Here, as elsewhere in this chapter, J. M. G. van der Poel provided helpful clarification concerning European usage. "The word 'pik'/(sith) (hence the verb 'pecken' or more recently 'pikken') has only been used in our province of Zeeland and in Belgium, not elsewhere in Europe. In the other provinces of my country this implement had the name 'zicht' or 'sicht(e).' The hook was called in Zeeland and Belgium 'pikhaak' and elsewhere in the Netherlands 'mathaak' or 'welhaak.' In Germany [the terms are] respectively 'Sichte' or 'Kniesense,' and 'Matthaken.' The American word 'sith' is obviously a corruption of 'zicht' or 'sicht(e),' and 'mathook' or 'mattock' of 'mathaak' or 'Matthaken.'" (Private correspondence, February 25, 1994.)
36. For British descriptions of the implement, see W. L. Rham 1841: 60; C. W. Johnson 1848, Vol. 2, 190–192; H. Stephens 1855, Vol. 2, 337–338; J. C. Loudon 1839, Vol. 1, 372–373. The classic American study of the Flemish scythe is P. H. Cousins 1973. The features noted result in the implement being "balanced with its fulcrum at a point behind the hand-grip" (P. H. Cousins 1973: 12). For a comparison of contemporary blades of scythes with what is now usually called a "scythette," see H. J. Hopfen 1960: 102.
37. W. L. Rham 1841: 61. The reader may find this abbreviated description difficult to follow. A more detailed description of the same operation is provided by Henry Stephens. "The reaping is done by pressing the back of the hook with the left hand, against the standing corn, in the direction of the wind, and by cutting with the scythe close to the ground against the standing corn, with a free swing of the right arm—less by force than by the impetus of the blade—till in three or more strokes, according to the thickness of the crop, a sufficiency is severed, which, when caught in the hook, with a portion of the standing corn against which it rests, is rolled into the form of a sheaf by the workman walking backwards, and cutting any of the standing corn caught by the hook with the point of the scythe, until he reaches the point he started from, where by gathering and keeping the heads in a line by means of the hook, closes together the but-end of the sheaf with the scythe, and then, with both, by a little adroitness, and the assistance of the foot, a perfect sheaf is lifted from the ground, and placed in the band ready for binding." (H. Stephens 1855, Vol. 2, 338.)
38. See Lindemans 1952, Vol. 2, Plates VIII, IX; A. Verhulst 1990: 128; G. R. Quick and W. F. Buchele 1978: 8; M. Partridge 1973: 135–136; J. M. G. van der Poel 1983: 32; P. H. Cousins 1973: 13; J. F. Niermeyer 1968: 224; H. J. Smit 1929: Vol. II, 363; M. de Meyer 1946: 145–153; J. J. Voskuil 1972: 12–22; E. Thoen 1988: 776–779. For many helpful suggestions concerning the agricultural history of the Low Countries, I am

indebted to J. M. G. van der Poel, Peter Hoppenbrouwers, and Jan Bieleman, including the last five sources noted (all of which are in Dutch).

39. For illustrations of the Flemish scythe in use in premodern Europe, see J. Longnon 1969: Plate 7V; F. M. Kelly 1936: Plate VII; P. Lindemans 1952, Vol. 2, Plate VII; G. R. Quick and W. F. Buchele 1978: 7; J. Meurers-Balke and C. Loennecken 1984: 40.

40. A. F. M. Willich 1803: 458–459.

41. As early as 1632, Kiliaen van Rensselaer was sending "19 Hainault scythes for grain," and by 1643, "Flemish scythes" can be found in farm inventories. (R. H. Blackburn and N. A. Kelley, eds. 1987: 195.) See also R. S. Gottesman 1938: 20 (which reproduces an advertisement from *The New-York Weekly Journal* of April 23, 1739); P. H. Cousins 1973: 13–14; D. S. Cohen 1992: 127.

42. Jesse Buel in *American Farmer*, 8 (April 21, 1826), 33.

43. This assumes that the supply is not infinitely elastic, which seems reasonable enough, given the requisite skills.

44. Thomas B. McElwee, in a "Sketch of Bedford County, Pennsylvania" in *American Farmer*, 11 (June 26, 1829), 115.

45. Committee of the Highland Society of Scotland 1829: 244–249.

46. What follows is a brief review of only the more important costs and benefits that ignores a number of relevant considerations, particularly those specific to different crops, whose importance seems to have changed over time. For American discussions of the relative merits of different premodern reaping implements, see *The Colombian Magazine*, 3 (April 1789), 217; A. Coventry 1978: 116; S. Deane 1790: 64; F. Knight 1847: 84; *American Farmer*, 2 (May 12, 1820), 54; ibid., 8 (April 21, 1826), 33; ibid., 9 (August 10, 1827), 162; ibid., 11 (June 26, 1829), 115; ibid., 12 (August 20, 1830), 178; *The Plough Boy*, 4 (August 27, 1822), 99; *New England Farmer*, 1 (July 19, 1823), 405; ibid., 4 (August 12, 1825), 21; E. Ruffin 1857: 12–13; L. Rogin 1931: 69–72. For British discussions of costs and benefits, see A. F. M. Willich 1803: 116–117, 458–459; A. Rees 1810–1824: "Reaping"; Committee of the Highland Society of Scotland 1829: 244–249; J. C. Loudon 1839, Vol. 2, 817–827; C. W. Johnson 1848, Vol. 2, 188–193; H. Stephens 1855, Vol. 2, 330–347; J. C. Morton 1856: Vol. 1, 189, Vol. 2, 10–11, 505, 777–778.

47. J. M. G. van der Poel pointed out that, in the Low Countries, beaten-down crops were also harvested with the Flemish scythe, using the mathook to lift the stalks. (Private correspondence, February 25, 1994.)

48. Where stubble was common property, institutional barriers inhibited the introduction of the scythe (G. R. Quick and W. F. Buchele 1978: 10). Cutting high had the additional advantage of gathering fewer weeds with the harvested crop, a feature particularly important for barley destined for brewing. The presence of green weeds in the harvested stack was commonly considered to encourage "sweating," "and by that means, the smut, which is light, will be carried with the steam through the whole mow, or stack." (*The Plough Boy*, 4 [August 27, 1822], 99.)

49. For a comparison of "smooth-edged" and "toothed" sickles, including the more violent stroke required with the former and the difficulty of designing the right curvature for a smooth-edged blade, see H. Stephens 1855, Vol. 2, 330–331; A. Rees 1810–1824: "Sickle."

50. Heaviness also affected the choice among scythe-like implements, if one of these was to be used. To swing a cradle with long teeth through a dense stand was extremely difficult, the cutting motion necessarily pulling the severed stalks away from the uncut grain. Both the Flemish scythe and the scythe-plus-bow had the advantage in such situations of cutting to (rather than away from) the standing crop. Alternatively, in light stands where grain would "not fall evenly over the scythe," a cradle was particularly well suited for cutting and gathering the stalks. (G. H. Andrews 1853: 373.) See also J. J. Voskuil 1972: 16–17.

51. H. Stephens 1855, Vol. 2, 337.
52. The British claim that the Hainault scythe was "in general use among the women and girls in Flanders" (Committee of the Highland Society of Scotland 1829: 247) is challenged by Dutch historians who argue that this implement "required skilled manpower" (J. M. G. van der Poel 1983: 32). All of the premodern pictures the author could find showed only males wielding Flemish scythes. This implement was also more dangerous to use than the sickle.
53. Further considerations include (but are not limited to) the following. Harvesting by the sickle had the disadvantage of compressing stalks held in the left hand (while cutting with the right), necessitating a longer period for the cut crop to dry. Serrated sickles required little if any sharpening in the field, whereas all scythe blades had to be resharpened as harvesting proceeded. Reapers with sickles commonly produced neater results in terms of the heads of severed stalks all aligned the same way, thereby making easier the subsequent task of threshing. Both J. M. G. van der Poel and Peter Hoppenbrouwers pointed out to me that harvests cut with the sickle also took less barn space than those cut with any of the scythe-like implements. (Private correspondence, February 21, 1994 and February 25, 1994.)
54. See, for example, A. F. M. Willich 1803: 116; P. W. Blandford 1976: 115; J. A. Perkins 1976: 47–58, and by the same author, 1977: 125–135; E. J. T. Collins 1972: 16–33. For an indication of how unresolved the question is as to when and why scythe-like implements replaced the sickle in harvesting small grains in Britain, compare the very detailed and region-specific arguments of Perkins with Collins's emphasis upon changing relative factor prices during the Napoleonic Wars.
55. Also delaying a British change in reaping implements was "the judicial problem . . . where cereal stubble was common property," and consequently "the low cutting scythe was forbidden." (G. R. Quick and W. F. Buchele 1978: 10.)
56. Whether British crops were, on the average, less dense than those of, say, Flanders is difficult to establish. But the inappropriateness of cradles for reaping heavy crops was a commonplace among contemporary observers on both sides of the Atlantic. See, for example, S. Deane 1790: 64; *American Farmer*, 12 (August 20, 1830), 178; C. W. Johnson 1848, Vol. 2, 189; also L. Rogin 1931: 72.
57. Comparing the British grain cradle with the scythe-plus-bow, a British writer noted that "to mow up against the standing crop" required "merely a loop or hoop attached to the [scythe's] handle and heel of the blade, rather than a cradle, which can be employed only when the grain is cut from the crop." (J. C. Morton 1856, Vol. 2, 10–11.) In short, the former implement is totally inappropriate in light stands. An American observer agreed. "The [American] manner of reaping is cradling by the scythe. . . . But the cause is that the crops are so thin and short, that many could not be reaped at all by any other means, except at more expence than they are worth. But where there is any thing like our crops in England, they cannot cradle at all, but are compelled to reap in the same manner as we do." (Richard Parkinson 1805: 203.) See also B. Romans 1775: 121.
58. H. Stephens 1855, Vol. 2, 338; P. H. Cousins 1973: 12.
59. J. C. Loudon 1839: Vol. 1, 373.
60. Wayne Randolph of Colonial Williamsburg finds that fragmented evidence from the colonial period suggests that sickles of the eighteenth century may have been narrower and more elliptical in shape than those of the seventeenth century. (Private communication, November 7, 1995.)
61. J. L. Bishop 1868, Vol. 1, 477. See also B. P. Poore 1867: 503; V. S. Clark 1929, Vol. 1, 48; H. S. Russell 1976: 83.
62. A theodolite, patented in 1735 for seven years by Rowland Houghton of Boston. (V. S. Clark 1929, Vol. 1, 48.)
63. A. F. M. Willich 1803: 459. What is not clear is how extensive American modifica-

tions were to the European models they began with. For example, French cradles at this time are pictured with fewer teeth (four, rather than five), but of similar length. See Denis Diderot 1763: Agriculture, Semoir, Plate I (reproduced almost half a century later in Louis Liger 1805, Vol. 2, 508). Miller is among those who would minimize American achievements, arguing (with no supporting evidence) that "doubtless all the European types had been brought over by the colonists," and subsequent American designs were "simply an improvement upon some of these earlier forms." (M. F. Miller 1902: 10.)

64. The list of those receiving patents for reaping machines between 1800 and 1815 were Richard French and J. T. Hawkins of New Jersey (1803), Samuel Adams of New Jersey (1805), J. Comfort of Pennsylvania and William P. Claiborne of Virginia (1811), Peter Gaillard of Pennsylvania (1812), and Peter Baker of New York (1814). (U.S. Bureau of the Census, *Eighth Census of the United States.* 1860. Agriculture 1864: xx.) One other historical curiosity of the 1787–1815 period should perhaps be noted. In 1788 John Beale Bordley reported on "a new method of reaping wheat," which amounted to little more than division of labor in the field and the use of bags to gather the severed heads. An early report appeared in *The Colombian Magazine,* 2 (September 1788), 510–512. The same report with a clearer illustration appears in Philadelphia Society for Promoting Agriculture, *Memoirs,* 4 (1939), 122–125. Bordley reports he witnessed this "new method" at the farm of an "ingenious Gentleman" in Talbot County, Maryland. The reaping implements used were sickles and scythes.

65. *American Farmer,* 5 (December 1823), 307.

66. Quoted in F. H. Higgins 1958: 15.

67. *American Agriculturist,* 2 (June 1843), 83.

68. J. C. Morton 1856, Vol. 2, 744.

69. About 1850 McCormick acquired the patent rights to the Hussey cutter-bar (J. T. Schlebecker 1975: 114). Prior to that design improvement, the McCormick reaper had the added defect of serving poorly as a mower, and prospective buyers, not surprisingly, wanted a single machine to perform both functions. For a discussion of the evolution of mowers, and the associated problems of reaper modification, see J. J. Thomas 1869: 158–165; L. Rogin 1931: 90–91; J. T. Schlebecker 1975: 114–115; C. H. Danhof 1969: 232–233; M. F. Miller 1902: 18–28.

70. For discussions of other differences between McCormick and Hussey reapers, see R. D. Hurt 1982: 42–43; R. L. Ardrey 1894: 45–46; L. Rogin 1931: 85–90; J. C. Morton 1856, Vol. 2, 743–747. Some of the better illustrations of these machines can be found in G. H. Andrews 1853: 364–367, and *The Southern Planter,* 3 (March, 1843), 68; M. F. Miller 1902: Plates IV, V.

71. Quoted in L. Rogin 1931: 87.

72. The additional problem, as cutting surfaces were dulled, was the resulting need "that the knives should move with great rapidity, and the gear for getting up this speed inevitably produces heavy draught." (J. C. Morton 1856, Vol. 2, 745.)

73. This McCormick defect, reported in the Royal Agricultural Society's *Journal* of 1851, was soon remedied by further design modifications, as outlined in J. C. Morton 1856, Vol. 2, 745–746.

74. Quoted in C. H. Danhof 1969: 230. In the same year (1853), a committee appointed to evaluate mechanical reapers in Virginia reported favorably on the performance of both the McCormick and the Hussey reaper, while (in the same periodical as this report) another Virginian judged McCormick's reaper to be "the most magnificent and costly humbug in its line." (Virginia State Agricultural Society, *Journal of Transactions* [1853], 146–148, 51.) The tendency of metal products in general to be less than reliable is evident in the 1859 complaint of another farmer: "The implements for the most part now made, are so slight, and in too many cases of such poor

material, that the use of them for one season, is about as long as they will last in credit; the bars, bolts, and screw-heads, after that time, will need constant repairing, and this is the case with nearly all the farming implements now made." (Ibid., 230, n. 44.)

75. E. Ruffin 1851: 104. For commentary on the extent to which the cradle had displaced the sickle in the South at this time, see *The Southern Cultivator*, 3 (July 1845), 100. For a recent study of mechanization efforts made by family farms at mid-century, see Sue Headlee, 1991.

76. For a visual review of many of these different rotary devices, see M. F. Miller 1902: Plate II. The rotary type of cutting device dominated British inventive efforts until mid-century. Reciprocating devices were proposed as early as 1807 (by Robert Salmon in Britain), but they were not incorporated into machines destined to be successful until William Manning's mower of 1831 and the Hussey and McCormick reapers originating in that decade. A summary of the characteristics of the more important reapers invented between 1786 and 1831 in England, Scotland, and America can be found in William T. Hutchinson, 1930: 58.

77. The cutting device is described by the inventor as "a circular horizontal frame work, on the circumference of which is screwed the scythes in six parts, with the edges turned outward, so as to form a complete circle." (Three of these six blades are clearly visible in Figure 7.20). To ensure the blades remain sharp, "the edge of the scythe in its revolution passes under a whetstone fixed on an axis and revolving with the scythe." As with the Hussey and McCormick reapers, the horse was positioned to the front and side, not to the rear as in many British mechanical reapers of this era. (*American Farmer*, 5 [December 12, 1823], 199.)

78. This patent was granted to Richard French and T. J. Hawkins of New Jersey. Little is known about this machine, but the "cutters consisted of a series of scythe-like knives which revolved on a vertical spindle." (R. D. Hurt 1982: 41.)

79. *American Farmer*, 5 (December 12, 1823), 199.

80. Ibid., 8 (March 24, 1826), 2. By one estimate, "fifty or more of these machines were made," although the author provides no supporting evidence. (Gilbert Cope 1881: 218.)

81. The history of American and British mechanical reapers is reviewed in many of the standard sources on agricultural development, including J. C. Loudon 1839, Vol. 1, 421–427; U.S. Bureau of the Census, *Eighth Census of the United States*. 1860. Agriculture 1864: xx–xxii; R. L. Ardrey 1894: 40–52; G. E. Fussell 1952: 115–139; M. Partridge 1973: 126–133; G. R. Quick and W. F. Buchele 1978: 17–38; and P. W. Blandford 1976: 113–133.

82. For a list of patents granted for mechanical reapers between 1803 and 1830, see U.S. Bureau of the Census, *Eighth Census of the United States*. 1860. Agriculture 1864: xx–xxi. Symptomatic of the limited progress in British reapers in these early years was the fact that while the Royal Society of Arts first offered a reward "for a machine capable of reaping corn [grain]" in 1774, in the next forty-six years "not a single award appears to have been made." (H. T. Wood 1913: 128.) For a review of reaper defects by contemporaries, see A. Rees 1810–1824: "Reaping Machine," and *American Farmer*, 12 (July 23, 1830), 147.

83. See J. L. Bishop 1868, Vol. 2, 205–206; *Niles' Register*, 1 (January 25, 1812), 390; ibid., 10 (August 10, 1816), 399; *New England Farmer*, 4 (August 26, 1825), 39.

84. *American Farmer*, 5 (December 19, 1823), 307.

85. *New England Farmer*, 4 (April 7, 1826), 296.

86. For advertisements of the 1820s featuring "best warranted" and "cast steel scythes" as well as "superior grain cradles," see *American Farmer*, 2 (June 23, 1820), 104; ibid., 10 (June 20, 1828), 112; ibid., 12 (May 14, 1830), 72; *New England Farmer*, 1 (January 25, 1823), 207; ibid., 2 (September 13, 1823), 55; ibid., 2 (April 3, 1824), 287.

87. J. T. Schlebecker 1975: 113.
88. *American Farmer*, 6 (January 28, 1825), 353. J. M. G. van der Poel pointed out that by 1800 a similar implement—a cradle-like harvesting device with "coarse cloth" linking the teeth—was used in the province of Groningen and the adjoining region of the province of Drenthe for reaping buckwheat.
89. The third condition, in the jargon of economics, implies a supply curve of labor that is extremely inelastic. For a review of the history of this Roman harvesting machine, including evidence that some of the constraints noted above did prevail on latifundia in Gallic provinces, see K. D. White 1967: 157–173; and White 1984: 60–62.
90. *American Farmer*, 1 (November 26, 1819), 279. A similar stripping mechanism also surfaced in Britain for reaping grain during the late eighteenth century, and in the British case, the inspiration for the development was knowledge of the Roman mechanism. See P. T. Dondlinger 1919: 81–82; G. R. Quick and W. F. Buchele 1978: 13–18.
91. As illustrated in Figure 7.22, this device was "to be drawn by a horse attached with chains or ropes, which are secured at the ends of the axle." The cutting surface consisted of "43 teeth 9 inches in length, flat in the top, tapered at the ends, and turned up below, resembling the fingers of a cradle." (*American Farmer*, 1 [September 5, 1819], 254.)
92. *American Farmer*, 1 (September 5, 1819), 254; ibid., 1 (November 26, 1819), 279; ibid., 2 (May 19, 1820), 60 (response to a query about agricultural practices in North Carolina); *New England Farmer*, 4 (March 3, 1826), 253; L. C. Gray 1933, Vol. 2, 800.
93. *The Plough Boy*, 1 (February 19, 1820), 30.
94. See *American Farmer*, 1 (November 26, 1819), 279; *New England Farmer*, 4 (March 3, 1826), 253; J. C. Loudon 1839, Vol. 1, 427. A version of the clover header was evidently in use in Britain as early as 1807. (R. L. Ardrey 1894: 54.) Whether the early British design influenced subsequent American development of the clover header is not known. But certainly by the 1830's, the machines in use in the two countries were strikingly similar in appearance. Compare, for example, the clover headers depicted in the *New England Farmer*, 4 (March 3, 1826), 253, with those depicted in J. C. Loudon 1839, Vol. 1, 427.
95. For more detailed descriptions of this earlier form of hay rake, see *American Farmer*, 3 (June 22, 1821), 104; ibid., 3 (July 20, 1821), 135. Danhof characterizes this implement as "merely an enlarged version of the common hand rake, equipped with shafts and a handle thereby serving as a drag rake." (C. H. Danhof 1969: 219.) He notes "the drag rake was considered to have some advantage in gleaning grain and hay fields after normal hand raking," but adds "there was little interest in such operations." The observations of a farmer from this era help to explain that lack of interest. "In gleaning the stubble of grain too, which it is often used for; and in the damp of the morning (which is the best time to save the straggled grain, being less apt to shell out,) this kind of rake drags much dirt with the rakings, and the grain is so affected with it, as to be unfit for use until cleaned in some way better than a barn-fan will do it." (*American Farmer*, 3 [July 20, 1821], 135.)
96. As illustrated in Figure 7.23, these early versions of the improved horse rake also had a number of "small pins set in perpendicular on the top of the head, to prevent the hay or grain from falling over when full." (*American Farmer* 3 [July 20, 1821], 135.) These pins are often omitted in later illustrations. See, for example, J. J. Thomas 1855: 153; R. D. Hurt 1982: 85; P. Bidwell and J. I. Falconer 1925: 213.
97. P. Bidwell and J. I. Falconer 1925: 213. See also the various citations from newspapers of the early 1820s in subsequent footnotes.
98. Testimonials from this early period about the efficacy of this hay rake for gathering

hay on level ground may be found in *The Plough Boy*, 2 (February 24, 1821), 310; *American Farmer*, 2 (December 21, 1820), 312; ibid., 3 (July 20, 1821), 135; ibid., 3 (September 14, 1821), 199; ibid., 4 (May 10, 1822), 49; *New England Farmer*, 1 (July 5, 1823), 389; ibid., 2 (July 10, 1824), 394; ibid., 3 (June 10, 1825), 361; ibid., 3 (July 8, 1825), 398. In addition to gathering hay, some farmers used the horse rake to gather cut grain. While this evidently worked tolerably well for oats and barley, it did "not usually answer so well for cradled wheat and rye, as the motion of the rake is apt to stir up dust, that injures the grain for bread stuff." (*New England Farmer*, 2 [July 10, 1824], 394; see also *American Farmer*, 3 [September 14, 1821], 199.)

99. "When small obstructions occur, the handles are depressed, and the points of the teeth rise and pass freely. Over large obstructions, the rake must be lifted." (J. J. Thomas 1855: 153.) Aside from the difficulty of effecting the second, both corrections for obstructions interrupt the raking and presumably result in uncollected hay. Obstacles not spotted in advance could break the wooden teeth.

100. Clarence Danhof claims the implement was invented "shortly before 1820," and presents a variety of claims from a variety of sources that date the origins of the horse rake from 1811, or alternatively, "about 1814." (C. H. Danhof 1956: 170, n. 8, and Danhof 1969: 220.) Butterworth suggests the mechanism was patented in 1822. (B. Butterworth 1892: "Horse Rakes.") By the latter date they were available for $10 from a Baltimore manufacturer. (*American Farmer*, 4 [June 28, 1822], 112.) The simple horse rake cost about $2. (*New England Farmer* 1 [July 5, 1823], 389.) For descriptions and endorsements of the revolving horse rake from the 1820s, see *American Farmer*, 4 (June 28, 1822), 112; *Niles' Register*, 28 (July 9, 1825), 298; *New England Farmer*, 1 (July 12, 1823), 398; ibid., 3 (February 25, 1825), 245; ibid., 6 (July 4, 1828), 399; ibid., 7 (October 24, 1828), 106. By the late 1830s, the *New England Farmer* reported that this implement was in general use in most parts of Pennsylvania and New Jersey, and was gaining popularity in other parts of the country (16 [May 29, 1839], 375).

101. See C. H. Danhof 1969: 220.

102. See for example *American Farmer*, 3 (September 14, 1821), 199; ibid., 10 (June 27, 1828), 113; *New England Farmer*, 3 (February 25, 1825), 245.

103. *American Farmer*, 4 (November 1, 1822), 253.

104. See *American Farmer*, 2 (June 23, 1820), 104.

105. See *American Farmer*, 9 (November 16, 1827), 275.

106. *New England Farmer*, 8 (April 16, 1830), 306.

107. See, for example, *American Magazine and Monthly Chronicle*, 1 (July, 1758), 492; J. Eliot 1935: 40; S. Deane 1790: 233–234; J. C. Fitzpatrick 1931–1944, Vol. 55, 504; *Agricultural Museum*, 2 (January 1811), 195; E. M. Betts 1953: 236; J. Buel 1819: 55, 60–61; *American Farmer*, 1 (May 21, 1819), 58; *New England Farmer*, 3 (February 25, 1825), 245; P. L. Haworth 1925: 95, 110.

8. Threshing

1. John C. Fitzpatrick 1925, Vol. 29, 455 (letter from Washington to Samuel Chamberline dated April 3, 1788).

2. Examples can be found of colonists using unjointed poles or clubs to thresh instead of flails, but these appear to be very much the exception, not the rule. See, for example, B. Romans 1775: 121; L. C. Gray 1958, Vol. 1, 171; L. Rogin 1931: 157n.

3. According to one colonist, the best material for the thong was either squirrel skin "raw or one-half tanned" or the skin of an eel (C. R. Woodward 1941: 369).

4. I am indebted to Bruce Craven of Old Sturbridge Village for demonstrating the ease and effectiveness of this latter type of flailing motion.

5. See, for example, the clubs depicted in G. R. Quick and W. F. Buchele 1978: 11.
6. For a discussion of the Roman evidence, see E. M. Jope 1956a: 97; H. S. Harrison 1954: 70; J. M. G. van der Poel 1983: 34. For early depictions of the flail in European art (other than those appearing in Figure 8.1), see James Carson Webster 1938: Plates XVIII, XX, and LIX; Elizabeth Hallam 1986: 49; Franco Cardini 1989: 74, 148.
7. Egyptian depictions of donkeys used for this purpose date from about 2500 B.C. (F. E. Zeuner 1954b: 343.) The Old Testament includes a number of references to livestock treading and threshing floors, including Judges 6:39 and Deuteronomy 25:4.
8. See, for example, Jack P. Greene 1965, Vol. 2, 855 (Landon Carter's 1774 observations), and John Beale Bordley's 1786 observations on "Treading out Wheat" in Philadelphia Society for Promoting Agriculture, *Memoirs*, 6 (1939), 117. For an example of how oxen were used, see H. Bonebright-Closz 1921: 55.
9. For comments by contemporaries, see S. Deane 1790: 283; D. M. Matteson 1931: 13; J. B. Bordley 1939: 116; J. C. Fitzpatrick 1931–1944, Vol. 29, 455; *The Columbian Magazine*, 2 (September 1788), 511; *The Plough Boy*, 2 (June 3, 1820), 5; W. Strickland 1801: 47. For summary judgments about dominant threshing technique by geographic region, see U.S. Bureau of the Census, *Preliminary Report of the Census of 1860* (Washington, D.C.: U.S. Government Printing Office, 1862), 95; L. Rogin 1931: 157–160; P. Bidwell and J. I. Falconer 1941: 126. For an exception to the reputed rule that the flail dominated in New England, see "Notes on Compton, Rhode Island," Massachusetts Historical Society *Collections*, 1st series, Vol. 9 (1804), 200–201.
10. Problems associated with dung removal were seldom mentioned by agricultural writers. One exception was Nicholas Ridgely, who discussed both the problem and a common solution. "In the winter, the dung is hard and of little moment. In the summer the horses may be confined a day or two in the stable; and fed on dry food, which will harden their dung, and put them in excellent condition for treading; and they will perform the work with better spirits, than when taken with a full belly from grass. This should always be done when they are employed in any kind of labour, in the summer. My horses when used in ploughing or harrowing Indian corn, turning fallow, ploughing-in wheat, etc. are constantly kept on dry food, except, perhaps, once a week, they are turned to grass, for the sake of opening their bowels, and keeping them in good health." (N. Ridgely 1818: 40–41.) For comments on surfaces used to tread out grain in the South, see J. P. Greene 1965, Vol. 1, 366 (Landon Carter); *The Columbian Magazine*, 1 (October 1786), 80; J. B. Bordley 1939: 121; D. M. Matteson 1931: 13 (George Washington); E. M. Betts 1953: 77 (Thomas Jefferson).
11. J. B. Bordley 1799: 249. See also Washington's comments in D. M. Matteson 1931: 27, and *American Farmer*, 2 (June 30, 1820), 109. When done in the field (rather than an enclosed area), treading was preceded by "smoothing down" a circular dirt surface. "Then the grain was hauled to it on wooden sleds, arranged in a certain way upon this circle and then four, six or even eight horses were driven two abreast around in a circle, over the grain, till at least most of the grain was tramped out. Sometimes a man stood in the middle of the circle, and with lines and whip kept the horses in the right place and on the move. More frequently, however, a small boy was put on each saddle horse, and when a householder did not have enough boys of his own, he borrowed the required remainder from the progeny of his neighbors. . . . During dry weather this method of threshing went on fairly well, but it happened quite frequently that heavy showers of rain fell upon the half-threshed wheat, in which case serious losses were unavoidable." (W. G. Bek 1922: 298.) J. M. G. van der Poel pointed out that in Europe, to reduce bruising of the grain, horseshoes were sometimes removed. (Private correspondence, December 17, 1994.) No references were found to a comparable practice in America, although its merits seem obvious.

12. *The Columbian Magazine,* 3 (April 1789), 217.
13. See U. S. Bureau of the Census, *Preliminary Report of the Census for 1860* (Washington, D.C.: U.S. Government Printing Office, 1862), 91. An Ohio observer reported the same division in that state about 1820, large farms using livestock and small farms using the flail (L. Rogin 1931: 159).
14. The economic theorist will be puzzled by this explanation. Why should small farmers not pool their livestock, or alternatively, why should itinerant threshers with teams of horses not rent their services to small farmers, as evidently happened in Wisconsin in the 1840s (L. Rogin 1931: 181n)? The answer appears to be, in part, the presence of imperfect markets and the importance of other factors discussed below.
15. Norman E. Lee also emphasizes the importance of climate as a determinant of threshing technique chosen, but for somewhat different reasons. In Europe, treading was prevalent in warm climates, "but in the colder climates of central and northern Europe, where the weather was uncertain, threshing had to be done by hand in barns" (N. E. Lee 1960: 83). This explanation would seem suspect on two counts. The problem of uncertain weather could be solved by building sheltered threshing floors, as sometimes happened in America. More important, the climate of the Middle States (including Delaware) is not more "uncertain" than that of Northwest Europe, and yet treading was far from absent in this American region.
16. By one popular estimate, the last quart of grain took as much labor to thresh out as the first half-bushel. See L. Rogin 1931: 178n, 179n.
17. If the horses were not driven in ranks but, as noted previously, allowed to run "promiscuously and loose," even less slave labor was required.
18. This contrast assumes that the opportunity cost for these inputs already available on the plantation (horses, slaves, and overseer) were comparatively low, an assumption that does not seem unrealistic, given the alternative tasks available to factors immediately after the grain harvest.
19. Leo Rogin emphasizes the role of the "moth-fly" as an explanation for the "universality" of livestock treading in Maryland and Virginia, the faster method enabling wheat farmers to escape much of the prospective damage from this insect by more prompt marketing. (L. Rogin 1931: 159n.) The difficulty with this explanation is that "the efficacy of early threshing in preventing much damage by insects was accidentally discovered in 1772" (ibid.), but the Southern preference for livestock treading was well established long before that date. (See also ibid., 182.)
20. The colonist who drew this machine to the attention of the *Pennsylvania Magazine* readily conceded that "the first hint he had of it, was from a model showed by the ingenious and worthy Mr. Ferguson, in his lectures in London." (*Pennsylvania Magazine,* 1 [February 1775], 71.)
21. The illustration dates from an 1892 work by Benjamin Butterworth prepared under the auspices of the U.S. Patent Office. It seems to be a variant of a machine that Butterworth drew in an 1888 edition, which Quick and Buchele reproduce and attribute to "Wardrop of New Jersey" (G. R. Quick and W. F. Buchele 1978: 45). A patent was awarded in 1794 to James Wardrop of Virginia for a machine "made with flails or elastic rods twelve feet in length, of which twelve were attached in a series having each a spring requiring a power of twenty pounds to raise it three feet high at the point. A wallower shaft with catches or teeth, in its revolution successively lifted each flail in alternate movements, so that three of the flails were operated upon by the whole power, viz, twenty pounds. The whole weight to be overcome was one hundred and twenty pounds, and the machine was worked by two men. The flails beat upon a grating, to which the corn to be threshed was fed by hand." (U.S. Bureau of the Census, *Preliminary Report on the Eighth Census* 1862: 96.) The machine depicted in Figure 8.3 has eight flails and is beating grain on a barn floor (not a grate), although the revolving mechanism appears to be similar.

22. This difficulty was evident from the outset. The pioneering British design based upon the beating principle—the machine of Michael Menzies, with "flails fixed in a beam and worked by a water-wheel"—was a momentary success, but "the flails . . . soon broke, and the invention fell into disgrace." (G. E. Fussell 1952: 152–153.)
23. R. H. Gabriel 1926: 230. The British pioneer of this type of machine is generally acknowledged to be Andrew Meikle, although his initial thresher used longitudinal bars (not pegs) rotating inside a fixed curved surface (that had no pegs). Meikle's description is difficult to follow (see, for example, G. R. Quick and W. F. Buchele 1978: 45–46), but the basic principles of his machine have been expertly distilled and illustrated by P. W. Blandford 1974: 128, Figure 28A. See also M. Partridge 1973: 163.
24. *Genesee Farmer*, 1 (April 30, 1831), 132.
25. In the judgment of one historian, "in the late eighteenth and early nineteenth centuries, [the threshing machine] was easily the most complex and most expensive piece of agricultural equipment in existence." (Stuart Macdonald 1978: 168.)
26. *Genesee Farmer*, 3 (June 8, 1833), 175.
27. *Genesee Farmer*, 1 (April 30, 1831), 132.
28. *Genesee Farmer*, 3 (June 8, 1833), 178.
29. Ibid.
30. *Genesee Farmer*, 1 (April 30, 1831), 132. For evidence that reservations persisted into the next decade, see *The Farmers' Register*, 10 (July 1842), 338.
31. See James T. Lemon 1972: 178; Jared Eliot 1935: 251; *Virginia Gazette*, November 19, 1772. In 1774, the Virginia Society for the Advancement of Useful Knowledge gave "a pecuniary reward and medal" to "Mr. Holiday for his model of a very ingenious and useful machine for threshing out wheat." (*Virginia Historical Register and Literary Companion*, 6 [October, 1853], 218.) The names are different—"Holiday" getting the medal, and "Hobday" advertising in the *Virginia Gazette*—but other evidence suggests that the 1853 periodical got the spelling wrong, and Hobday received the medal. See John Pope 1792: 6 (who spelled the name Hobdy).
32. The American who reported the details of the British machine in Figure 8.3 to the *Pennsylvania Magazine* claimed to have "heard of machines for threshing of grain erected in America, but never saw or heard a description of one of them." (Ibid., 71.)
33. See John C. Fitzpatrick 1925, Vol. 29, 456; Jared Sparks 1837, Vol. 12, 293; D. M. Matteson 1931: 12; Donald Jackson 1976: xxxii–xxxiii; E. M. Betts 1953: 68, 72; Philadelphia Society for Promoting Agriculture, *Minutes*, May 6, 1788, 40; *The Columbian Magazine*, 2 (September, 1788), 512.
34. See, for example, *The Columbian Magazine*, 2 (September 1788), 576–578 (on the Winlaw machine); *Pennsylvania Packet, and Daily Advertiser*, October 9, 1788 (on a model brought to New York by Baron Polinitz); Philadelphia Society for Promoting Agriculture, *Minutes*, August 17, 1790, 61 (on a model by Samuel Mulliken); ibid., November 8, 1791, 71; S. W. Fletcher 1959: 26–27; John Beale Bordley 1799: 586 (all on the machines of Colonel Alexander Anderson); Samuel Deane 1790: 283 (describing "an engine . . . once made by a gentleman in the State of New Hampshire"); *Boston Weekly Magazine*, 36 (July 2, 1803), 147 (on a machine for threshing clover seed by Timothy Kirk); A. F. M. Willich 1803: 114 (an editorial aside by James Mease on "a few contrivances for threshing" designed in America); *Literary Magazine and American Register*, 1 (January 1804), 318 (on Jedediah Turner's patent); ibid., 3 (March 1805), 236 (on a machine of David Prentice); ibid., 5 (February, 1806), 109 (report on a British machine); Philadelphia Society for Promoting Agriculture, *Minutes*, May 12, 1807, 101 (on a model exhibited by Napp of New York); *American Farmer*, 9 (August 10, 1827), 162 (reproducing an 1807 document that includes mention of a new threshing machine).

35. Editorial aside by James Mease in A. F. M. Willich 1803: 114.
36. D. M. Matteson 1931: 12; P. L. Haworth 1925: 126–127; U.S. Bureau of the Census, *Preliminary Report on the Eighth Census* 1862: 96; James Mease 1818: xxxiv; *Literary Magazine and American Register*, 1 (January, 1804), 318.
37. W. C. Ford 1891, Vol. 12, 341. Washington did add that once he could personally supervise the activity, he would consider experimenting again. (Ibid.) But one year later his correspondence evidences a concern about how to build a better threshing floor, with no mention of mechanical alternatives. (J. C. Fitzpatrick 1940, Vol. 33, 296.)
38. Franklin Knight 1847: 33–34.
39. J. B. Bordley 1939: 116–121; and by the same author, 1799: 245–257. Bordley was well aware of both British and American machines. See for example 1799: 586–591, and his 1803: 318–319 (in which he surveys British machines but offers no personal endorsement of any of them). One curiosity in Bordley's footnote is his assertion that "the thrashing-mill certainly gives this [speedy threshing] method; and in every respect is a superior instrument for getting out wheat from straw." The puzzle is why, if he truly believed in this superiority, he did not advocate acquiring machines instead of improving treading procedures, and why, in his review of British machines, he did not single out those he deemed preferable for his American readers. One possible answer is that like most agricultural writers of his day, Bordley wanted to make clear he was abreast of technological developments, but could not bring himself to recommend machines whose superior efficiency was still in doubt. Consider, for example, the uncertainty apparent in his discussion of Anderson's machine. "In 1782 Colonel Anderson then of Philadelphia, now residing on the Susquehanna, near Lancaster, invented a mill moved by horses, for thrashing wheat and other small grain out from its straw. . . . In 1791 he built one of full size; which I saw work to advantage, though as Colonel Anderson well observed, it was capable of considerable improvement. But having since invented a thrashing mill, on different principles, a model of which I saw work admirably well, he probably has not further attended to the first; and I wait to hear of his ordinary business admitting him to build one of full size, on his new invention of *rubbing*, instead of *striking* out the grain." (Bordley 1799: 586.)
40. Massachusetts Society for Promoting Agriculture, *Repository and Journal*, 1 (1801), 3.
41. Ibid., 3 (1815), 55–67.
42. Ibid., 1 (1801), 6–7.
43. S. W. Fletcher 1959: 21–23. The list remained substantially the same until 1793. (Ibid.)
44. Philadelphia Society for Promoting Agriculture, *Memoirs*, 1 (1815), xxxii–xlii.
45. U.S. Bureau of the Census, *Preliminary Report on the Eighth Census* 1862: 97.
46. U.S. Bureau of the Census, *Preliminary Report on the Eighth Census* 1862: 97.
47. Massachusetts Agricultural Society, *Repository and Journal*, 5 (1819), 16.
48. Ibid., 259–260.
49. Jesse Buel 1819: 61. (Italics in the original.)
50. William Tilghman, in an address to the Society on January 18, 1820; reported in *American Farmer*, 2 (July 28, 1820), 138.
51. *Genesee Farmer*, 1 (April 30, 1831), 132; ibid., 3 (June 8, 1833), 175, 178; U.S. Bureau of the Census, *Preliminary Report on the Eighth Census* 1862: 97. (In 1830, thirty-four threshing machines were patented; in 1831, a record thirty-eight.)
52. W. H. Brewer 1880: 520.
53. L. Rogin 1931: 164; see also J. T. Schlebecker 1975: 118, who endorses Rogin's modification of Brewer's original assertion.
54. *American Farmer*, 11 (May 8, 1829), 58, cited in L. Rogin 1931: 164, n. 62.
55. Most price data must be approached with considerable caution. Prices could vary depending upon whether the threshing machine was new or used, stationary or por-

table, or the quote included or excluded "horse power." Often reported prices do not specify such details.

56. *The Plough Boy,* 1 (September 18, 1819), 122–123.

57. For price quotations from the 1820s, see *American Farmer,* 2 (August 18, 1820), 168; ibid., 3 (June 29, 1821), 106; ibid., 3 (December 7, 1821), 296; ibid., 4 (August 2, 1822), 152; ibid., 4 (November 22, 1822), 280; ibid., 5 (November 21, 1823), 275; ibid., 6 (January 23, 1824), 352; ibid., 6 (September 3, 1824), 192; ibid., 6 (December 31, 1824), 323; ibid., 7 (May 13, 1825), 60; ibid., 10 (June 20, 1828), 107; ibid., 11 (December 4, 1829), 299; ibid., 11 (March 5, 1830), 408; ibid., 12 (December 3, 1830), 297; *New England Farmer,* 1 (June 14, 1823), 363; ibid., 2 (January 17, 1824), 197; ibid., 5 (October 6, 1826), 84; *New York Farmer and Horticultural Repository,* 2 (July 1829), 163.

58. Recall the complaints noted earlier, dating from the early 1830s. For descriptions of, and commentary on, some of the threshing machines developed in the 1820s, see *American Farmer,* 2 (July 7, 1820), 120; ibid., 5 (November 21, 1823), 275; ibid., 5 (January 2, 1824) 327; ibid., 6 (September 3, 1824), 192; ibid., 6 (October 22, 1824), 248; ibid., 6 (December 3, 1824), 294; ibid., 6 (December 31, 1824), 323; ibid., 6 (January 28, 1825), 354; ibid., 7 (May 13, 1825), 60; ibid., 8 (May 12, 1826), 62–63; ibid., 9 (January 18, 1828), 345; ibid., 10 (May 9, 1828), 57; ibid., 10 (June 20, 1828), 107; ibid., 10 (October 31, 1828), 258; ibid., 11 (May 8, 1829), 58; *New England Farmer,* 1 (October 26, 1822), 99; ibid., 1 (June 14, 1823), 363; ibid., 2 (October 25, 1823), 98; ibid., 5 (October 6, 1826), 84; ibid., 6 (October 27, 1827), 97.

59. *American Farmer,* 5 (January 23, 1824), 352. Working this machine by hand, however, was extremely hard work. See S. L. Boardman 1892: 15–16.

60. Occasionally water power was used, but most threshing machines not using hand power seem to have been driven by horse power in the 1820s. While both sweep and treadmill devices were available during that decade, the latter were significantly improved by a treadmill device patented in 1830 by H. A. Pitts. (See R. L. Ardrey 1894: 105, 112.)

61. See R. L. Ardrey 1894: 105–112; C. H. Danhof 1969: 222–228; J. T. Schlebecker 1975: 118–120.

62. For comments on, and descriptions of, rollers and sledges used in ancient times for threshing purposes, see K. D. White 1984: 30, 62; G. R. Quick and W. F. Buchele 1978: 39; E. C. Curwen and G. Hatt 1953: 120; R. L. Ardrey 1894: 103–107; Sian E. Rees 1979: 485–486.

63. B. Romans 1775: 121.

64. According to Nicolas Ridgely, this type of surface was first designed by Benjamin Sylvester of Maryland. (Nicholas Ridgely 1818: 43.) It was subsequently introduced into Kent County, Delaware, by John Clayton. (U.S. Bureau of the Census, *Preliminary Report on the Eighth Census* 1862: 95.)

65. See R. S. Elliot 1883: 46; John Nicholson 1814: 263–264; W. H. Brewer 1880: 520; G. R. Quick and W. F. Buchele 1978: 42–43. For a description and picture of another roller from this period designed for threshing but little used, see C. Churchman and G. Martin 1814: 400–402.

9. Winnowing and Straw Cutting

1. S. Deane 1790: 326.

2. In premodern America, these sieves tended to be made of splint. For pictures and descriptions, see Mary Earle Gould 1962: 15, 212, 214, 224–227; J. Ritchie Garrison 1991: 48. For a description of how this secondary separation was achieved prior to

the use of sieves, see G. R. Quick and W. F. Buchele 1978: 50. A variant of the ancient Egyptian method evidently also persisted in early America: "on a windy day the big barn doors were opened and threshed grain was tossed in the air by clean wooden scoop shovels, and the lighter chaff blown away." (Charles S. Phelps 1917: 46–47.)

3. Dublin Society 1740: 163.
4. S. Deane 1790: 295.
5. One British description of winnowing with the aid of a crude fan was as follows. "The earliest methods of performing this operation was by merely allowing it to fall from a shovel through a current of air; the wind carried off much of the lighter material, and the [grain] fell down much cleaner. This had to be repeated several times; a large wicker fan was at first used to create a draught, and afterwards a sort of reel upon which was fastened, by one edge, pieces of canvas;–these, as they revolve, cause considerable draught, the [grain] during this time being sifted through wide-meshed sieves called riddles, and during its passage to the ground the lighter particles are blown away; this operation takes place between the two doors upon the thrashing floor of the barn." (G. H. Andrews 1853: 126.) For illustrations of these primitive (unenclosed) fans, see E. J. T. Collins 1969c: 38, and M. and C. H. B. Quennell 1961: 131.
6. Dublin Society 1740: 164.
7. B. Butterworth 1892: "Winnowing and Sifting Grain."
8. Thus, for example, an American inventor in 1845 advertised an "improved" winnowing machine "which consists, in addition to the screen and other parts in general use in fan-mills, an additional screen, and what is denominated a chess board, which are arranged in such a manner as to cause a much stronger blast of wind to act upon the grain at the lower part of the shaking sieves or screens than at the upper, and thereby to aid the action of said sieves in effecting the screening, by which means the chaffing and screening are performed simultaneously." (*American Agriculturist*, 4 [November 1845], 344.)
9. See for example Gösta Berg 1976: 25–46; S. Deane 1822: 515; E. M. Jope 1956a: 98; G. R. Quick and W. F. Buchele 1978: 48.
10. R. Bradley, 1725: 343.
11. Dublin Society 1740: 165. See also Gösta Berg 1976: 34. For a good illustration of an early British winnowing fan "with its sides laid open," see A. M. Bailey 1782, Vol. 1, Plate XXI. The latter model included an "upper riddle-box," a "fixed riddle," and an "under riddle." Another illustration (not laid open) of a winnowing fan that included a system for separating "sticks or stones . . . into different drawers" before "the clean wheat comes out at the screen in front" may be found in *The Farmer's Magazine and Useful Family Companion*, 2 (1777), plate opposite 311.
12. Examples can be found of a "Fan" or "Dutch Fan" in farm inventories in 1766 and 1773 (Gilbert Cope 1881: 219), and "Dutch Fans" were advertised in 1774 as "made and sold by Adam Ekart, in Market Street, Philadelphia" (*Virginia Gazette*, May 26, 1774, 1). In his editorial aside on fans in 1803, Mease suggests that they were first introduced into Philadelphia "forty or fifty years since" (that is, between 1753 and 1763) "by Adam Echard . . . from a Holland model." (A. F. M. Willich 1803: 421.)
13. After noting the availability of Dutch fans for "Cleaning wheat or any other kind of grain," the 1774 advertisement continued: "Likewise rolling screens, sieves for sifting iron ore, etc., warranted of the best make; also all sorts of wire work, for cleaning wheat, barley, rye, flax seed, Indian corn, oats, or any other kind of grain, and wire short-cloths for millers."
14. For the American description, see S. Deane 1790: 295–296; for the 1738 description, see Dublin Society 1740: 164–165.
15. The 1774 model bears a striking resemblance to pictures of British fans published

at about this time. (See, for example, *The Farmer's Magazine and Useful Family Companion*, 2 [1777], plate opposite 311.) By the early 1830s, the position of the fan in some machines had shifted from a horizontal axis to a vertical axis, as depicted in the bottom diagram of Figure 9.4.

16. See Gösta Berg 1976: 34–36.
17. S. Deane 1790: 87.
18. William C. Howells 1963: 63. Manning the sheet had a reputation for being one of the more demanding agrarian tasks. According to John Reynolds, "About the hardest work I ever performed was winnowing the wheat with a sheet." (John Reynolds 1879: 90.) See also David E. Schob, 1975: 105.
19. For references to winnowing fans in use in America during the period 1783–1810, see J. F. D. Smyth 1784, Vol. 2, 119; J. C. Fitzpatrick 1924, Vol. 4, 73; *New York Magazine*, 3 (June 1792), 357; *Massachusetts Magazine*, 5 (April 1793), 254; Mease's editorial aside in A. F. M. Willich 1803: 421.
20. See Gilbert Cope 1881: 219; L. Rogin 1931: 154, 156, 176; U.S. Department of Labor 1898, Vol. 2, 470–471; Ohio Department of Agriculture 1904: 10; John B. Conner 1893: 8.
21. As described by one observer, the product of defective fans was well calculated to prompt consideration of alternatives: "The end-product of this process of fanning often represented a gruesome mixture of good and bad wheat, chaff, remnants of straw, and the seed of weeds." (W. G. Bek 1922: 298.)
22. See for example *American Farmer*, 2 (May 19, 1820), 64; ibid., 2 (June 23, 1820), 104; ibid., 4 (July 5, 1822), 120; ibid., 6 (June 18, 1824), 104; ibid., 7 (April 1, 1825), 16; ibid., 8 (May 5, 1826), 55; ibid., 8 (June 23, 1826), 112; ibid., 8 (October 27, 1826), 256; ibid., 9 (August 3, 1827), 160; ibid., 11 (June 5, 1829), 96; ibid., 11 (September 11, 1829), 208; ibid., 11 (October 9, 1829), 240; *New England Farmer*, 5 (January 12, 1827), 194.
23. *American Farmer*, 3 (June 22, 1821), 99.
24. *American Farmer*, 5 (February 13, 1824), 371.
25. *New England Farmer*, 3 (December 3, 1824), 146.
26. *American Farmer*, 6 (December 31, 1824), 323.
27. *American Farmer*, 8 (June 9, 1826), 92.
28. U.S. Bureau of the Census, *Historical Statistics* 1975, Part 1, 201, series E52.
29. Price quotations can be found in *American Farmer*, 1 (June 25, 1819), 103–104; ibid., 2 (June 23, 1820), 104; ibid., 4 (July 5, 1822), 120; ibid., 4 (May 24, 1822), 72; ibid., 6 (October 8, 1824), 232; ibid., 6 (December 31, 1824), 323; ibid., 8 (May 5, 1826), 55; ibid., 8 (June 23, 1826), 112; ibid., 9 (August 3, 1827), 160; *The Plough Boy*, 1 (September 18, 1819), 123; *New England Farmer*, 3 (December 3, 1824), 146; ibid., 5 (January 12, 1827), 194.
30. This assertion is contradicted by Rogin's suggestion "that the fanning mills in use in Virginia in the early forties were cheaper but no better than the earlier ones; indeed, some who had new ones came to borrow one purchased thirty years earlier, for it cleaned faster than the new ones." (L. Rogin 1931: 157, n. 17; see also ibid., 178.) But Rogin's suggestion of a static (or deteriorating) technology is based upon a single observation in 1842 by one individual about the superiority of his old fan when compared with his neighbors' newer fans. (Virginia Board of Agriculture, *Report*, Richmond, 1842, 8; in *Journal* of the House of Delegates of Virginia, 1842–1843, Doc. 12.) Clarence Danhof would seem closer to the truth when he argues (without supporting footnotes) that prior to 1820 winnowing fans were "little used", and that as the machines improved, their use increased. Without supporting evidence, Danhof asserts that the main period of improvement and diffusion was the 1830s and 1840s. (C. H. Danhof 1969: 222.) The evidence presented above suggests that the initial great surge in improvement and use occurred during the first half of the 1820s.

31. *American Farmer*, 8 (May 5, 1826), 55.
32. *American Farmer*, 1 (October 29, 1819), 241.
33. Thus, for example, a New England farmer claimed that when he switched to chopped feed, his cows gave more milk, adding the vague observation "and likewise something for the improvement of the condition of my whole stock." (*New England Farmer*, 12 [February 5, 1834], 133.)
34. *American Farmer*, 1 (June 18, 1819), 90; ibid., 11 (August 14, 1829), 169. A typical explanation was that chopped food exposed "more points for the extraction of nutriment; to the maceration of the liquids in, and the action of, the stomach, or stomachs, of animals. And no provender is wasted, as it is by feeding entire; either by negligence in servants, or uselessly passing through the viscera." (Richard Peters 1817: 103, 105.) See also J. D. Schoepf 1911, Vol. 2, 89; *Massachusetts Agricultural Repository and Journal*, Vol. 2, 12 (Letters and Extracts Communicated to the Society [Boston: Russell and Cutler, 1809], a reprint from the *Kennebec Gazette* of June 1807); ibid., 4 (1816), 138–140 (reproduction from the Bath Society's Papers of "On The Use of Chaff, Compared With Hay, For Horses"); *American Farmer*, 1 (October 29, 1819), 241; *New England Farmer*, 2 (February 28, 1824), 246; letter from William Phillips in New York Board of Agriculture, *Memoirs*, 3 (1826), 374.
35. As of the 1760s, this cutting device was something of a novelty in Britain. (See *Museum Rusticum*, 1 [1764], 259; ibid., 5 [1765], 208–213; ibid., 6 [1766], 8–16.) These boxes soon included a treadle device to keep the straw pressed down and in place as the cutting stroke was administered. For a picture of a primitive cutting device, see H. J. Hopfen 1960: 124. For pictures and descriptions of the cutting box plus treadle, see *Museum Rusticum*, 6 (1766), 8–11; John Vince 1983: 66–67; and P. A. Wright 1961: 51–52. In a discussion about cutting boxes in 1777, James Sharp claimed that James Edgill of London designed "the first machine for cutting chaff by a single spiral knife" for which he received a prize of twenty guineas in 1768 from the Society of Arts in London, in whose opinion "the application of its spiral cutter seems to be new." One of Sharp's workmen added a second knife "whereby [the straw cutter] cuts twice in each round." (*The Farmer's Magazine and Useful Family Companion*, 2 [1777], 311.) J. M. G. van der Poel has pointed out that European illustrations of simple cutting boxes can be found as early as the sixteenth century, suggesting that the basic design originated on the Continent and then spread to Britain some time in the eighteenth century, where the design was significantly improved by adding spiral cutters, as noted. (Private communication, December 17, 1994.) Early European illustrations may be found in W. Hansen 1982: 244.
36. This 1803 design (appearing in a source published in America) is not significantly different from the design illustrated twenty-six years earlier in *The Farmer's Magazine and Useful Family Companion*, 2 (1777), plate opposite 391. Curiously, while the basic design appears to have originated on the Continent and migrated to Britain (see previous footnote), this improved British design migrated back, but by accident. From the cargo of an English ship, wrecked on the coast of Friesland in the first decade of the nineteenth century, a British straw cutter of the sort illustrated in Figure 9.6 was salvaged. Sold at public auction, it was publicized in *Magazijn van Vaderlandschen Landbouw* (2 [1805], Figure III), the first agricultural periodical of the Netherlands, complete with illustrations of the design, and was subsequently copied by Friesland artisans. (J. M. G. van der Poel, private communication, December 17, 1994.)
37. The British tended to justify their terminology by focusing on the output, not the input, as, for example, "Chaff-Cutter: in Rural Economy, an implement constructed for the purpose of cutting straw, and other substances into chaff." (Abraham Rees 1810–1824: "Chaff-Cutter.")
38. For a review of some of the major British developments, see Henry Trueman Wood

1913: 133–136; O. Beaumont and J. W. Y. Higgs 1958: 10–12. Illustrations of some of the early British machines may be found in A. M. Bailey 1782, Vol. I, Plate XV, and G. E. Fussell 1952: illustrations 90, 92, 93, 94. For a description of how some of the better known of these machines operated, see Abraham Rees 1810–1824: "Chaff-Cutter."

39. John C. Fitzpatrick 1931–1944, Vol. III, 3; John Taylor 1811: 53.

40. Hotchkiss's response to the prospect of a premium is described in an 1807 letter from Thomas Gold to G. W. Jeffreys of North Carolina, reproduced in *American Farmer*, 9 (August 10, 1827), 161–162. The feeding apparatus, apparently consisting of a revolving band, had a number of defects. "First, the band would always stretch or shrink according to the weather—again, the straw was constantly pressing against the knife, even when it was cutting—thereby occasioned great friction and labour, and caused the knife to spring from the frame, against which it cut, which prevented it after a short time from cutting clean. Neither did it cut any regular precise length." (*American Farmer*, 2 [June 9, 1820], 87.)

41. S. Deane 1790 and 1797.

42. Article from the *Kennebec Gazette* of January 1807, reproduced in *Massachusetts Agricultural Repository and Journal*, 2 (1809), 12.

43. In a letter to Dr. James Mease, dated January 30, 1809. (John Taylor 1811: 53.)

44. Letter also addressed to Dr. Mease, dated March 15, 1809. (Richard Peters 1811b: 64.)

45. For discussions of Hotchkiss's various straw cutters, see Thomas Gold 1817: 352; *Massachusetts Agricultural Repository and Journal*, 4 (1816/17), 400–401; *American Farmer*, 2 (June 9, 1820), 87.

46. In a letter dated April 8, 1817, Richard Peters noted the defects of the flywheel, as well as the tendency to the machine to produce "slow and irregular protrusions of the hay, or straw, for the operation of the knife." (Richard Peters 1818c: 104.) But the cutting defects were more fundamental than this. Reviewing this mechanism in 1820 the editor of the *American Farmer* pointed out that "the cast steel knife cut against cast iron, but the inequality in the texture of the two metals soon put the knife out of order." (*American Farmer*, 2 [June 9, 1820], 87.)

47. For a detailed description of the various parts of this straw cutter, see Eastman's explanation in *American Farmer*, 3 (January 4, 1822), 326–327.

48. Board of Agriculture of the State of New York, *Memoirs*, 3 (1826), 372–373.

49. Ibid.

50. The British influence on some American designs was a commonplace at the time. For an example of prizes withheld from American inventors of straw cutters because their machines were judged "similar to the English," and thus not innovative enough to warrant a prize, see *Massachusetts Agricultural Repository and Journal*, 5 (1818), 17. In 1826, according to one of the larger dealers in agricultural implements, "the chaff cutters used in our country . . . are copied, with slight alterations, from the English ones." (*American Farmer*, 8 [May 5, 1826], 55.)

51. See *American Farmer*, 4 (January 29, 1823), 350; *New England Farmer*, 4 (February 3, 1826), 217.

52. See *American Farmer*, 1 (January 7, 1820), 328; *Massachusetts Agricultural Repository and Journal*, 6 (1820), 234; *American Farmer*, 3 (December 14, 1821), 304; ibid., 4 (July 5, 1822), 115–116; ibid., 4 (July 25, 1822), 144; ibid., 4 (January 24, 1823), 350; *New England Farmer*, 3 (January 14, 1825), 194; ibid., 3 (February 25, 1825), 245; ibid., 4 (February 3, 1826), 217; *American Farmer*, 8 (June 9, 1826), 93.

53. See, for example, B. R. Carroll 1836, Vol. 2, 482; L. C. Gray 1933: Vol. 1, 201–202; Vol. 2, 835, 844; P. Bidwell and J. I. Falconer 1925: chapter XVII. Also increasing demand for straw cutters was the rise in the practice of penning livestock to secure manure. See, for example, *Southern Planter*, 2 (February 1842), 32–33.

54. See, for example, *New England Farmer*, 12 (February 5, 1834), 133; ibid., 14 (September 23, 1835), 87; ibid., 16 (January 3, 1838), 207; ibid., 16, Supplement, 2–3.
55. Virginia Board of Agriculture, *Report*, Richmond, 1842 (in the *Journal* of the House of Delegates of Virginia, 1842–1843, Document 12, 8).
56. Thomas Gold 1816/17: 352.

10. *Novus Ordo Seclorum?*

1. Quoted in F. A. Updike 1915: 369.
2. The news arrived with Henry Carroll, Secretary of the American Legation at Ghent, aboard a British sloop of war, which reached New York harbor on Saturday, February 11, 1815. Accounts of subsequent celebrations may be found in the *Freeman's Journal* (Philadelphia), Monday, February 13, 1815, special supplement; *New-York Herald*, Wednesday, February 15, 1815; ibid., Saturday, February 18, 1815; *New-York Spectator*, Wednesday, February 15, 1815; ibid., Saturday, February 18, 1815.
3. Edward Channing 1917, Vol. 4, 56.
4. Edward Channing 1917, Vol. 4, 564.
5. Arthur Ekirch, for example, argues that with the cessation of hostilities with Britain in 1814, the American faith in the progress of their nation "became a dogma of widespread mass appeal" (Ekirch 1969: 32). In support of this contention he offers a minimum of evidence: a few optimistic quotations from Thomas Jefferson in 1816 and 1818, and the summary judgment of Henry Adams (made at the end of the nineteenth century) that "In 1815 for the first time Americans ceased to doubt the path they were to follow. Not only was the unity of their nation established, but its probable divergence from older societies also well defined." (Ekirch 1969: 32–33; Henry Adams 1889–1891, Vol. 9, 220.)
6. James Mease 1818: vii.
7. Quoted in H. S. Russell 1976: 274.
8. *The Agricultural Museum*, "designed to be a repository of valuable information to the farmer and manufacturer," was published in Georgetown, D.C., between July 1810, and June 1812. For a discussion of this periodical's brief history, see C. R. Barnett 1928.
9. See C. S. Brigham 1925: 199.
10. *American Farmer*, initially published in Baltimore, lasted until 1897. *The Plough Boy*, published in Albany, was less successful, lasting only until 1821. The goal of all of these newspapers was well summarized by the editor of the *American Farmer*. "The great aim, and the chief pride, of the *American Farmer* will be, to collect information from every source, on every branch of Husbandry, thus to enable the reader to study the various systems which experience has proved to be the best, under given circumstances; and in short, to put him in possession of that knowledge and skill in the exercise of his means, without which the best farm and the most ample materials, will remain but as so much *dead capital* in the hands of their proprietor." (*American Farmer*, 1 [April 2, 1819], 6.) For a discussion of the agricultural press in America at this time, see A. L. Demaree 1941; P. W. Gates 1960, chapter XVI; H. T. Pinkett 1950.
11. See, for example, the premiums initially offered by the Philadelphia Society for Promoting Agriculture (*Boston Magazine*, 3 [March 1786], 123–126; *Universal Asylum*, 4 [March 1790], 163–166), by the New-York Agricultural Society (*New-York Magazine*, 6 [June 1795], 371–373), and by the Massachusetts Society for Promoting Agriculture (*Massachusetts Magazine*, 5 [June 1793], 358–360; *Massachusetts Agricultural Repository and Journal*, 1 [1801], 6–8). Only the Philadelphia Society list mentioned implements, offering one premium for sowing wheat "with a machine." This same

society featured on its original emblem an illustration of oxen pulling a plow, not because of a special concern with plow improvements, but to encourage the use of oxen as draught animals.

12. See, for example, surveys by the Philadelphia Society for Promoting Agriculture (*The Columbian Magazine*, 3 [February 1789], 87–90), by the New-York Society for the Promotion of Agriculture, Arts, and Manufactures, in their *Transactions*, 1 (1792), Part 1, viii–xiii, and 1 (1794), Part 2, xxix–xxxii, by the Massachusetts Agricultural Society (*Massachusetts Agricultural Repository and Journal*, 1 [1801], 3–29), by the Society for Promoting Agriculture in the State of Connecticut in their *Transactions* (New Haven: William W. Morse, 1802), 4–6, and by the Richmond Agricultural Society (*The Agricultural Museum*, 1 [May 8, 1811], 329–335). Only the last of these asked about new farm implements, with two questions in a total of eighty-four queries asking (about plows [#78] and threshing machines [#79])what kinds were in use and whether "there are lately introduced that have peculiar advantages, and what are they?"
13. S. Deane 1790: 85–86; 1797: 99–100. The two editions differ by a single sentence appended at the end of the 1797 analysis: "To prevent these evils [useful discoveries dying with the discoverer], the forming of societies in various parts of the country *might* be of great use." (Italics added; Deane evidently was not certain of the usefulness of societies for this purpose.)
14. Jefferson 1799: 313.
15. See, for example, *Universal Asylum*, 6 (January 1791), 27–31; New-York Society for the Promotion of Agriculture, Arts, and Manufactures, *Transactions*, 1 (1792), Part 1, 1–24; ibid., Vol. 1, Part 4, 1799, 1–20; *Massachusetts Agricultural Repository and Journal*, 2 (1809), 9–14 (a review of Maine agriculture by the Kennebec Agricultural Society, reproduced from the *Kennebec Gazette* of June 19 and 26, 1807); *The Agricultural Museum*, 1 (December 7, 1810), 177–178.
16. The articles originally appeared in a recently established newspaper (*The Spirit of Seventy-Six*) three or four years before reappearing in *Arator*. (*The Farmers' Register*, 8 [December 31, 1840], 704; see also John Taylor 1817: 217.) For commentary on their impact, see *The Farmers' Register* source noted. Edmund Ruffin, the editor of *The Farmers' Register*, judged *Arator* to be "the first original agricultural work (worthy to be so called) which had ever been published in Virginia, or in the southern states." (Ibid.) See also *American Farmer*, 10 (November 14, 1828), 273.
17. Livingston's article originally appeared in David Brewster, ed., *New Edinburgh Encyclopaedia* (New York: Samuel Whiting and John L. Tiffany, 1813–1831), 332–341 (the first American edition); and was reprinted in the 1832 edition (with identical pagination). For a discussion of the dating of this article to "about 1813," see L. Rogin 1931: 5, n. 8, and P. Bidwell and J. I. Falconer 1925: 465.
18. *Massachusetts Agricultural Repository and Journal*, 4 (1817), 303. Symptomatic of how broadly this society was prepared to cast its net in search of novel implements was the prize of $20 for "the person who shall produce at the Show [the Brighton Cattle Show and Exhibition] any other Agricultural Implement of his own invention, which shall, in the opinion of the Trustees, deserve a reward." (Ibid., 5 [1818], 100.)
19. *Massachusetts Agricultural Repository and Journal*, 3 (1815), 338–349.
20. *The Plough Boy*, 1 (September 18, 1819), 122–123. Richard N. Harrison, of 211 Front-Street, New-York, announced he had "established a Factory," and could "furnish Agricultural Machines of every kind" or "make them to order."
21. Jesse Buel 1819: 56–62.
22. Philadelphia Society for Promoting Agriculture, *Memoirs*, 4 (1818), v–xxviii. Mease was the second vice-president of this society. His announced topic was "On the Progress of Agriculture, with Hints for Its Improvement in the United States."
23. Webster's address was reproduced in four successive editions of *The Rural Magazine*

and Farmer's Monthly Magazine, 1 (March 1819), 50–52; (April 1819), 83–86; (May 1819), 114–115; (June 1819), 152–155. According to the editors, Webster (the vice-president of the society) furnished in this address "an outline for that which is so much needed in New-England: *A system of Agriculture.*" ([March 1819)], 49, italics in original.)

24. *American Farmer,* 1 (August 20, 1819), 161–163; (August 27, 1819), 169–171; (September 3, 1819), 177–179. Madison was president of the society.

25. See, for example, *The Plough Boy,* 2 (October 7, 1820), 146; ibid., 2 (March 17, 1821), 331; ibid., 3 (May 18, 1822), 404; *American Farmer,* 2 (January 12, 1821), 329; ibid., 2 (July 28, 1820), 138; ibid., 2 (September 15, 1820), 194; ibid., 3 (June 8, 1821), 85; ibid., 3 (August 24, 1821), 170; ibid., 4 (May 3, 1822), 42; ibid., 5 (December 19, 1823), 307; ibid., 12 (December 31, 1830), 330; *New England Farmer,* 1 (November 30, 1822), 137; ibid., 1 (January 18, 1823), 196; ibid., 2 (January 17, 1824), 194; ibid., 3 (March 25, 1825), 276; ibid., 7 (November 28, 1828), 145.

26. In the 1830s and beyond, similar publications "continued to spring up like mushrooms," until by 1860 they numbered close to 100 with an estimated circulation in excess of 250,000. (L. C. Gray 1958, Vol. 2, 788.)

27. See, for example, *Massachusetts Agricultural Repository and Journal,* 4 (1816/17), 284; ibid., 5 (1819), i, 325; ibid., 6 (1821), 162; *The Plough Boy,* 2 (October 14, 1820), 157; ibid., 2 (February 3, 1821), 281; ibid., 4 (August 13, 1822), 86; *New England Farmer,* 1 (January 4, 1823), 180; ibid., 1 (January 18, 1823), 194; ibid., 5 (February 16, 1827), 233.

28. See, for example, *The Plough Boy,* 1 (June 12, 1819), 13; ibid., 2 (March 10, 1821), 324; ibid., 4 (December 17, 1822), 222; *American Farmer,* 3 (July 6, 1821), 114; ibid., 3 (November 16, 1821), 288; ibid., 4 (May 10, 1822), 49; ibid., 4 (January 24, 1823), 347; *New England Farmer,* 1 (August 10, 1822), 10; ibid., 1 (January 18, 1823), 194, 197; ibid., 1 (February 22, 1823), 236–237; ibid., 1 (May 31, 1823), 345; ibid., 4 (March 17, 1826), 267.

29. See, for example, *American Farmer,* 1 (July 2, 1819), 105–106; ibid., 4 (March 29, 1822), 12; ibid., 4 (May 31, 1822), 77; ibid., 7 (October 14, 1825), 233–234; ibid., 10 (February 13, 1829), 378–379; ibid., 10 (March 6, 1829), 403; *New England Farmer,* 1 (December 14, 1822), 157–158; ibid., 3 (October 9, 1824), 84–85; ibid., 3 (July 22, 1825), 412; *The Plough Boy,* 4 (December 17, 1822), 222–223.

30. In the calculation of relevant costs, the concept of depreciation is noticeably absent in the 1820s in the estimation of expenses associated with new farm implements, but on occasion it does appear in estimates associated with livestock. See, for example, *American Farmer,* 1 (December 10, 1819), 291; ibid., 4 (May 10, 1822), 53; ibid., 9 (July 6, 1827), 122; ibid., 11 (March 20, 1829), 3.

31. *American Farmer,* 3 (July 6, 1821), 113; italics in original.

32. Jesse Buel, in an address to the Albany County Agricultural Society, yreported in the *American Farmer,* 3 (November 30, 1821), 285. The books were "the *Practical Farmer* by Gen. Armstrong, Taylor's *Arrator* [sic], [and] Coxe on Fruit Trees," the newspapers were the only two then available (*The Plough Boy* and the *American Farmer*), plus the *Memoirs* of the Philadelphia Society for the Promotion of Agriculture and the papers of the Massachusetts Agricultural Society (available in the *Massachusetts Agricultural Repository and Journal*). (John Armstrong's book, it should perhaps be noted, was published in Albany by Jesse Buel.) Even this short list was not acceptable to all. *The New England Farmer,* for example, characterized the publications of the Massachusetts Agricultural Society as "little known and less regarded" (1 [December 14, 1822], 156.)

33. One potential source for transmitting agricultural information and transforming practice in the period under scrutiny was unambiguously unimportant. About the turn of the century, Columbia, Yale, and Harvard established professorships in "Natural History," and in 1821, "the first exclusively agricultural and industrial school" in

the nation was established at Gardiner, Maine. (H. S. Russell 1976: 268; H. T. Pinkett 1950: 149; S. L. Boardman 1892: 17; P. Bidwell and J. I. Falconer 1925: 194.) By the end of the 1820s, a few other agricultural schools had been founded, but most were short-lived, and the influence of the survivors on farming practice appears to have been of little consequence in antebellum America. See *The Plough Boy,* 4 (February 4, 1823), 292; *New England Farmer,* 2 (April 17, 1824), 303; ibid., 4 (August 19, 1825), 30–31; *American Farmer,* 10 (August 15, 1828), 176; *Southern Agriculturist,* 2 (November, 1829), 522n; Bidwell and Falconer 1925: 194–195; L. C. Gray 1958, Vol. 2, 789–792.

34. W. C. Ford 1891: Vol. 13, 348. For other contemporary views on the expected role that agricultural societies could play, or should play, in reforming agriculture, see *American Magazine and Monthly Chronicle,* 1 (July 1758), 489; *Universal Asylum,* 4 (March 1790), 156; *American Farmer,* 3 (May 25, 1821), 67; *New England Farmer,* 1 (April 5, 1823), 284.

35. Quoted in Earl Cathcart 1891: 15.

36. *American Farmer,* 3 (July 6, 1821), 114.

37. *The Plough Boy,* 4 (April 22, 1823), 374.

38. In 1823, for example, one critic of the cattle shown at Brighton noted that "plain practical farmers, while they admire at the overgrown, pampered beauty of *such* animals, would no more think of taking them for their own farm stock than they would go to a turtle feast." (*New England Farmer,* 1 [March 22, 1823], 266.)

39. Reported in *American Farmer,* 4 (January 17, 1823), 340. For similar criticisms, see *American Farmer,* 3 (November 16, 1821), 268; *New England Farmer,* 1 (April 5, 1823), 284; ibid., 3 (October 29, 1824), 108; ibid., 4 (December 9, 1825), 157.

40. A proposal to establish "a society . . . under the patronage of the general government," to be called the American Society of Agriculture, was sent to the House in 1797, but never acted upon (*New-York Magazine,* 2 [April 1797], 174–176), and it was never raised again. In the late colonial period, John Adams had anticipated Washington's concerns by urging each colony to establish "a society for the improvement of agriculture, arts, manufactures, and commerce," but this plea also fell largely upon deaf ears. (Brooke Hindle 1981: 16.)

41. Philadelphia Society for Promoting Agriculture 1939: 18–19.

42. *The Plough Boy,* 2 (October 14, 1820), 155; ibid., 3 (April 6, 1822), 356; *American Farmer,* 1 (July 16, 1819), 121; ibid., 4 (June 14, 1822), 90; ibid., 6 (June 4, 1824), 81; *New England Farmer,* 2 (November 15, 1823), 123; ibid., 2 (December 13, 1823), 156; ibid., 2 (June 26, 1824), 380; ibid., 3 (March 11, 1825), 260; ibid., 4 (December 9, 1825), 157; ibid., 6 (April 18, 1828), 307.

43. New-York Society for the Promotion of Agriculture, Arts and Manufactures, *Transactions,* 2 (1807), iii.

44. See, for example, L. C. Gray 1958, Vol. 2, 784; P. Bidwell and J. I. Falconer 1925: 187, n. 8; W. C. Neely 1935: 46; C. M. Andrews 1919: 120–122.

45. One indicator of prevailing conservatism at this time was the difficulty Watson had in achieving even this modest second step. The farmers in the Pittsfield area "held back their animals . . . for fear of being laughed at, which compelled me to lead the way with several prime animals." (Watson 1820: 119n.)

46. James Mease, for example, reports in 1812 on the sixth cattle show held at Bush Hill, Pennsylvania, which evidently had the twin objectives of "improving the breed of cattle" and providing a cattle market. (Mease 1812, Vol. 2, 286–290.) About the same time the Columbian Agricultural Society for the Promotion of Rural and Domestic Economy was holding livestock exhibits at Georgetown in the District of Columbia. (*Agricultural Museum,* 1 [May 22, 1811], 366–369; C. L. Flint 1874: 118–119.) See also W. C. Neely 1935: 46–50; C. W. Wright 1910: 25.

47. P. Bidwell and J. I. Falconer 1925: 188.

48. 5 (October 23, 1819), 113. See also ibid., 15 (October 31, 1818), 151. Much of the rapid spread of the Berkshire-type fairs to 1825 can be attributed to the receipt of state aid, and much of their rapid demise thereafter, to the withdrawal of that aid. For a description of this history plus the subsequent development of fairs in antebellum America, see W. C. Neely 1935; P. Bidwell and J. I. Falconer 1925: 189–193; L. C. Gray 1958, Vol. 2, 784–788. At the core of these early fiscal difficulties would seem to be a public good problem: that a benefit (an annual cattle show and fair) once created could not be withheld from others, but only because the sponsors of fairs refused to charge admission and entry fees to cover related costs. The latter solution would be implemented when the sponsorship of fairs changed later in the nineteenth century.
49. See, for example, *Niles' Weekly Register,* 1 (October 19, 1811), 118–119; ibid., 9 (October 14, 1815), 111–112.
50. *Niles' Weekly Register,* 9 (October 14, 1815), 111.
51. Commissioner of Patents 1860: 24.
52. Massachusetts Board of Agriculture 1854: 11. This focus upon "the strength and docility" of the oxen, rather than the implement the oxen were pulling, according to this source is to be explained primarily by the fact that "the plough, at that time, was nearly the same as that which had been used a century before." (Ibid.) For evidence that the initiation of fairs by the Massachusetts Agricultural Society was influenced by the success of the Berkshire Society fairs, see R. S. Bates 1965: 58–59; E. Watson 1857: 499.
53. See above, page 206.
54. See, for example, *Niles' Weekly Register,* 13 (November 1, 1817), 147–148; *American Farmer,* 3 (June 22, 1821), 99; ibid., 4 (July 5, 1822), 115–116; ibid., 4 (August 23, 1822), 173; ibid., 4 (November 29, 1822), 283; ibid., 5 (December 26, 1823), 313; ibid., 5 (February 13, 1824), 370–371; ibid., 6 (May 2, 1824), 11; ibid., 7 (April 29, 1825), 43; *New England Farmer,* 1 (August 24, 1822), 26; ibid., 3 (February 18, 1825), 237; ibid., 4 (February 24, 1826), 245.
55. *American Farmer,* 5 (May 25, 1823), 70.
56. *American Farmer,* 8 (October 6, 1826), 226.
57. Tench Coxe 1794: 272.
58. Tench Coxe 1814: 687.
59. R. Dickinson 1813: 63–75.
60. The relevant sources are those listed in Table 10.5 and U.S. Congress, House, *Documents Relative to Manufacturing in the United States,* 22nd Cong., 1st Sess., 1833, Doc. No. 308.
61. "A Gentleman" in the *American Magazine and Monthly Chronicle,* 1 (February 1758), 235.
62. 1 (May 28, 1819), 68.
63. *New England Farmer,* 1 (August 10, 1822), 15.
64. Ibid., italics in the original.
65. *New England Farmer,* 1 (January 4, 1823), 180.
66. *New-York Farmer and Horticultural Repository,* 2 (December 1829), 278. (Italics in original.)
67. The complete list is as follows: "distillation by steam, the pendulum and lever mill, the machine for splitting skins, the pressed nail mill, the great increase of chemical preparations for dyers, colormen, and manufacturers, the conversion of fossil coal into a pigment, the cask for preserving fermented liquors on tap in sound condition, weaving machinery in several new forms, the manufacture of edge tools from rolled steel, various improvements to save fuel, the variations and extensions of the application of steam, the manufacture of opium from the common red poppy and from the lettuce, the increase of the pharmaceutic preparations to the number of seventy, the

division of labor in the cultivation of the cane, activity in the manufacture of the currant wine, the tanning of deer skins, activity and ingenuity in the substitutions for wool, by the manufacturing of thick and warm cotton goods, and by cotton warps under woollen woofs, the machine for manufacturing dipt candles, the activity, extension, and improvement of the sheep breeding and farming, the new employment of the children in the cities, boroughs, and villages, and the active employment of the females in general in manufactures, the extension and facilitation of communication between the producers and importers and the manufacturing citizens, by the various and unprecedented improvements in the post office department, the extension of the funds of the manufacturers by many of the banks, which are solidly and rigidly founded, constituted, and administered, the introduction of new exotic raw materials, by means of commerce, and of laborers, artisans, and manufacturers and processes of every branch from various foreign nations." (Tench Coxe 1814: 688.)

68. T. G. Fessenden 1835: 3.

69. Ibid., 7.

70. Ibid., 330.

71. Ibid., 330. Contrast this characterization with a later description of earlier American prejudice. "Obstinate adherence to prejudice of any kind, is now generally regarded as a mark of ignorance or stupidity. A century ago, the reverse was the case. In many a small country town, a greater degree of intelligence than was possessed by his neighbors brought down upon the farmer the ridicule of the whole community. If he ventured to make experiments, to strike out new paths of practice, and adopt new modes of culture; or if he did not plant just as many acres of corn as his fathers did, and that too in 'the old of the moon;' if he did not sow just as much rye to the acre, use the same number of oxen to plough, and get in his crops on the same day, or hoe as many times as his father and grandfather did; . . . he was shunned in company by the old and young, and looked upon as a mere visionary." (Massachusetts Board of Agriculture 1854: 9.)

72. See, for example, Paul Bairoch 1981; Jan de Vries 1972; F. Dovring 1966; David Grigg 1982: chapter 13; E. Kerridge 1967 and 1969; G. E. Mingay 1963 and 1969; B. H. Slicher van Bath 1960 and 1969; Joan Thirsk 1987: 56–61.

73. Lest any be in doubt that the "New Order" had divine sponsorship, the framers of the Great Seal added the words *Annuit Coeptis* (He [God] has favored our undertakings).

Appendix A. Remedies for Soil Exhaustion

1. P. L. Haworth 1915: 53.

2. See, for example, Jared Eliot 1934: 29–30, 207; P. L. Haworth 1925: 92–95; *Royal American*, 2 (January 1775), 29–30.

3. New-York Society for the Promotion of Agriculture, Arts, and Manufactures, *Transactions*, 1 (1792), Part 1, 7.

4. *The Colombian Magazine*, 3 (March 1789), 156.

5. *Universal Asylum*, 4 (March, 1790), 163. The society's original list of premiums, published in 1785, included this award plus another that was dropped from the 1790 list ("for the greatest quantity and variety of manure collected in one year").

6. S. Deane 1790: 162–166; see also *Boston Magazine*, 1 (April 1789), 239–242; *The Colombian Magazine*, 4 (January 1790), 9–12; P. L. Haworth 1925: 104–105; A. O. Craven 1926: 88, 94–96.

7. S. Deane 1790: 163–166; J. B. Bordley 1803: 199–209. In 1811 a commentator in *The Agricultural Museum* emphasized only fourteen "main" types, including ashes, lime, gypsum, seaweed, and animal manures.

8. *The Plough Boy*, 2 (February 10, 1821), 292.
9. *New England Farmer*, 2 (December 6, 1823), 145.
10. See, for example, *Universal Asylum*, 6 (January 1791), 27–31; ibid., 6 (March 1791), 162–168; *New-York Magazine*, 1 (November 1796), 569; ibid., 2 (August 1797), 418–420; New-York Society for the Promotion of Agriculture, Arts, and Manufactures, *Transactions*, 1 (1792), Part 1, 26; ibid., I (1801), 34–56; ibid., III (1814), 56; W. Strickland 1801: 44; George Logan 1797: 24; Philadelphia Society for Promoting Agriculture, *Memoirs*, 1 (1808), 162–175; H. G. Spafford 1813: 18; G. Cope 1881: 221; Connecticut Society for Promoting Agriculture, *Transactions*, 1802, 6–9; P. L. Haworth 1925: 102; *New England Farmer*, 7 (July 10, 1829), 405; *American Farmer*, 3 (July 27, 1821), 14. For efforts to date the earliest experiments with gypsum, particularly those of Richard Peters (circa 1770), see Lyman Carrier 1923: 270–271; G. Cope 1881: 219–220; *American Farmer*, 7 (April 8, 1825), 20; *New England Farmer*, 7 (December 19, 1828), 172.
11. See *Niles' Weekly Register*, 6 (April 30, 1814), 152; Timothy Dwight 1969: 352; *Massachusetts Agricultural Repository and Journal*, 3 (1815), 56, 64; ibid., 6 (1821), 155; *American Farmer*, 2 (April 7, 1820), 14; ibid., 2 (March 2, 1821), 391; *The Plough Boy*, 2 (December 9, 1820), 220; ibid., 3 (March 9, 1822), 323; *New England Farmer*, 2 (February 21, 1824), 234; ibid., 3 (March 18, 1825), 268; *New-York Farmer and Horticultural Repository*, 1 (November 1828), 258.
12. See W. Strickland 1801: 44; John Binns 1803: 2; John Taylor 1813: 94; *Massachusetts Agricultural Repository and Journal*, 6 (1821), 386–387; *American Farmer*, 1 (March 10, 1820), 394; *The Plough Boy*, 3 (March 9, 1822), 323.
13. See the interchange between John Taylor and Richard Peters in Philadelphia Society for Promoting Agriculture, *Memoirs*, 2 (1811), 51–61, 63–66; also *Niles' Weekly Register*, 12 (March 29, 1817), 79; ibid., 15 (October 10, 1818), 106; New-York Society for the Promotion of Agriculture, Arts, and Manufactures, *Transactions*, 1 (1799), Part 4, 62; *Massachusetts Agricultural Repository and Journal*, 4 (1817), 285; *American Farmer*, 1 (June 4, 1819), 74.
14. See, for example, John Taylor 1813: 81.
15. *American Farmer* 1 (July 16, 1819), 121–123; ibid., 2 (September 1, 1820), 182; ibid., 3 (July 27, 1821), 14; ibid., 5 (December 19, 1823), 307; ibid., 6 (April 23, 1824), 34; ibid., 10 (January 2, 1829), 329–330; *The Plough Boy*, 3 (October 13, 1821), 156; *New England Farmer*, 2 (December 6, 1823), 145; ibid., 3 (December 17, 1824), 163; ibid., 4 (August 19, 1825), 29; ibid., 4 (January 6, 1826), 185; ibid., 5 (July 28, 1826), 6; ibid., 8 (February 12, 1830), 238; *New-York Farmer and Horticultural Repository*, 1 (November 1828), 259.
16. *The Rural Magazine and Farmer's Monthly Museum*, 1 (April 1819), 84.
17. *American Farmer*, 9 (December 14, 1827), 306.
18. *New England Farmer*, 4 (August 19, 1825), 29.
19. See, for example, Lyman Carrier 1923: 269; *American Magazine and Monthly Chronicle*, 1 (July 1758), 490.
20. For examples of early commentary about lime as a fertilizer, see *The Columbian Magazine*, 1 (September 1787), 632–633; *Universal Asylum*, 4 (March 1790), 167–169; ibid., 6 (April 1791), 247; ibid., 7 (October 1791), 236–237; *New England Farmer*, 5 (November 24, 1826), 141.
21. *Universal Magazine*, 4 (May 1790), 298; J. B. Bordley 1797; query #20; *Massachusetts Agricultural Repository and Journal*, "Answers to Agricultural Queries," Vols. 1–2 (1807), 45 (in continuous pagination of Vols. I-II, 233).
22. New-York Society for the Promotion of Agriculture, Arts, and Manufactures, *Transactions*, 1 (1799), Part 4, 64.
23. John Taylor 1813: 81.
24. Quoted in J. T. Lemon 1972: 173.

25. *The Plough Boy*, 2 (December 9, 1820), 220; *Massachusetts Agricultural Repository and Journal*, 3 (1815), 65.
26. *The Plough Boy*, 1 (November 20, 1819), 195; see also *New England Farmer*, 2 (February 21, 1824), 233–234.
27. *The Rural Magazine and Farmer's Monthly Museum*, 1 (April 1819), 84.
28. The Ruffin record is summarized in L. C. Gray 1958, Vol. 2, 780–782. For contemporary commentary upon the rising fashionability of manure after 1810, see *American Farmer*, 6 (August 13, 1824), 163; *Southern Agriculturist*, 1 (February 1828), 75.
29. *American Farmer*, 2 (January 26, 1821), 346.
30. John Taylor 1813: 117.
31. See, for example, B. H. Slicher van Bath 1960; David Grigg 1982: chapter 13; Joan Thirsk 1985: 533–589.
32. See, for example, H. J. Carman 1939: 39, 56, 58, 79, 122, 123, 125, 257, 317, 321.
33. See J. B. Bordley 1784; *The Columbian Magazine*, 1 (May 1787), 369–371; P. A. Haworth 1925: 52–53, 120–124; S. Deane 1790: 235–237; *Universal Asylum*, 6 (March 1791), 162–168; ibid., 9 (October 1792), 257–261; New-York Society for the Promotion of Agriculture, Arts and Manufactures, *Transactions*, 1 (1799), Part 4, 6–8; A. O. Craven 1926: 89, 97–100.
34. The list is reproduced in *Universal Asylum*, 4 (March 1790), 163.
35. *Universal Asylum*, 6 (June 1791), 408.
36. S. Deane 1790: 236.
37. *American Farmer*, 2 (June 16, 1820), 92. Consider, for example, the classification scheme used by John Binns in 1803, simply to describe the soils on his own farm: (1) soil with a sandy top and clay bottom, (2) a yellow stiff clay soil, (3) a red clay soil, (4) a grey sandy soil, (5) a loose red soil, (6) a mulatto soil, (7) a black soil, and (8) a white livery soil. (J. Binns 1803: 37–38.)
38. *Massachusetts Agricultural Repository and Journal*, 1 (1807), 217; *The Plough Boy*, 1 (October 23, 1819), 164; *American Farmer*, 2 (January 26, 1821), 346; *Southern Agriculturist*, 2 (October 1829), 462; *Farmers' Register*, 7 (October 1839), 609–610. Typical of generalizations about crop rotation in the 1820s and 1830s was that of agrarian expert Jesse Buel. "Without investigating the physiology of plants, it is enough for the present to observe, that they are furnished with different systems of roots, through which the plants are principally supplied with nourishment; that some of these search for food near the surface, and that others penetrate for it deep in the soil; that some render the soil hard and compact—others loose and friable;—that some plants exhaust the soil, while others fertilize it; and that the object of a rotation is, to make such a selection from different classes, as shall produce the greatest profit to the cultivator, without impoverishing his soil." (*New England Farmer*, 2 [January 3, 1824], 178.)
39. Sources indicating early interest have been cited above. For evidence of growing interest in the first few decades of the nineteenth century, see *Agricultural Museum*, 2 (August 1811), 733–738; *Niles' Weekly Register*, 11 (November 2, 1816), 148; *The Plough Boy*, 2 (December 23, 1820), 234–235; ibid., 3 (January 19, 1822), 269–270; *New England Farmer*, 2 (January 3, 1824), 177–179; *American Farmer*, 1 (July 23, 1819), 132–133; ibid., 1 (January 21, 1820), 344; ibid., 1 (February 11, 1820), 363–365; ibid., 1 (March 17, 1820), 401; ibid., 2 (April 14, 1820), 17–19; ibid., 2 (May 17, 1820), 61; ibid., 2 (June 16, 1820), 92–93; ibid., 3 (March 8, 1822), 399–400; ibid., 6 (May 28, 1824), 78–79; ibid., 6 (August 24, 1824), 171–172; ibid., 7 (December 30, 1825), 321–323; ibid., 9 (May 4, 1827), 49–50.
40. P. Bidwell and J. I. Falconer 1925: 20–21, 104; L. C. Gray 1958, Vol. 1, 177; P. L. Haworth 1925: 91–92; H. S. Russell 1976: 130–133; J. P. Greene, ed. 1965, Vol. 2, 670, 693; *Massachusetts Agricultural Repository and Journal*, 4 (1817), 389; H. J. Carman 1939: 42, 94, 255; G. Cope 1881: 212; W. B. Weeden 1963: 689, 735; *American*

Magazine and Monthly Chronicle, 1 (July 1758), 491; *American Farmer*, 2 (July 28, 1820), 137; ibid., 7 (April 8, 1825), 20; H. L. Kerr 1964: 87, 89.

41. W. Strickland 1801: 39.
42. L. C. Gray 1958, Vol. 2, 780–781, 803; P. Bidwell and J. I. Falconer 1925: 104, 233–235; C. R. Woodward 1927: 289; W. Strickland 1801: 39; *American Farmer*, 9 (August 10, 1827), 161; *New-York Magazine*, 1 (February 1796), 191; *The Columbian Magazine*, 1 (February 1787), 311; Jacquelin Ambler 1800; H. L. Kerr 1964: 87; A. O. Craven 1926: 88, 97.

Appendix B. Breeding and Raising Livestock

1. Horses will not be considered, largely because of the complexity of tracing the development of what for some was primarily a draught animal, for others, a means of transportation, and for upper-income groups, a luxury good acquired for reasons quite divorced from raising the net profits of a farm. Poultry developments are also ignored, largely because most of the early experimentation of consequence appears to have occurred after 1830.
2. See, for example, H. Carman 1939: 250, 255; P. Bidwell and J. I. Falconer 1925: 107; Adolph Benson 1937, Vol. 1, 55–56; Richard Parkinson 1805: 54; L. G. Connor 1918: 94, 100; S. L. Stover 1962: 103; H. L. Kerr 1964: 87.
3. Bakewell began his experiments "about 1760," and they became widely publicized when G. Culley's essay describing them was published in 1786. (C. W. Johnson 1848, Vol. 1, 28.) Curiously, the issue of improving breeds was almost completely ignored by the Royal Society of Arts in subsequent decades, despite the growing importance of this topic to the agrarian community. (H. T. Wood 1913: 139.)
4. See T. C. Byerly 1976: 259; *Massachusetts Magazine*, 5 (December 1793), 713–714; New-York Society for the Promotion of Agriculture, Arts, and Manufactures, *Transactions*, 1 (1799), Part 4, 12. The latter includes an address by Simeon DeWitt, Surveyor-General, who notes in 1799 concerning the improvement of cattle breeds, "The manner in which this may be effected . . . [is] so well known and have become topics so familiar to every one, that it would seem quite superfluous to mention them on such an occasion as this." Noah Webster did use the occasion of his 1819 address to a local agricultural society to labor the obvious: "the improvement of domestic animals" could be "effected . . . by selecting for propagation those of the best shape and qualities from the species we now possess." (*The Rural Magazine and Farmer's Monthly Museum*, 1 [May 1819], 114.)
5. *Eighty Years' Progress* 1864: 41.
6. See, for example, H. Carman 1939: 190; P. Bidwell and J. I. Falconer 1925, chapter VIII; L. C. Gray 1958, Vol. 1, Chapter IX.
7. *Universal Asylum*, 4 (March 1790), 164.
8. P. Bidwell and J. I. Falconer 1925: 229, G. F. Lemmer 1943: 79–80; L. G. Connor 1918: 100–101; C. W. Pursell, Jr. 1959: 86–87; *Eighty Years' Progress* 1864: 59; Charles Plumb 1906: 344–345.
9. A pioneer in the introduction of Merinos to America, Robert Livingston, upon returning to this country in 1805, was "chagrined to find that the efforts of DuPont, Humphreys, and himself had brought so little result." (A. H. Cole 1926, Vol. 1, 75.) One indicator of that absence of results was the treatment of some of the offspring of "Don Pedro," a prize ram imported by E. I. DuPont in 1801. One farmer who purchased the ram's services subsequently sold his stock to neighboring farmers, and the latter "neglected those valuable animals, great numbers of which . . . perished in their hands or were sold to butchers." (C. W. Pursell, Jr. 1959: 88.) See also C. W. Wright 1910: 10–12.

10. See L. G. Connor 1918: 102–103; A. H. Cole 1926, Vol. 1, 75.
11. E. A. Carman, H. A. Heath, and J. Minto 1892: 197; see also A. H. Cole 1926, Vol. 1, 76.
12. *Niles' Weekly Register*, 1 (October 26, 1811), 133–134; ibid., 6 (July 16, 1814), 333. See also C. W. Wright 1910: 16–34.
13. C. W. Pursell, Jr. 1962: 99.
14. See, for example, A. H. Cole 1926, Vol. 1, 73; C. W. Pursell, Jr. 1962: 91.
15. E. M. Betts 1953: 131.
16. C. S. Plumb 1906: 183; A. H. Saunders 1914: 275; and Saunders 1918: 158, 162; G. F. Lemmer 1947: 79–80; L. C. Gray 1958, Vol. 1, 204; P. Bidwell and J. I. Falconer 1925: 221. See also *Massachusetts Agricultural Repository and Journal*, 2 (1809), 281.
17. A. H. Saunders 1914: 264n.
18. The words "Shorthorn" and "longhorn" are encountered in the historical literature with and without hyphens, and with and without the capitalization of the first letter. The spellings used here are consistent with that of modern agricultural dictionaries.
19. Charles Plumb 1906: 205, 252, 277, 289, 356, 381, 425, 442, 474, 522; A. H. Saunders 1918: 164–177; and Saunders 1914: 260–278; A. H. Cole 1926, Vol. 1, 81; L. G. Connor 1918: 106–107; C. T. Leavitt 1933: 58–59; U.S. Bureau of the Census 1864: cxxxii.
20. See *American Farmer*, 6 (June 3, 1825), 81–82; ibid., 6 (June 24, 1825), 105–109; ibid., 6 (July 15, 1825), 129–130; ibid., 6 (September 9, 1825), 193–194.
21. Timothy Pickering in an address to the Massachusetts Agricultural Society, quoted in *American Farmer*, 5 (June 27, 1823), 106. See also *New England Farmer*, 3 (January 14, 1825), 194; H. T. Pinkett 1950: 149.
22. *American Farmer*, 11 (October 9, 1829), 240.
23. Ibid.
24. The rise in American interest in developing better seeds, including improvement through importations or by the seed equivalent of selective breeding, seems to follow this same general pattern. For evidence of rising interest after 1783, see P. L. Haworth 1925: 105; New-York Society for the Promotion of Agriculture, Arts, and Manufactures, *Transactions*, 1 (1794), Part 2, xxix–xxxi; *Massachusetts Magazine*, 4 (May 1792), 288; *Boston Weekly Magazine*, 3 (December 22, 1804), 35; *Agricultural Museum*, 1 (September 12, 1810), 81; *Massachusetts Agricultural Repository and Journal*, 1 (1801), 3–4; *New England Farmer*, 3 (September 18, 1824), 60. The evidence for a quickening interest after 1815 includes (1) rising commentary on how to choose the best seeds for future crops, (2) signs of increasing specialization in stores that marketed seeds, and (3) more systematic approaches to canvassing foreign possibilities. See *American Farmer*, 1 (December 31, 1819), 316; ibid., 2 (February 2, 1821), 360; ibid., 4 (March 29, 1822), 8; ibid., 1 (March 17, 1820), 405; ibid., 5 (September 5, 1823), 187–188; *New England Farmer*, 3 (December 3, 1824), 146; ibid., 5 (October 13, 1826), 94; ibid., 2 (April 3, 1824), 288; ibid., 4 (April 21, 1826), 312; ibid., 5 (February 9, 1827), 227; ibid., 6 (December 21, 1827), 172–173; *Southern Agriculturist*, 1 (September 1828), 423.
25. There were no herd books for cattle in either America or Britain until 1822, when the British initiated such records for their herds.
26. See, for example, P. Bidwell and J. I. Falconer 1925: 223; C. T. Leavitt 1933: 51–52.

Bibliography

Acrelius, Israel. 1876. Description of the Former and present condition of the Swedish Churches in what was called New Sweden. . . . Stockholm: Harberg & Hasselberg, 1759; reprinted in Memoirs of the Historical Society of Pennsylvania, Vol. 11.

Adams, Charles Francis. 1850–1856. The Works of John Adams. Boston, Mass.: Little, Brown.

Adams, Daniel, ed. 1808. Medical and agricultural register for the years, 1806–1807; containing practical information on husbandry; cautions and directions for the preservation of health, management of the sick etc. designed for families. Boston, Mass.: Manning & Loring.

Adams, Henry. 1889–1891. History of the United States of America. New York: Charles Scribner's Sons.

Adams, James Truslow, ed. 1961. Album of American History. New York: Charles Scribner's Sons.

Adams, John. Works. See Adams, Charles Francis.

——. 1923. Revolutionary New England, 1691–1776. Boston, Mass.: Atlantic Monthly Press.

Adams, John Quincy. 1821. An Address delivered at the request of a Committee of Citizens of Washington on the occasion of reading The Declaration of Independence on the Fourth of July, 1821. Washington, D.C.: Davis and Force.

——. 1976. "Address delivered at ceremonies commencing work on the Chesapeake and Ohio Canal, held at Little Falls near Georgetown, District of Columbia, July 4, 1828." Reprinted in Henry A. Hawken, Trumpets of Glory: Fourth of July Orations, 1786–1861. Granby, Conn.: Salmon Brook Historical Society, 115–118.

Adams, Rufus W. 1814. The Farmer's Assistant, containing a correct description of the best methods of raising and keeping cows, and making butter and cheese. Marietta, Oh.: S. Fairlamb.

Affleck, Thomas. 1844. "Southern Agricultural Implements." American Agriculturalist, 3, 247–248.

Allen, A. B. 1846. "The Improvement of the Plow in the United States." Transactions of the New-York State Agricultural Society, 6, 229–238.

Allen, James T. 1872. History of the Short-Horn Cattle. New York: the author.

——. 1879. Digest of Seeding Machines and Implements Patented in the United States

from the Year 1800 up to, and including, June, 1878, With a Supplement to January, 1879. Washington, D.C.: U.S. Patent Office.
——. 1883. Allen's Digest of Plows, with Attachments, Patented in the United States From A.D. 1789 to January, 1883. Washington, D.C.: J. Bart.
Allen, Lewis F. 1868. American Cattle: Their History, Breeding and Management. New York: Taintor.
Allen, R. L. 1849. The American Farm Book; or Compendium of American Agriculture. New York: Orange Judd.
——. 1886. Digest of Agricultural Implements Patented in the United States from A.D. 1789 to July, 1881. New York: J. C. Von Arx.
Allen, Robert C. 1991. "The Two English Agricultural Revolutions, 1450–1850." In Bruce M. S. Campbell and Mark Overton, eds. Land, Labour and Livestock. Manchester: Manchester University Press, 236–254.
——. 1992. Enclosure and the Yeoman. Oxford: Clarendon Press.
Alvord, Henry E. 1899. "Dairy Development in the United States." U.S. Department of Agriculture Yearbook, 381–402.
Ambler, Jacquelin. 1800. Treatise on the Culture of Lucerne. Richmond: T. Nicolson.
American Husbandry. See Carman, Harry J.
"American Ploughs." Philadelphia Society for Promoting Agriculture, Memoirs, 4, 1818, 160–161.
Amos, William. 1810. Minutes in agriculture and planting . . . illustrated with specimens of eight sorts of the best, and two sorts of the worst natural grasses. . . . Boston, Mass.: J. Hellaby.
Anburey, Thomas. 1923. Travel through the Interior Parts of America. Boston, Mass.: Houghton Mifflin. (Original 1789.)
Anderson, B. G. 1907. "Improvement of Virginia Fire-Cured Tobacco." U.S. Department of Agriculture, Bureau of Soils, Bulletin 46, 1–40.
Anderson, Russell H. 1936. "Grain Drills through Thirty-Nine Centuries." Agricultural History, 10 (October), 157–205.
Andrews, Charles M. 1919. Colonial Folkways. New Haven, Conn.: Yale University Press.
Andrews, G. H. 1853. Modern Husbandry. London: Nathaniel Cooke.
Appelbaum, Diana Karter. 1989. The Glorious Fourth: An American Holiday, An American History. New York: Facts on File.
Aptheker, Herbert. 1960. The American Revolution. New York: International.
Arber, Edward. 1910. Travels and Works of Captain John Smith. Edinburgh: John Grant.
Ardrey, R. L. 1894. American Agricultural Implements. Chicago, Ill.: The author.
Arendt, Hannah. 1968. "Constitution Libertatis." In Jack P. Greene, ed. The Reinterpretation of the American Revolution, 1763–1789. New York: Harper & Row, 579–609.
Armstrong, John. 1819. A Treatise on Agriculture . . . by a Practical Farmer. Albany: J. Buel.
Arnold, B. W., Jr. 1897. History of the Tobacco Industry in Virginia from 1860 to 1894. Baltimore, Md.: Johns Hopkins University Press.
Ashmead, Henry Graham. 1884. History of Delaware County, Pennsylvania. Philadelphia, Penn.: L. H. Everts.
Ashton, T. S. 1948. The Industrial Revolution, 1760–1830. New York: Oxford University Press.
Atack, Jeremy, and Fred Bateman. 1987. To Their Own Soil: Agriculture in the Antebellum North. Ames, Ia.: Iowa State University Press.
Austin, Ivers James. 1839. An Oration delivered by request of the City Authorities before the Citizens of Boston on the Sixty-Third Anniversary of American Independence, July 4, 1839. Boston: John H. Eastburn.

Austin, Samuel. 1798. An Oration Pronounced at Worcester on the Fourth of July, 1798, the Anniversary of the Independence of the United States of America. Worcester: Leonard Worcester.
Axelrod, Alan, ed. 1985. The Colonial Revival in America. New York: Norton.
Baatz, Simon. 1985. "Venerate the Plough": A History of the Philadelphia Society for Promoting Agriculture, 1785–1985. Philadelphia, Penn.: Philadelphia Society for Promoting Agriculture.
Bailey, Alexander M. 1782. One Hundred and Six Copper Plates of Mechanical Machines, and Implements of Husbandry. . . . London: Benjamin White.
Bailey, L. H., ed. 1907a. Cyclopedia of American Agriculture. New York: Macmillan.
——. 1907b. "Treatment of Soil By Means of Tillage." In L. H. Bailey, ed. Cyclopedia of American Agriculture. New York: Macmillan, Vol. 1, 372–378.
Bailey, Robert C. 1973. Farm Tools and Implements before 1850. Spring City, Tenn.: Hillcrest Books.
Baillie, John. 1951. The Belief in Progress. New York: Charles Scribner's Sons.
——. 1965. Pamphlets of the American Revolution, 1750–1776. Cambridge, Mass.: Belknap.
——. 1968a. "The Logic of Rebellion." In Jack P. Greene, ed. The Reinterpretation of the American Revolution, 1763–1789. New York: Harper & Row, 208–235.
——. 1968b. "Political Experience and Enlightenment Ideas in Eighteenth-Century America." In Jack P. Greene, ed. The Reinterpretation of the American Revolution, 1763–1789. New York: Harper & Row, 277–291.
Bailyn, Bernard. 1960. Education in the Formation of American Society: Needs and Opportunities for Study. Chapel Hill, N.C.: University of North Carolina Press.
Bainer, Roy. 1975. "Science and Technology in Western Agriculture." Agricultural History, 49 (January), 56–72.
Baird, Thomas. 1793. General View of Agriculture in the County of Middlesex. London: Board of Agriculture.
Bairoch, Paul. 1973. "Agriculture and the Industrial Revolution." In Carlo M. Cipolla, ed. The Fontana Economic History of Europe, Vol. 2: The Industrial Revolution. London: Collins, 452–506.
Baker, Andrew, and Frank White. 1994. "The Impact of Changing Plow Technology in Rural New England in the Early 19th Century." Old Sturbridge Village, Sturbridge, Mass., typescript.
Baker, Andrew H., and Holly Izard Paterson. 1988. "Farmers' Adaptations to Markets in Early Nineteenth-Century Massachusetts." In Peter Benes, ed. The Farm. Boston, Mass.: Boston University, 95–108.
Bakewell, Robert. 1814. Observations on the influence of soil and climate upon wool. . . . Philadelphia, Penn.: Kimber and Conrad.
Balassa, Iván. 1975. "The Earliest Ploughshares in Central Europe." Tools and Tillage, 2, No. 4, 242–255.
Ball, D. E., and Gary M. Walton. 1976. "Agricultural Productivity Change in Eighteenth-Century Pennsylvania." Journal of Economic History, 36 (March), 102–117.
Bancroft, George. 1854. History of the United States. Boston, Mass.: Little, Brown.
——. 1976. "An Oration delivered on the Fourth of July, 1825, at Northampton, Massachusetts." Reprinted in Henry A. Hawken, ed. Trumpets of Glory: Fourth of July Orations, 1786–1861. Granby, Conn.: Salmon Brook Historical Society, 81–93.
Banister, John. 1970. John Banister and His Natural History of Virginia: 1678–1692, ed. Joseph Ewan and Nesta Ewan. Urbana, Ill.: University of Illinois.
Barber, John W., and Henry Howe. 1975. Early Woodcut Views of New York and New Jersey: 304 illustrations from "Historical Collections." New York: Dover.
Barlow, Joel. 1809. Oration delivered at Washington, July Fourth, 1809, at the request

of the Democratic Citizens of the District of Columbia. Washington, D.C.: R. C. Weightman.

Barnett, Claribel R. 1928. "The Agricultural Museum: An Early American Agricultural Periodical." Agricultural History, 2 (April), 99–102.

Baron, Robert C., ed. 1987. The Garden and Farm Books of Thomas Jefferson. Golden, Colo.: Fulcrum.

Baron, William. 1864. History of the Agricultural Associations of New York, from 1791 to 1862. Albany: Van Benthuysen's.

Barrus, Clara. 1914. Our Friend John Burroughs. Boston, Mass.: Houghton Mifflin.

Bateman, Fred. 1983. "Research Developments in American Agricultural History." Agricultural History Review, 21, Part II, 132–148.

Bates, Ralph S. 1965. Scientific Societies in the United States. Cambridge, Mass.: M.I.T. Press.

Baumer, Franklin L. 1960. Religion and the Rise of Skepticism. New York: Harcourt, Brace.

Beard, Charles A. 1982. "Introduction." In J. B. Bury. The Idea of Progress. Westport, Conn.: Greenwood, ix–xi.

Beard, Charles A., and Mary R. Beard. 1934. The Rise of American Civilization. New York: Macmillan.

Beatson, Major-General Alexander. 1820. A New System of Cultivation, without Lime, or Dung, or Summer Fallows, as practiced at Knowle-Farm, in the County of Sussex. London: W. Bulmer and W. Nicol.

Beaumont, Olga, and J. Geraint Jenkins. 1957. "Farm Tools, Vehicles, and Harnesses, 1500–1900." In Charles Singer, E. J. Holmyard, and A. R. Hall, eds. A History of Technology. Oxford: Clarendon, Vol. 3, 134–150.

Beaumont, Olga, and J. W. Y. Higgs. 1958. "Primary Production: Part I. Agriculture: Farm Implements." In Charles Singer, E. J. Holmyard, and A. R. Hall, eds. A History of Technology. Oxford: Clarendon, Vol. 4, 1–12.

Becker, Carl. 1910. "Kansas." In P. Smith, ed. Essays in American History Dedicated to Frederick Jackson Turner. New York: Henry Holt.

——. 1934. "Progress" in Edwin R. A. Seligman, ed. Encyclopedia of the Social Sciences. New York: Macmillan, Vol. 12, 495–499.

Becker, Raymond B. 1973. Dairy Cattle Breeds: Origin and Development. Gainesville, Fla.: University of Florida Press.

Behrens, June, and Pauline Brower. 1976. Colonial Farm. Chicago, Ill.: Childrens Press.

Bek, William G. 1922. "The Followers of Duden." Missouri Historical Review, 16 (January), 289–307.

Bell, Jonathan. 1983. "Harrows Used In Ireland." Tools and Tillage, 4, No. 4, 195–205.

Belloc, Hillaire. 1914. The Book of the Bayeux Tapestry. London: Chatto & Windus.

Benes, Peter, ed. 1988. The Farm. Dublin Seminar for New England Folklife: Annual Proceedings, 1986. Boston, Mass.: Boston University.

Bennett, Hugh H. 1944. Thomas Jefferson: Soil Conservationist. Washington, D.C.: U.S. Department of Agriculture Miscellaneous Publication 548.

Benson, Adolph B. 1937. Peter Kalm's Travels in North America. New York: Wilson-Erickson.

Benson, Albert Emerson. 1929. History of the Massachusetts Horticultural Society. Norwood, Mass.: Plimpton.

Bentzien, Ulrich. 1968. "The Dabergotz Ard." Tools and Tillage, 1, No. 1, 50–55.

Berg, Gösta. 1976. "The Introduction of the Winnowing-Machine in Europe in the 18th Century." Tools and Tillage, 3, No. 1, 25–46.

Berglung, Berndt. 1976. Wilderness Living: A Complete Handbook and Guide to Pioneering in North America. Toronto: Pagurian Press.

Berkeley, Edmund, and Dorothy Smith Berkeley. 1965. The Reverend John Clayton: A Parson with a Scientific Mind. Charlottesville, Virg.: University Press of Virginia.

Betts, Edwin Morris. 1944. Thomas Jefferson's Garden Book 1766–1824 with Relevant Abstracts from His Other Writing. Philadelphia, Penn.: American Philosophical Society.

———. 1953. Thomas Jefferson's Farm Book. Princeton, N.J.: Princeton University.

Beverley, Robert. 1705. The history and present state of Virginia, in four parts. . . . London: R. Parker.

Biddle, Nicholas. 1939. "Address to the Philadelphia Society for Promoting Agriculture, January 15, 1822." Reprinted in Philadelphia Society for Promoting Agriculture, Memoirs, 6, 31–56.

Bidwell, Percy W. 1916. Rural Economy in New England at the Beginning of the nineteenth Century. New Haven, Conn.: Academy of the Arts and Sciences of Connecticut.

———. 1921. "The Agricultural Revolution in New England." American Historical Review, 26 (July), 683–702.

Bidwell, Percy W., and John I. Falconer. 1925. History of American Agriculture in the Northern United States, 1620–1860. Washington, D.C.: The Carnegie Institution.

Bieleman, Jan. 1992. Geschiedenis van de Landbouw in Nederland, 1500–1950. Amsterdam: Boom Meppel.

Bigelow, Andrew. 1815. An Oration delivered before the Washington Benevolent Society at Cambridge, July 4, 1815. Cambridge, Mass.: Hilliard and Metcalf.

Bigelow, Jacob. 1831. Elements of Technology. Boston, Mass.: Hilliard, Gray, Little and Wilkins.

Billings, E. R. 1875. Tobacco: Its History, Varieties, Culture, Manufacture and Commerce. Hartford, Conn.: American Publishing Company.

Binns, John A. 1803. Treatise on Practical Farming. Fredrick-town, Md.: John B. Colvin.

Birbeck, Morris. 1818. Notes on a Journey in America, from the Coast of Virginia to the Territory of Illinois. London: Severn.

Bishop, J. Leander. 1868. A History of American Manufactures from 1608 to 1860. Philadelphia, Penn.: Edward Young.

Blackburn, Roderic H., and Nancy A. Kelley, eds. 1987. New World Dutch Studies: Dutch Arts and Culture in Colonial America, 1609–1776. Albany: Albany Institute of History and Art.

Blandford, Percy W. 1974. Country Craft Tools. London: David & Charles.

———. 1976. Old Farm Tools and Machinery: An Illustrated History. Fort Lauderdale, Fla.: Gale Research.

Blith, Walter. 1652. The English Improver Improved. London: John Wright.

Blodget, Samuel. 1806. Economica: A Statistical Manual for the United States of America. Washington, D.C.: Samuel Blodget.

Blome, Richard. 1687. The present state of His Majesties isles and territories in America. . . . London: H. Clark.

Bloomfield, Robert. 1798. The Farmer's Boy. London: Sampson Low, Marston, Low and Searle.

Blum, Jerome, ed. 1982. Our Forgotten Past: Seven Centuries of Life and Land. New York: Thames and Hudson.

Boardman, Samuel Lane. 1892. History of the Agriculture of Kennebec County, Maine. New York: H. W. Blake.

Bogart, Ernest L. 1923. Economic History of American Agriculture. New York: Longmans, Green and Co.

Bonebright-Closz, Harriet. 1921. Reminiscences of Newcastle, Iowa. Des Moines, Ia.: History Department of the University of Iowa.

Bonner, James C. 1964. A History of Georgia Agriculture, 1732–1860. Athens: University of Georgia Press.

Boodin, John Elof. 1939a. "The Idea of Progress." Journal of Social Philosophy, 4 (January), 101–120.

——. 1939b. The Social Mind: Foundations of Social Philosophy. New York: Macmillan.

Boorstin, Daniel J. 1958. The Americans: The Colonial Experience. New York: Random House.

Bordley, John Beale. 1784. A Summary View of the Course of Crops in the Husbandry of England and Maryland, with a comparison of their Products, and a system of improved courses, proposed for Farms in America. Philadelphia, Penn.: Charles Cist.

——. 1797a. Queries Selected from a paper of the Board of Agriculture in London, on the nature and principles of vegetation: with answers and observations by J. B. B. Philadelphia, Penn.: Charles Cist.

——. 1797b. Sketches on Rotations of Crops, and Other Rural Matters. . . . Philadelphia, Penn.: C. Cist.

——. 1799. Essays and Notes on Husbandry and Rural Affairs. Philadelphia, Penn.: Budd and Bartram.

——. 1803. Gleanings from the most Celebrated Books on Husbandry, Gardening, and Rural Affairs. Philadelphia, Penn.: James Humphreys.

——. 1939a. "Observations on an intended Drill for Clustering Wheat." February 16, 1786. Reprinted in Philadelphia Society for Promoting Agriculture, Memoirs, 6, 112–114.

——. 1939b. "Some Account of Treading out Wheat." Philadelphia Society for the Promotion of Agriculture, Memoirs, 6 (1939), 116–121. (Original report dated October 1786.)

Borgstrom, Georg. 1967. "Food and Agriculture in the Nineteenth Century." In Melvin Kranzberg and Carroll W. Pursell, Jr., eds. Technology in Western Civilization. New York: Oxford University Press, Vol. 1, pp. 408–424.

Boserup, Ester. 1987. "Agricultural Development and Demographic Growth: A Conclusion." In Antoinette Fauve-Chamoux, ed. Évolution Agraire & Croissance Démographique. Liege: Derovaux Ordina, 385–389.

Botke, I. J., ed. 1988. Het 'Schrijfboek' van Marten Aedsges (1742–1806). Landbouwer te Zuurdijk. Historia Agriculturae, Vol. 17. Groningen: Nederland Agronomisch-Historisch Instituut.

Bowler, Metcalf. 1786. A Treatise on Agriculture and Practical Husbandry. . . . Providence, R.I.: Bennett Wheeler.

Bowman, Allen. 1943. The Morale of the American Revolutionary Army. Washington, D.C.: American Council on Public Affairs.

Boyd, Julian P., ed. 1956. The Papers of Thomas Jefferson. Princeton, N.J.: Princeton University Press.

Bradley, Richard. 1725. A Survey of the Ancient Husbandry and Gardening. London: B. Motte.

Brady, Niall D. K. 1988. "The Plough Pebbles of Ireland." Tools and Tillage, 6, No. 1, 47–60.

Brand, Donald D. 1939. "The Origin and Early Distribution of New World Cultivated Plants." Agricultural History, 13 (April), 109–117.

Brasch, Frederick E. 1931. "The Royal Society of London and Its Influence upon Scientific Thought in the American Colonies." Scientific Monthly, 33 (October), 448–469.

——. 1939. "The Newtonian Epoch in the American Colonies: 1680–1783." American Antiquarian Society, 49 (October), 314–332.

Bray, Francesca. 1979. "The Evolution of the Mouldboard Plough in China." Tools and Tillage, 3, No. 4, 227–239.

Breasted, James Henry. 1916. Ancient Times: A History of the Early World. Boston, Mass.: Ginn.

Bremer, Frederika. 1853. The Homes of the New World. London: Arthur Hall, Virtue, and Co.
Bressler, Leo A. 1955. "Agriculture among the Germans in Pennsylvania during the Eighteenth Century." Pennsylvania History, 22 (April), 102–133.
Brewer, William H. 1880. "Cereal Report." In U.S. Bureau of the Census, Tenth Census of the United States. Agriculture.
Brewster, David, ed. 1832. The Edinburgh Encyclopaedia. Philadelphia, Penn.: Joseph and Edward Parker.
Brickell, John. 1737. A Natural History of North-Carolina. . . . Dublin: J. Carson.
Bridenbaugh, Carl. 1955. Cities in Revolt: Urban Life in America, 1743–1776. New York: Alfred A. Knopf.
Brigham, Clarence S. 1925. "An Account of American Almanacs and Their Value for Historical Study." Proceedings of the American Antiquarian Society, 35 (April–October), 195–218.
——. 1976. History and Bibliography of American Newspapers, 1690–1820. Westport, Conn.: Greenwood Press.
Brissot de Warville, Jacques Pierre. 1970. New Travels in the United States of America, Performed in 1788. New York: Augustus M. Kelley. (Original 1792.)
Bristed, John. 1818. The Resources of the United States of America. New York: James Eastburn.
Brock, R. A. 1883. "A Succinct Account of Tobacco in Virginia." In J. B. Killebrew, ed. "Report on the Culture and Curing of Tobacco in the United States." U.S. Bureau of the Census, Tenth Census, 1880, Report on the Production of Agriculture, 212–225.
Broehl, Wayne G., Jr. 1984. John Deere's Company: A History of Deere & Company and its Times. New York: Doubleday.
Brooke, Walter Edwin. 1919. The Agricultural Papers of George Washington. Boston, Mass.: Richard G. Badger.
Brooks, Jerome E. 1952. The Mighty Leaf: Tobacco through the Centuries. Boston, Mass.: Little, Brown.
Brown, P. E. 1935. "The Beginnings and Development of Soil Microbiology in the United States." Soil Science, 40 (June–December), 49–58.
Brown, Ralph M. 1939. "Agricultural Science and Education in Virginia Before 1860." William and Mary College Quarterly Historical Magazine, 19 (April), 197–213.
Brown, Richard D. 1989. Knowledge Is Power: The Diffusion of Information in Early America, 1700–1865. New York: Oxford University Press.
Brown, Theo. 1925. "Some Fundamentals of Plow Design." Agricultural Engineering, 6 (June), 124–129.
Browne, Charles A. 1927. "Some Historical Relations of Agriculture in the West Indies to that of the United States." Agricultural History, 1 (July), 23–33.
——. 1994. "Thomas Jefferson and the Scientific Trends of His Time." Chronica Botanica, 8, No. 3 (Summer), 363–423.
Browne, Charles A., and Joseph S. Chamberlain. 1926. Chemistry in Agriculture: A Cooperative Work Intended to Give Examples of the Contribution Made to Agriculture by Chemistry. New York: The Chemical Foundation.
Brownell, Thomas Church. 1815. Address on the Theory of Agriculture. . . . Albany: Webster & Skinner.
Bruce, K. 1932. "Virginia Agricultural Decline to 1860: A Fallacy." Agricultural History, 6 (January), 3–13.
Bruce, Philip A. 1935. Economic History of Virginia in the Seventeenth Century. New York: P. Smith.
Bruchey, Stuart. 1972. "The Business Economy of Marketing Change, 1790–1840: A Study of Sources of Efficiency." Agricultural History, 46 (January), 211–226.

Bryan, A. B. 1949. "Some Farming Societies and Farming Science." Better Crops with Plant Food, 33 (April), 21–24, 38–42.

Buel, Jesse. 1819. A Treatise on Agriculture, Comprising a Concise History of its Origins and Progress, the Present Condition of the Art, Abroad and at Home, and the Theory and Practice of Husbandry. Albany, N.Y.: Jesse Buel.

——. 1823. "On the Sowing of Clover Seed, and the Construction of Harrows." Board of Agriculture of the State of New-York, Memoirs, Vol. 2, 247–249.

——. 1840. The Farmer's Companion, or Essays on the Principles and Practice of American Husbandry. Boston, Mass.: Marsh, Capen, Lyon and Webb.

Buel, Richard, Jr. 1969. "Democracy and the American Revolution: A Frame of Reference." In Jack P. Greene, ed., The Reinterpretation of the American Revolution, 1763–1789. New York: Harper & Row, 122–149.

Bullard, John. 1804. A Discourse on Agriculture. . . . Cambridge, Mass.: W. Hillard.

Bullion, Brenda. 1988. "The Agricultural Press: 'To Improve the Soil and the Mind'." In Peter Benes, ed. The Farm. Boston, Mass.: Boston University, 74–94.

Bullock, William. 1649. Virginia Impartially Examined. . . . London: John Hammond.

Bureau of Agricultural Economics. 1937. Washington, Jefferson and Lincoln and Agriculture. Washington, D.C.: United States Department of Agriculture.

Burke, Edmund. 1758. An Account of the European Settlements in America. . . . London: R. and J. Dodsley.

Burkett, Charles William. 1900. History of Ohio Agriculture. Concord, N.H.: Runford Press.

Burnaby, Andrew. 1819. Travels through the Middle Settlements of North America, 1759–1760. In John Pinkerton, ed. General Collection of the Best and Most Interesting Voyages and Travels. London: Longman, Hurst, Rees, and Orme, Vol. 13.

Burrage, Walter L. 1923. A History of the Massachusetts Medical Society. Norwood, Mass.: Plimpton Press.

Bury, J. B. 1982. The Idea of Progress. Westport, Conn.: Greenwood.

Butterfield, L. H., ed. 1963. Adams Family Correspondence. Cambridge, Mass.: Belknap.

Butterworth, Benjamin. 1892. The Growth of Industrial Art. Washington, D.C.: U.S. Patent Office.

Byerly, T. C. 1976. "Changes in Animal Science." Agricultural History, 50 (January), 258–274.

Byrd, William. 1940. William Byrd's Natural History of Virginia: or the Newly Discovered Eden, ed. Richard Croom Beatty and William J. Mulley. Richmond, Virg.: Dietz Press.

——. 1942. Another Secret Diary of William Byrd of Westover, 1739–1741, ed. Maude H. Woodfin and Marion Tinling. Richmond, Virg.: Dietz Press.

Cabell, Nathaniel F. 1800. Early History of Agriculture in Virginia. Washington, D.C.: Lemuel Towers.

Campbell, Bruce M. S., and Mark Overton, eds. 1991a. Land, Labour and Livestock: Historical Studies in European Agricultural Productivity. Manchester: Manchester University Press.

——. 1991b. "Productivity Change in European Agricultural Development." In their Land, Labour and Livestock. Manchester: Manchester University Press, 1–50.

Campbell, Charles, ed. 1840. The Bland Papers. Petersburg: Edmund & Julian C. Ruffin.

Campbell, H. W. 1907. "A System of Scientific Soil Culture for Semi-Arid Regions." In L. H. Bailey, ed., Cyclopedia of American Agriculture. New York: Macmillan, Vol. I, 398–402.

Campbell, Patrick. 1937. Travels in the Interior Parts of North America in the Years 1791–1792, ed. H. H. Langton. Toronto: Champlain Society.

Cantor, Jay E. 1976. The Landscape of Change: Views of Rural New England, 1790–1865. Sturbridge, Mass.: Old Sturbridge Village.
Cardini, Franco. 1989. Europe 1492: Portrait of a Continent Five Hundred Years Ago. New York: Facts on File.
Carman, Ezra A., H. A. Heath, and John Minto. 1892. Special Report on the History and Present Condition of the Sheep Industry of the United States, prepared for U.S. Department of Agriculture, Bureau of Animal Industry. Washington, D.C.: Government Printing Office.
Carman, Harry J., ed. 1939. American Husbandry. New York: Columbia University Press. (Original 1775.)
Carp, E. Wayne. 1984. To Starve the Army at Pleasure: Continental Army Administration and American Political Culture, 1775–1785. Chapel Hill, N.C.: University of North Carolina Press.
Carpenter, Edward. 1954. "The Groundhog Thresher: An Enigma." Wisconsin Magazine of History, 37 (Summer), 217–218.
Carrier, Lyman. 1923. The Beginnings of Agriculture in America. New York: McGraw-Hill.
——. 1957. Agriculture in Virginia, 1607–99. Williamsburg, Virg.: 350th Anniversary Celebration Corporation.
Carroll, B. R. 1836. Historical Collections of South Carolina. New York: Harper & Brothers.
Carter, Robert. 1940. Letters of Robert Carter, 1720–1727: The Commercial Interests of a Virginia Gentleman, ed. Louis B. Wright. San Marino, Calif.: Huntington Library.
Carver, Jonathan. 1779. A Treatise on the Culture of the Tobacco Plant. London: Jonathan Carver.
Cassara, Ernest. 1975. The Enlightenment in America. Boston, Mass.: Twayne.
Cassedy, James H. 1976. "Medicine and the Learned Society in the United States, 1600–1850." In Alexandra Oleson and Sanborn C. Brown, eds. The Pursuit of Knowledge in the Early American Republic. Baltimore, Md.: John Hopkins University Press, 261–278.
Cathcart, Earl. 1891. "Jethro Tull: His Life, Times and Teaching." The Journal of the Royal Agricultural Society of England, 2 (Third Series), 1–40.
Cathey, Cornelius Oliver. 1956. Agricultural Developments in North Carolina, 1783–1860. James Sprout Studies in History and Political Science, Vol. 38. Chapel Hill, N.C.: University of North Carolina Press.
——. 1966. Agriculture in North Carolina before the Civil War. Raleigh, N.C.: State Department of Archives and History.
Cazenove, Theophile. 1922. Cazenove Journal, 1794; A Record of a Journey of Theophile Cazenove through New Jersey and Pennsylvania, ed. Rayner Wickersham Kelsey. Haverford, Penn.: Pennsylvania History Press.
Chambers, J. D., and G. E. Mingay. 1975. The Agricultural Revolution, 1750–1880. London: B. T. Batsford.
Channing, Edward. 1917. A History of the United States. New York: Macmillan.
Chase, Benjamin. 1869. History of Old Chester, from 1719 to 1869. Auborn, N.H.: John B. Clarke.
Chastellux, Marquis de. 1963. Travels in North America in the Years 1780, 1781, and 1782, ed. Howard C. Rice, Jr. Chapel Hill, N.C.: University of North Carolina Press.
Childe, V. Gordon. 1951. "The Balanced Sickle." In W. F. Grimes, ed. Aspects of Archeology in Britain and Beyond. London: H. W. Edwards, 39–48.
Chinard, Gilbert, ed. 1934. A Huguenot Exile in Virginia. New York: Press of the Pioneers.
——. 1947. "Eighteenth Century Theories on America as a Human Habitat." American Philosophical Society Proceedings, 91 (February), 27–57.

Chivers, Keith. 1976. The Shire Horse: A History of the Breed, the Society and the Men. London: J. A. Allen.

Chorley, G. P. H. 1981. "The Agricultural Revolution in Northern Europe, 1750–1880: Nitrogen, Legumes, and Crop Productivity." Economic History Review, 34 (February), 71–93.

Church, Lillian M. 1935. Bibliography on the Development of Farm Machinery. Washington, D.C.: U.S. Department of Agriculture, Information Series 14.

——. 1939. Partial History of the Development of Grain Threshing Implements and Machines. Washington, D.C.: U.S. Department of Agriculture Information Series 73.

Churchman, Caleb, and George Martin, Jr. 1814. "A Description of the Improved Patent Bark Mill, or Machine for Threshing Grain, Called the Pennsylvania Rubber, with Directions for Using." Philadelphia Society for Promoting Agriculture, Memoirs, Vol. 3, 400–402.

Clark, Christopher. 1990. The Roots of Rural Capitalism: Western Massachusetts, 1780–1860. Ithaca, N.Y.: Cornell University Press.

Clark, Gregory. 1991. "Labour Productivity in English Agriculture, 1300–1860." In Bruce M. S. Campbell and Mark Overton, eds., Land, Labor and Livestock. Manchester: Manchester University Press, 211–235.

Clark, Victor S. 1929. History of Manufactures in the United States. New York: McGraw-Hill.

Clarke, William H. 1945. Farms and Farmers: The Story of American Agriculture. Freeport, N.Y.: Books for Libraries Press.

Clayton, John. 1965. The Reverend John Clayton: A Parson with a Scientific Mind: His Scientific Writings and Other Related Papers, ed. Edmond Berkeley and Dorothy Smith Berkeley, eds. Charlottesville, Virg.: University Press of Virginia.

Clemen, Rudolf. 1923. The American Livestock And Meat Industry. New York: Ronald Press Co.

Clemens, Paul G. E. 1975. "The Operation of an Eighteenth-Century Chesapeake Tobacco Plantation." Agricultural History, 49 (July), 517–531.

Clinton, DeWitt. 1814. "A Discourse delivered before the Literary Philosophical Society of New York, July 4, 1814." North American Review, 8, 157–168.

——. 1976. "An Address delivered at ceremonies breaking ground on the Ohio Canal, performed at Licking Summit near Newark, Ohio, July 4, 1825." Reprinted in Henry A. Hawken, ed. Trumpets of Glory: Fourth of July Orations, 1786–1861. Granby, Conn.: Salmon Brook Historical Society, 119–121.

Clonts, F. W. 1926. "Travel and Transportation in Colonial North Carolina." North Carolina Historical Review, 3 (January), 16–35.

Cluny, Alexander. 1769. The American Traveller: or, Observations on the Present State, Culture, and Commerce of the British Colonies in America. . . . London: E. and C. Dilly.

Cobb, N. A. 1907. "Agriculture in Hawaii." In L. H. Bailey, ed. Cyclopedia of American Agriculture. New York: Macmillan, Vol. 1, 114–121.

Cobbett, William. 1818. A Year's Residence in the United States of America. Belfast: Ulster Register Office.

Cochrane, Willard W. 1976. The Development of American Agriculture: A Historical Analysis. Minneapolis, Minn.: University of Minnesota Press.

Coclanis, Peter A. 1989. The Shadow of a Dream: Economic Life and Death in the South Carolina Low Country, 1670–1920. New York: Oxford University Press.

Cohen, David Steven. 1992. The Dutch-American Farm. New York: New York University Press.

Cohen, I. Bernard, ed. 1941. Benjamin Franklin's Experiments. Cambridge, Mass.: Harvard University Press.

Cole, Arthur H. 1926a. "Agricultural Crazes." American Economic Review, 16 (December), 622–639.
——. 1926b. The American Wool Manufacture. Cambridge, Mass.: Harvard University Press.
Cole, S. M. 1954. "Differentiation of Non-Metallic Tools." In Charles Singer, E. J. Holmyard, and A. R. Hall, eds., A History of Technology. Oxford: Clarendon, Vol. 1, 495–519.
Coleman, D. C. 1992. Myth, History and the Industrial Revolution. London: Hambledon Press.
Coleman, Kenneth. 1958. The American Revolution in Georgia: 1763–1789. Athens, Ga.: University of Georgia Press.
——. 1959. "Agricultural Practices in Georgia's First Decade." Agricultural History, 33 (October), 196–199.
Collins, E. J. T. 1969a. "Harvest Technology and Labour Supply in Britain, 1790–1870." Economic History Review, 22 (December), 453–473.
——. 1969b. "Labour Supply and Demand in European Agriculture, 1800–1880." In E. L. Jones and S. L. Woolf, eds. Agrarian Change and Economic Development. London: Methuen, 61–94.
——. 1969c. Sickle to Combine: A Review of Harvest Techniques from 1800 to the Present Day. Oxford: Alden & Mowbray.
——. 1972. "The Diffusion of the Threshing Machine in Britain, 1790–1880." Tools and Tillage, 2, No. 1, 16–33.
Collinson, Peter, and John Custis. 1948. "Brothers of the Spade: Correspondence of Peter Collinson, of London, and of John Custis, of Williamsburg, Virginia, 1734–1776," ed. Earl G. Swan. Proceedings of the American Antiquarian Society, 58 (April), 17–190.
Colman, Gould P. 1968. "Innovation and Diffusion in Agriculture." Agricultural History, 42 (July), 173–187.
Combe, George. 1841. Notes on the United States of North America During a Phrenological Visit, in 1838–9–40. Philadelphia, Penn.: Carey & Hart.
Commager, Henry Steele, and Richard B. Morris, eds. 1967. The Spirit of Seventy-Six: The Story of the American Revolution as Told by Participants. New York: Harper & Row.
Commissioner of Agriculture. See U.S. Commissioner of Agriculture.
Commissioner of Patents. See U.S. Commissioner of Patents.
Committee of the Highland Society of Scotland. 1829. "Report in regard to the Experiments made in Scotland, in Harvest 1825, with the Flemish Scythe." Highland Society of Scotland, Prize-Essays and Transactions, Vol. 7, 244–249.
Conard, John. 1826. "On a New Corn Planter." Philadelphia Society for Promoting Agriculture, Memoirs, Vol. 5, 99–101.
Condy, Thomas D. 1819. An Oration delivered before an assemblage of the inhabitants of Charleston, S.C. on the 5th day of July, 1819 in commemoration of American Independence. Charleston, S.C.: A. E. Miller.
Connecticut State Agricultural Society. 1802. Transactions of the Society, for promoting agriculture in the state of Connecticut. . . . New Haven, Conn.: William W. Morse.
Conner, John B. 1893. Indiana Agriculture: Agricultural Resources and Development of the State: Struggles of Pioneer Life Compared with Present Conditions. Indianapolis, Ind.: W. B. Burford.
Connor, L. G. 1918. "A Brief History of the Sheep Industry in the United States." Agricultural History Association Annual Report, Vol. 1, 93–197.
Cook, Peter W. 1988. "Domestic Livestock of Massachusetts Bay, 1620–1775." In Peter Benes, ed. The Farm. Boston, Mass.: Boston University, 109–125.
Cooper, Thomas. 1794. Some Information Respecting America. Dublin: William Porter.

Cope, Gilbert. 1881. "Historical Sketch of Chester County Agriculture." In Pennsylvania State Board of Agriculture, Fourth Annual Report, 1880. Harrisburg, Penn.: Lane S. Hart, 208–225.

Copeland, Peter F. 1977. Working Dress in Colonial and Revolutionary America. Westport, Conn.: Greenwood Press.

Cornell, E. 1843. "Southern Plows and Plowing." The American Agriculturist, 2 (April), 49–50.

Cousins, Peter H. 1973. Hog Plow and Sith: Cultural Aspects of Early Agricultural Technology. Dearborn, Mich.: Edison Institute.

Coventry, Alexander. 1978. Memoirs of an emigrant: The journal of Alexander Coventry, Md., in Scotland, the United States and Canada during the period 1783–1831. Albany, N.Y.: Albany Institute of History and Art.

Coxe, Tench. 1814. "Digest of Manufactures." American State Papers, Finance, Vol. 2, 13th Cong., 2nd Sess., Doc. 407.

——. 1965. A View of the United States of America, in a series of papers written at various times, in the years between 1787 and 1794. New York: Sentry.

Coxe, William. 1817. A View of the Cultivation of Fruit Trees. Philadelphia, Penn.: M. Carey and Son.

Crafts, N. F. R., and T. C. Mills. 1994. "The Industrial Revolution as a Macroeconomic Epoch: An Alternative View." Economic History Review, 47 (November), 769–775.

Crallé, Richard K., ed. 1856. Reports and Public Letters of John C. Calhoun. New York: D. Appleton.

Craven, Avery. 1926. Soil Exhaustion as a Factor in the Agricultural History of Virginia and Maryland, 1606–1860. University of Illinois Study in the Social Sciences, 13, No. 1, Urbana, Ill.

——. 1932. Edmond Ruffin, Southerner: A Study in Secession. New York: D. Appleton.

Cresswell, Donald H. 1975. The American Revolution in Drawings and Prints. Washington, D.C.: Library of Congress.

Crèvecoeur, St. John de. 1961. Crèvecoeur's Eighteenth-Century Travels in Pennsylvania and New York, ed. Percy G. Adams. Louisville, Ky.: University of Kentucky Press. (Original 1801.)

Crittenden, Charles Christopher. 1931. "Overland Travel and Transportation in North Carolina, 1763–1789." North Carolina Historical Review, 8 (July), 239–257.

Cronon, William. 1983. Changes in the Land: Indians, Colonists, and the Ecology of New England. New York: Hill and Wang.

Crouthers, David D. 1962. Flags of American History. Maplewood, N.J.: C. S. Hammond.

Crowley, J. E. 1974. This Sheba Self: The Conceptualization of Economic Life in Eighteenth-Century America. Baltimore, Md.: Johns Hopkins University Press.

Culley, George. 1804. Observations on live stock: containing hints for choosing and improving the best breeds. . . . New York: D. Longworth.

Curwen, E. Cecil. 1946. Plough and Pasture. London: Cobbett.

Curwen, E. Cecil, and Gudmund Hatt. 1953. Plough and Pasture: The Early History of Farming. New York: Henry Schuman.

Cushing, Caleb. 1832. An Oration delivered before the Citizens of Newburyport on the Fifty-Sixth Anniversary of American Independence. Newburyport, Mass.: T. B. and E. L. White.

——. 1833. Oration pronounced at Boston before the Colonization Society of Massachusetts on the Anniversary of American Independence, July 4, 1833. Boston, Mass.: G. W. Light.

Dabney, John. 1796. An Address to Farmers. . . . Salem, Mass.: J. Dabney.

Daggett, David. 1799. An Oration Pronounced on the Fourth of July, 1799, at the request of the Citizens of New-Haven. New Haven, Conn.: Thomas Green and Son.

Danckaerts, Jasper. 1913. Journal of Jasper Danckaerts, 1679–1680, ed. Bartlett Burleigh James and J. Franklin Jameson. New York: Charles Scribner's Sons.
——. 1969. Diary of our Second Trip from Holland to New Netherland, 1683, ed. Kenneth Scott. Upper Saddle River, N.J.: Gregg Press.
Danhof, Clarence H. 1956. "Gathering the Grass." Agricultural History, 30 (October), 169–173.
——. 1969. Change in Agriculture: The Northern United States, 1820–1870. Cambridge, Mass.: Harvard University Press.
Daniels, George H. 1972. Nineteenth-Century American Science: A Reappraisal. Evanston, Ill.: Northwestern University Press.
Davidson, J. B. 1907. "Tillage Machinery." In L. H. Bailey, ed. Cyclopedia of American Agriculture. New York: Macmillan, Vol. 1, 387–398.
Davidson, Marshall. 1951. Life in America. Boston, Mass.: Houghton Mifflin.
Davies, Norman DeGaris. 1917. The Tomb of Nakht at Thebes. New York: Metropolitan Museum of Art.
Davis, Joseph R., and Harvey S. Duncan. 1921. History of the Poland China Breed of Swine. Omaha, Neb.: Poland China History Association.
Day, Clarence A. 1954. A History of Maine Agriculture, 1604–1860. Orono, Me.: Maine University Press.
Deák, Gloria Gilda. 1988. Picturing America, 1497–1899. Princeton, N.J.: Princeton University Press.
Deane, Samuel. 1790. The New-England Farmer. Worcester, Mass.: Isaiah Thomas. 2d ed., same publisher, 1797; 3d ed., Boston: Wells and Lilly, 1822.
Dearborn, Henry Alexander. 1806. An Oration Delivered at Salem on the Fourth of July, 1806. Salem, Mass.: Register Office.
——. 1811. Oration pronounced at Boston, on the Fourth of July, 1811, before the Supreme Executive and in the presence of the Bunker-Hill Association. Boston, Mass.: Monroe and French.
De Coin, Colonel Robert L. 1864. History and Cultivation of Cotton and Tobacco. London: Chapman and Hall.
Deering Harvester Company. 1900. Official Retrospective Exhibition of Development of Harvesting Machinery for the Paris Exposition of 1900. Chicago, Ill.: Deering Harvester Co.
Delmage, Rutherford E. 1947. "The American Idea of Progress, 1750–1800." Proceedings of the American Philosophical Society, 91 (October), 307–314.
Demaree, Albert Lowther. 1941. The American Agricultural Press, 1819–1860. New York: Columbia University Press.
de Meyer, Maurits. 1946. "Sikkel, zichte, zeis en pik," Volkskunde, 5, 145–153.
Denton, Daniel. 1670. A brief description of New-York. . . . London: J. Hancock and W. Bradley.
Desaussure, Henry William. 1798. An Oration before the Inhabitants of Charleston, South Carolina, on the Fourth of July, 1798, in Commemoration of American Independence. Charleston, S.C.: W. P. Young.
De'Sigmond, Alexius A. J. 1935. "Development of Soil Science." Soil Science, 40 (June–December), 77–86.
de Vries, Jan. 1972. "Labor–Leisure Trade-off." Peasant Studies Newsletter, 1 (January), 45–50.
Dickinson, Rodolphus. 1813. A Geographical and Statistical View of Massachusetts Proper. Greenfield, Mass.: Denio and Phelps.
Diderot, Denis. 1762–1772. Encyclopédie Recueil de Planches, sur Sciences, les Arts Libéraux, et les Arts Méchaniques. Paris: Briasson, David, Le Breton, et Durand.
Dies, Edward Jerome. 1949. Titans of the Soil: Great Builders of Agriculture 1752–1940. Chapel Hill, N.C.: University of North Carolina Press.

Dodgshon, R. A., and C. A. Jewell. 1970. "Paring and Burning and Related Practices with particular references to the South-Western Counties of England." In Alan Gailey and Alexander Fenton, eds. The Spade in Northern and Atlantic Europe. Belfast: Ulster Folk Museum, 74–87.

Dondlinger, Peter Tracy. 1919. The Book of Wheat. New York: Orange Judd.

Doner, Ralph D., and M. L. Nichols. 1934. "The Dynamics of Soil on Plow Moldboard Surfaces Related to Scouring." Agricultural Engineering, 15 (January), 9–13.

Dosedla, H. C. 1984. "František Šach's Contribution towards Research on Pre-Industrial Tillage Implements in Austria." Tools and Tillage, 5, No. 1, 43–56.

Douglass, William. 1749. A Summary, Historical and Political, of the first Planting, Progressive Improvements, and Present State of the British Settlements in North America. Boston, Mass.: Rogers and Fowle.

Dovring, Folke. 1966. "The Transformation of European Agriculture." In H. J. Habakkuk and M. Postan, eds. The Cambridge Economic History of Europe, Vol. 6: The Industrial Revolutions and After. Cambridge, England: Cambridge University Press, 603–667.

——. 1969. "Eighteenth-Century Changes in European Agriculture: A Comment." Agricultural History, 43 (January), 181–186.

Dowdey, Clifford. 1957. The Great Plantation: A Profile of Berkeley Hundred and Plantation Virginia from Jamestown to Appomattox. New York: Rinehart.

Drepperd, Carl W. 1930. Early American Prints. New York: Century.

——. 1972. Pioneer America, Its First Three Centuries. New York: Cooper Square Publishers.

Drower, M. S. 1954. "Water-Supply, Irrigation, and Agriculture." In Charles Singer, E. J. Holmyard, and A. R. Hall, eds. A History of Technology. Oxford: Clarendon, Vol. 1, 520–557.

Drown, William. 1824. Compendium of Agriculture or the Farmer's Guide. Providence, R.I.: Field and Maxcy.

Dublin Society. 1740. Essays and Observations. Dublin: Dublin Society.

Duffy, John. 1951. "The Passage to the Colonies." Mississippi Valley Historical Review, 38 (June), 21–38.

Duke, J. S. 1846. "American Tobacco." De Bow's Review, 2 (October), 248–267.

Dumbauld, Edward. 1976. Thomas Jefferson, American Tourist. Norman, Okla.: University of Oklahoma Press.

Dunbar, James. 1780. Essays on the History of Mankind in Rude and Cultivated Ages. London: W. Strahan.

Dunbar, Seymour. 1915. A History of Travel in America. Indianapolis, Ind.: Bobbs-Merrill.

Dunbary, Gary S. 1961. "Colonial Carolina Cowpens." Agricultural History, 35 (July), 125–131.

Dupree, A. Hunter. 1976. "The National Pattern of American Learned Societies, 1709–1863." In Alexandra Oleson and Sanborn C. eds. Brown, The Pursuit of Knowledge in the Early American Republic. Baltimore, Md.: Johns Hopkins University Press, 21–32.

Dwight, Theodore. 1798. An Oration spoken at Hartford in the State of Connecticut on the Anniversary of American Independence, July 4th, 1798. Hartford, Conn.: Hudson and Goodwin.

Dwight, Timothy. 1969. Travels in New England and New York, ed. Barbara Miller Soloman and Patricia M. King. Cambridge, Mass.: Belknap.

Earle, Alice Morse. 1901. Stage-Coach and Tavern Days. New York: Macmillan.

Eaves, Charles Dudley. 1945. The Virginia Tobacco Industry, 1780–1860. Lubbock, Tx.: Texas Tech Press.

Ebeling, Walter. 1979. The Fruited Plain: The Story of American Agriculture. Berkeley, Calif.: University of California Press.

Edelstein, Ludwig. 1967. The Idea of Progress in Classical Antiquity. Baltimore, Md.: Johns Hopkins Press.

Edwards, Everett E., ed. 1937. Washington, Jefferson, Lincoln and Agriculture. Washington, D.C.: U.S. Department of Agriculture.

——. 1938. References on American Colonial Agriculture. Washington, D.C.: U.S. Department of Agriculture, Bibliographical Contributions 30.

——. 1940. "American Agriculture—The First 300 Years." U.S. Dept of Agriculture Yearbook, 71–276.

——. 1943. Jefferson and Agriculture. U.S. Department of Agriculture, Agricultural History Series 7. Washington, D.C.

——. 1949. "Europe's Contribution to the American Dairy Industry." Journal of Economic History, 9 (supplement), 72–84.

Edwards, Paul, ed. 1967. The Encyclopedia of Philosophy. New York: Macmillan.

Egloff, Keith. 1980. "Colonial Plantation Hoes of Tidewater Virginia." Virginia Historic Landmarks Commission. Research Report Series 1, Yorktown, Virginia, June.

Eighty Years' Progress of the United States. . . . New York: New National Publishing House, 1864.

Eisinger, Chester E. 1954. "The Farmer in the Eighteenth Century Almanac." Agricultural History, 28 (July), 107–112.

Ekirch, Arthur Alphonse, Jr. 1969. The Idea of Progress in America, 1815–1860. New York: Ames.

Eliot, Jared. 1934. Essays upon Field Husbandry in New England and Other Papers, 1748–1762, ed. Harry J. Carman and Rexford G. Tugwell. New York: Columbia University Press.

Ellet, Elizabeth Fries. 1850. Domestic History of the American Revolution. New York: Baker and Scribner.

Elliot, Richard Smith. 1883. Notes Taken in Sixty Years. St. Louis, Mo.: R. P. Studley.

Ellis, Franklin, and Samuel Evans. 1883. History of Lancaster County, Pennsylvania. Philadelphia, Penn.: Everts and Peck.

Ellis, William. 1742. The Modern Husbandman. London: T. Osborne.

Ellison, Rosemary. 1982. "The Agriculture of Mesopotamia c. 3000–600 B.C." Tools and Tillage, 4, No. 3, 173–184.

Ellsworth, Lucius F. 1968. "The Philadelphia Society for Promotion of Agriculture and Agriculture Reform, 1785–1793." Agricultural History, 42 (July), 189–199.

Emerson, George B., and Charles L. Flint. 1862. Manual of Agriculture. Boston, Mass.: Swan, Brewer and Tileston.

Emmet, William Temple. 1983. A Computer-Aided Analysis of the Performance of Coated Moldboard Plow Surfaces. Master of Science thesis, Cornell University (January).

Encyclopaedia Britannica; or, A Dictionary of Arts and Sciences. Edinburgh: A. Bell and C. Macfarquhar, 1771. 2d ed., Edinburgh: J. Balfour, 1778–83 3d ed., Philadelphia, Penn.: Budd and Bartram, 1803.

Engleman, Fred L. 1962. The Peace of Christmas Eve. New York: Harcourt, Brace and World.

Evans, E. Estyn. 1970. "Introduction." In Alan Gailey and Alexander Fenton, eds. The Spade in Northern and Atlantic Europe. Belfast: Ulster Folk Museum, 1–9.

Evans, Emory G. 1975. Thomas Nelson of Yorktown: Revolutionary Virginian. Charlottesville, Virg.: University Press of Virginia.

Everett, A. H. 1839. An Oration delivered at Holliston, Massachusetts, on the Fourth of July, 1839, at the request of the Democratic Citizens of the Ninth Congressional District. Boston, Mass.: Henry L. Devereux.

Everett, Edward. 1825. An Oration delivered at Plymouth, December 22, 1824. Boston, Mass.: Cummings, Hilliard.

——. 1826. An Oration delivered at Cambridge on the Fiftieth Anniversary of the Declaration of Independence. Boston, Mass.: Cummings, Hilliard.

——. 1835. Address delivered before the Literary Societies of Amherst College, August 25, 1835. Boston, Mass.: Russell, Shattuck, and Williams.

Ewan, Joseph. 1976. "The Growth of Learned and Scientific Societies in the Southern United States to 1860." In Alexandra Oleson and Sanborn C. Brown, eds. The Pursuit of Knowledge in the Early American Republic. Baltimore, Md.: Johns Hopkins University Press, 208–218.

Ewan, Joseph, and Nesta Ewan. 1970. John Banister and His National History of Virginia, 1678–1692. Urbana, Ill.: University of Illinois Press.

Fairholt, Frederick William. 1859. Tobacco: Its History and Associations. London: Chapman and Hall.

Farish, Hunter Dickinson, ed. 1943. Journal and Letters of Philip Vickers Fithian, 1773–1774: A Plantation Tutor of the Old Dominion. Williamsburg, Virg.: Colonial Williamsburg.

Faust, Drew Galpin. 1979. "The Rhetoric and Ritual of Agriculture in Antebellum South Carolina." Journal of Southern History, 4 (November), 541–568.

Fay, Bernard. 1932. "Learned Societies in Europe and America in the Eighteenth Century." American Historical Review, 37 (January), 255–266.

Fay, Sidney B. 1947. "The Idea of Progress." American Historical Review, 52 (January), 231–246.

Feller, Irwin 1962. "Inventive Activity in Agriculture 1837–1890." Journal of Economic History, 22 (December), 560–577.

Fenton, Alexander. 1970. "The Plough-Song: A Scottish Source for Medieval Plough History." Tools and Tillage, 1, No. 3, 175–191.

——. 1974. "The Cas-chrom: A Review of the Scottish Evidence." Tools and Tillage, 2, No. 3, 131–148.

Ferguson, Adam. 1792. Principles of Moral and Political Science. Edinburgh: W. Creech.

Ferleger, Lou. 1990. Agriculture and National Development: Views on the Nineteenth Century. Ames, Ia.: Iowa State University Press.

Fermigier, André. 1977. Jean-François Millet. Geneva: Albert Skira.

Fessenden, Thomas G. 1835. The Complete Farmer and Rural Economist. Boston, Mass.: Russell, Odiorne and Co.

Firor, John. 1990. The Changing Atmosphere: A Global Challenge. New Haven, Conn.: Yale University Press.

Fischer, David Hackett. 1962. "John Beale Bordley, Daniel Boorstin, and the American Enlightenment." Journal of Southern History, 28 (August), 327–342.

Fithian, Philip Vickers. 1965. Journal and Letters of Philip Vickers Fithian, 1773–1774: A Plantation Tutor of the Old Dominion, ed. Hunter Dickinson Farish. Williamsburg, Virg.: Colonial Williamsburg.

Fitzherbert, John. 1523. The Boke of Husbandrie. London: Rycharde Pynson. (Often mistakenly attributed to John's younger brother Anthony.)

Fitzpatrick, John C. 1925. The Diaries of George Washington, 1748–1799. Boston, Mass.: Houghton Mifflin.

Fitzpatrick, John C., ed. 1931–1944. The Writings of George Washington from the Original Manuscript Sources, 1745–1799. Washington, D.C.: U.S. Government Printing Office.

Fletcher, Stevenson W. 1947. "The Subsistence Farming Period in Pennsylvania Agriculture, 1640–1840." Pennsylvania History, 14 (July), 185–195.

——. 1950. Pennsylvania Agriculture and Country Life, 1640–1840. Harrisburg, Penn.: Pennsylvania Historical and Museum Commission.

——. 1959. The Philadelphia Society for Promoting Agriculture, 1785–1955. Philadelphia, Penn.: Philadelphia Society for Promoting Agriculture.

Flint, Charles L. 1864. "Progress in Agriculture." In Eighty Years' Progress of the United States. New York: L. Stebbins, 19–102.
——. 1874. "A Hundred Years' Progress of American Agriculture." In Maine Board of Agriculture, Annual Report for the Year 1874. Augusta, Ga.: Sprague, Owen and Nash, 105–153.
Flon, Christine. 1985. The World Atlas of Archeology. Boston, Mass.: G. K. Hall.
Fogel, Robert W. 1994. "Economic Growth, Population Theory, and Philosophy: The Bearing of the Long-Term Processes on the Making of Economic Policy." American Economic Review, 84 (June), 369–395.
Foner, Philip S., ed. 1945. The Complete Writings of Thomas Paine. New York: Citadel.
Forbes, R. J. 1958. "Power to 1850." In Charles Singer, E. J. Holmyard, and A. R. Hall, eds. A History of Technology. Oxford: Clarendon, Vol. 4, 148–167.
Ford, Paul Leicester, ed. 1892–1899. The Writings of Thomas Jefferson. New York: G. P. Putnam's Sons.
Ford, Worthington Chauncey, ed. 1891. The Writings of George Washington. New York: G. P. Putnam's Sons.
Forni, Gaetano. 1980. "Recent Archeological Findings of Tilling Tools and Fossil Ard Traces in Italy." Tools and Tillage, 4, No. 1, 60–64.
Forrest, Edwin. 1838. Oration delivered at the Democratic Republican Celebration of the Sixty-Second Anniversary of the Independence of the United States, in the City of New-York. New-York: J. W. Bell.
Forsyth, Robert. 1804. The Principles and Practice of Agriculture. . . . Edinburgh: A. Bell.
Fowler, Eldridge M. 1895. "Agricultural Machinery and Implements." In Chauncey M. Depew, ed. One Hundred Years of American Commerce. New York: D. O. Haynes, Vol. 2, 352–356.
Franklin, Benjamin. 1887. "Observations Concerning the Increase of Mankind and the Peopling of Countries." In John Bigelow, ed. The Complete Works of Benjamin Franklin. New York: G. P. Putnam's Sons, Vol. 2, 223–234.
Franklin, Ellis, and Samuel Evans. 1883. History of Lancaster County, Pennsylvania. Lancaster, Pa.: Lancaster County Heritage.
Franklin, W. Niel. 1926. "Agriculture in Colonial North Carolina." North Carolina Historical Review, 3 (October), 539–574.
Furlong, William Rea, and Byron McCandless. 1981. So Proudly We Hail: The History of the United States Flag. Washington D.C.: Smithsonian Institution.
Fussell, G. E. 1945. "History and Agricultural Science." Agricultural History, 19 (April), 126–127.
——. 1952. The Farmers Tools, 1500–1900. London: Andrew Melrose.
——. 1958. "Primary Production: Part II. Agriculture: Techniques of Farming." In Charles Singer, E. J. Holmyard, and A. R. Hall, eds. A History of Technology. Oxford: Clarendon, Vol. 4, 13–43.
——. 1959. "Low Countries' Influence on English Farming." English Historical Review, 74 (October), 611–622.
——. 1966. "Ploughs and Ploughing Before 1800." Agricultural History, 40 (July), 177–186.
——. 1973. Jethro Tull: His Influence on Mechanized Agriculture. Reading, England: Osprey.
——. 1976. "Agricultural Science and Experiment in the Eighteenth Century: An Attempt at a Definition." Agricultural History Review, 24, Part 1, 44–47.
Gabriel, Ralph Henry. 1940. Course of American Democratic Thought. New York: Ronald Press.
Gabriel, Ralph Henry, ed. 1926. The Pageant of America: A Pictorial History of the United States, Vol. 3. Toilers of Land and Sea. New Haven, Conn.: Yale University Press.

Gage, Charles E. 1937. "Historical Factors Affecting American Tobacco Types and Uses and the Evolution of the Auction Market." Agricultural History, 11 (January), 43–57.
Gagliardu, John G. 1959. "Germans and Agriculture in Colonial Pennsylvania." Pennsylvania Magazine of History and Biography, 83 (April), 192–218.
Gailey, Alan, and Alexander Fenton, eds. 1970a. The Spade in Northern and Atlantic Europe. Belfast: Ulster Folk Museum.
——. 1970b. "The Typology of the Irish Spade." In Alan Gailey and Alexander Fenton, eds. The Spade in Northern and Atlantic Europe. Belfast: Ulster Folk Museum, 35–48.
Gallatin, Albert. 1810. "Report on Manufactures." American State Papers, Finance, Vol. 2, 11th Cong., 2nd Sess., Doc. 325.
Gambrill, Olive Moore. 1942. "John Beale Bordley and the Early Years of the Philadelphia Agricultural Society." Pennsylvania Magazine, 66 (October), 410–439.
Gardner, W. H. 1977. "Historical Highlights in American Soil Physics, 1776–1976." Journal of the Soil Science Society of America, 41 (March–April), 221–229.
Garner, W. W. 1913. "Tobacco Curing." U.S. Department of Agriculture, Farmers' Bulletin 523, 1–24.
Garner, W. W., et al. 1922. "History and Status of Tobacco Culture." U.S. Department of Agriculture Yearbook, 395–468.
Garrison, J. Ritchie. 1991. Landscape and Material Life in Franklin County, Massachusetts, 1770–1860. Knoxville, Tenn.: University of Tennessee Press.
Gates, Paul W. 1960. The Farmer's Age: Agriculture 1815–1860. New York: Holt, Rinehart and Winston.
——. 1972. "Problems of Agricultural History, 1790–1840." Agricultural History, 46 (January), 33–58.
Gay, Peter. 1969. The Enlightenment: An Interpretation. New York: Alfred A. Knopf.
Gaylord, Willis, and Luther Tucker. 1840. American Husbandry. New York: Harper and Brothers.
Gehrke, William H. 1935. "The Antebellum Agriculture of the Germans in North Carolina." Agricultural History, 9 (July), 143–160.
Gerdts, William H. 1990. Art Across America: Two Centuries of Regional Painting, 1710–1920. New York: Abbeville.
Gibbon, David, and Ted Smart. 1979. Colonial Virginia. New York: Crescent.
Gibson, Elizabeth Bordley. 1865. Biographical Sketches of the Bordley Family of Maryland. Philadelphia, Penn.: H. B. Ashmead.
Gilbert, Frank. 1882. Jethro Wood: Inventor of the Modern Plow. Chicago, Ill.: Rhodes and McClure.
Gill, William R., and Glen E. Vanden Berg. 1967. Soil Dynamics in Tillage and Traction. Agricultural Handbook No. 316. Washington, D.C.: U.S. Government Printing Office.
Gilmore, E. H. 1878. History of Tobacco: Its Original Introduction and its Present Culture and Manufacture. Washington, D.C.: n.p..
Gilmore, John W. 1907. "The Utility of Farm Machinery." In L. H. Bailey, ed. Cyclopedia of American Agriculture. New York: Macmillan, Vol. 1, 203–207.
Glassie, Henry. 1968. The Pattern in the Material Folk Culture of the Eastern United States. Philadelphia, Penn.: University of Pennsylvania Press.
Glob, P. V. 1951. Ard and Plough in Prehistoric Scandanavia. Aarhus: Aarhus University Press.
Godwin, Peter. 1816/17. "Straw Cutter." Massachusetts Agricultural Repository and Journal, 4, 352–353.
——. 1839. "The Course of Civilization." Democratic Review, 6 (September), 208–217.
Gold, Thomas. 1817. Address . . . delivered before the Berkshire association of agriculture and manufactures, at Pittsfield, October 2, 1817. Pittsfield, Mass.: P. Allen.

Goode, George Brown. 1901. "The Beginnings of National History in America." Report of the U.S. National Museum; Annual Report of the Board of Regents of the Smithsonian Institution, Part 2. Washington, D.C.: Government Printing Office, 355–406.

Googe, Barnaby. 1614. The whole art and trade of husbandry. London: Richard More.

Gordon, Robert B., and Patrick M. Malone. 1994. The Texture of Industry: An Archeological View of the Industrialization of North America. New York: Oxford University Press.

Gottesman, Reta Susswein, ed. 1938. The Arts and Crafts in New York, 1726–1776. New York: New-York Historical Society.

Gould, Clarence P. 1915. Money and Transportation in Maryland, 1720–1765. Baltimore, Md.: Johns Hopkins Press.

Gould, Mary Earle. 1962. Early American Wooden Ware and Other Kitchen Utensils. Rutland, Vt.: Charles E. Tuttle.

Gras, N. S. B. 1940. History of Agriculture in Europe and America. New York: F. S. Crofts.

Gray, Lewis C. 1958. History of Agriculture in the Southern United States to 1860. Washington, D.C.: The Carnegie Institution.

Greasley, David, and Les Oxley. 1994. "Rehabilitation Sustained: The Industrial Revolution as a Macroeconomic Epoch." Economic History Review, 47 (November), 760–768.

Grece, Charles F. 1819. Facts and Observations respecting Canada and the United States of America. . . . London: J. Harding.

Green, C. E., and D. Young, eds. 1980. Encyclopedia of Agriculture. London: William Green & Sons.

Greenburg, David B., ed. 1946. Furrow's End: An Anthology of Great Farm Stories. New York: Greenberg.

Greene, Jack P. 1965. The Diary of Colonel Landon Carter of Sabine Hill, 1752–1778. Charlottesville, Virg.: University Press of Virginia.

——. 1968a. "Introduction." In Jack P. Greene, ed. The Reinterpretation of the American Revolution, 1763–1789. New York: Harper & Row, 2–75.

——. 1968c. "The Role of the Lower Houses of Assembly in Eighteenth-Century Politics." In Jack P. Greene, ed. The Reinterpretation of the American Revolution, 1763–1789. New York: Harper & Row, 86–109.

——. 1989. Selling a New World: Two Colonial South Carolina Promotional Pamphlets by Thomas Nairne and John Norris. Columbia, S.C.: University of South Carolina Press.

Greene, Jack P., ed. 1968b. The Reinterpretation of the American Revolution, 1763–1789. New York: Harper & Row.

Greeno, Follett L. 1912. Obed Hussey: Who, Of All Inventors, Made Bread Cheap. Rochester, New York: Rochester Herald Publishing Co.

Grevin, Phillip J. 1972. Four Generations: Population, Land, and Family in Colonial Andover, Mass. Ithaca, N.Y.: Cornell University Press.

Grigg, David. 1982. The Dynamics of Agricultural Change. New York: St. Martin's.

——. 1992. The Transformation of Agriculture in the West. Oxford: Blackwell.

Grimke, Thomas S. 1829. An Oration delivered in St. Philip's Church before the Inhabitants of Charleston on the Fourth of July, 1809. Charleston, S.C.: William Riley.

Groce, George C., Jr. 1937. "Benjamin Gale." New England Quarterly, 10 (December), 697–716.

Gross, Robert A. 1976. The Minutemen and Their World. New York: Hill and Wang.

Guetter, Fred J., and Albert E. McKinley. 1924. Statistical Tables Relating to the Economic Growth of the United States. Philadelphia, Penn.: McKinley.

Haberey, Waldemar. 1949. "Gravierte Glasschale und Sogenannte Mithrassymbole aus

einem spätrömischen Grabe von Rodenkirchen bei Köln." Bonner Jahrbücher, Vol. 149, 94–104.
Hagedoorn, A. L. 1948. Animal Breeding. London: Crosby Lockwood & Son.
Hall, A. R. 1957. "The Rise of the West." In Charles Singer, E. J. Holmyard, and A. R. Hall, eds. A History of Technology. Oxford: Clarendon, Vol. 3, 709–721.
Hallam, Elizabeth, ed. 1986. The Plantagenet Chronicles. New York: Weidenfeld & Nicolson.
Halsey, Francis W., ed. 1906. A Tour of Four Great Rivers . . . in 1769, being the Journal of Richard Smith. New York: Charles Scribner's Sons.
Hamilton, E. W. 1931. "George Washington, Farmer." American Thresherman, 33 (February), 5, 18.
Hamilton, Sinclair. 1958. Early American Book Illustrators and Wood Engravers, 1670–1870. Princeton, N.J.: Princeton University Press.
Hanke, Oscar A., and John L. Skinner, eds. 1974. American Poultry History, 1823–1973. Madison, Wisc.: American Poultry Historical Society.
Hansen, W. 1982. Hauswesen und Tagewerk in Alten Lippe. Münster: Aschendorff.
Harper, James. 1972. "The Tardy Domestication of the Duck." Agricultural History, 46 (July), 385–389.
Harris, L. E. 1957. "Land Drainage and Reclamation." In Charles Singer, E. J. Holmyard, and A. R. Hall, eds. A History of Technology. Oxford: Clarendon, Vol. 3, 300–323.
Harrison, H. S. 1954. "Discovery, Invention, and Diffusion." In Charles Singer, E. J. Holmyard, and A. R. Hall, eds. A History of Technology. Oxford: Clarendon Press, Vol. 1, 58–84.
Hart, Albert Bushnell. 1908. American History Told by Contemporaries. New York: Macmillan.
Hartlib, Samuel. 1652. Samuel Hartlib, his Legacy; or an enlargement of the Discourse of Husbandry used in Brabant and Flanders. London: R. and W. Leybourn.
Hassall, W. O., ed. 1970. The Holkham Library. Oxford: Oxford University Press.
Haudricourt, André G., and Mariel Jean-Brunhes Delamarre. 1955. L'Homme et la Charrue à Travers le Monde. Paris: Gallimard.
Hawken, Henry A. 1976. Trumpets of Glory: Fourth of July Orations, 1786–1861. Granby, Conn.: Salmon Brook Historical Society.
Haworth, Paul Leland. 1915. George Washington: Farmer. Indianapolis, Ind.: Bobbs-Merrill.
——. 1925. George Washington, Country Gentleman: Being an Account of His Home Life and Agricultural Activities. Indianapolis, Ind.: Bobbs-Merrill.
Hays, Willet M. 1901. Plant Breeding. U.S. Department of Agriculture, Division of Vegetable Physiology and Pathology, Bulletin 29.
Haywood, C. Robert. 1959. "Mercantilism and South Carolina Agriculture, 1700–1763." South Carolina Historical Magazine, 60 (January), 15–27.
Headlee, Sue. 1991. The Political Economy of the Family Farm: The Agrarian Roots of American Capitalism. New York: Praeger.
Hedrick, Ulysses P. 1933. A History of Agriculture in the State of New York. Albany, N.Y.: New York State Agricultural Society.
Henretta, James A. 1973. The Evolution of American Society, 1700–1815. Lexington, Mass.: D. C. Heath.
Herndon, G. Melvin. 1957. Tobacco in Colonial Virginia: The Sovereign Remedy. Williamsburg, Virg.: Virginia 350th Anniversary Celebration Corporation.
——. 1969. William Tatham and the Culture of Tobacco. Coral Gables, Fla.: University of Miami Press.
Hewitt, W. 1766. "Account of the Construction and Use of two Horse-hoes, or Hoe-ploughs, laid before the Society for the Encouragement of Arts, etc., by the Inventor." Museum Rusticum et Commerciale, 6, 397–403.

Hieronimus, Robert. 1989. America's Secret Destiny: Spiritual Vision and the Founding of a Nation. Rochester, N.Y.: Destiny Books.

Higgins, F. Hal. 1958. "John M. Horner and the Development of the Combined Harvester." Agricultural History, 32 (January), 14–24.

Hildebrand, George H. 1949. "Introduction." In Frederick J. Teggart. The Idea of Progress. Berkeley, Calif.: University of California Press, 3–30.

Hilldrup, Robert L. 1959. "A Campaign to Promote the Prosperity of Colonial Virginia." Virginia Magazine of History and Biography, 67 (October), 410–428.

Hindle, Brooke. 1956. The Pursuit of Science in Revolutionary America, 1735–1789. Chapel Hill, N.C.: University of North Carolina Press.

——. 1981. Emulation and Invention. New York: W. W. Norton.

Hinke, William J. 1916. "Report of the Journey of Francis Louis Michel from Berne, Switzerland to Virginia, October 2, 1701–December 1, 1702." Virginia Magazine, 24 (January), 1–43.

Hirsch, Arthur H. 1930. "French Influence on American Agriculture in the Colonial Period with Special Reference to Southern Provinces." Agricultural History, 4 (January), 1–9.

Hobi, Franz. 1926. Die Benennungen von Sichel und Sense in den Mundarten der Romanischen Schweiz. Heidelberg: Carl Winter's Universitätsbuchhandlung.

Hogner, Dorothy C. 1944. Our American Horse. New York: T. Nelson & Sons.

Hoke, Donald R. 1990. Ingenious Yankees: The Rise of the American System of Manufactures in the Private Sector. New York: Columbia University Press.

Holbrook, Stewart H. 1955. Machines of Plenty: Pioneering in American Agriculture. New York: Macmillan.

Holland, James W. 1983. "The Beginning of Public Agricultural Experimentation in America: The Trustee's Garden in Georgia." Agricultural History, 12 (July), 271–298.

Holland, Josiah Gilbert. 1855. History of Western Massachusetts. Springfield, Ill.: Samuel Bowles.

Holmes, George K. 1899. "Progress of Agriculture in the United States." U.S. Department of Agriculture Yearbook, 307–334.

——. 1912. Tobacco Crop of the United States, 1612–1911. U.S. Department of Agriculture, Bureau of Statistics, Circular 33. Washington, D.C.: Government Printing Office.

——. 1919. "Three Centuries of Tobacco." U.S. Department of Agriculture Yearbook, 151–175.

Hopfen, H. J. 1960. Farm Implements for Arid and Tropical Regions. FAO Agricultural Development Paper No. 67. Rome: Food and Agriculture Organization of the United Nations.

Hounshell, David A. 1984. The History of American Technology: Exhilaration or Discontent? Wilmington, Del.: Hagley Museum.

Howard, Robert W. 1965. The Horse in America. Chicago, Ill.: Follett.

Howells, William Cooper. 1963. Recollections of Life in Ohio, from 1813 to 1840. Gainesville, Fla.: Scholar's Facsimiles & Reprints.

Hulbert, Archer B. 1920. The Paths of Inland Commerce. New Haven, Conn.: Yale University Press.

Humphreys, David. 1816. A discourse on the agriculture of the state of Connecticut. New Haven, Conn.: W. W. Woodward.

Humphries, Walter R., and R. B. Gray. 1949. Partial History of Haying Equipment. Washington, D.C.: U.S. Department of Agriculture, Information Series 74.

Hunt, Frazier. 1949. Horses and Heroes: The Story of the Horse in America for 450 Years. New York: Charles Scribner's Sons.

Hurt, R. Douglas. 1982. American Farm Tools: From Hand-Power to Steam-Power. Manhattan, Kan.: Sunflower University Press.

Hutchinson, William T. 1930. Cyrus Hall McCormick: Seed-Time, 1809–1856. New York: Century.
——. 1935. Cyrus Hall McCormick: Harvest, 1856–1884. New York: D. Appleton-Century.
Iinuma, Jiro. 1969. "The Ne-No-Hi-Kara-Suki of Shosoin." Tools and Tillage, 1, No. 2, 105–117.
——. 1982. "The Development of Ploughs in Japan." Tools and Tillage, 4, No. 3, 139–154.
Inge, W. R. 1920. The Idea of Progress. Oxford: Clarendon Press.
International Harvester Company of America. 1913. Harvest Scenes of the World. Chicago, Ill.: International Harvester Company of America.
Jackson, Donald, ed. 1976. The Diaries of George Washington. Charlottesville, S.C.: University Press of Virginia.
Jacobstein, Meyer. 1907. The Tobacco Industry in the United States. New York: Columbia University Press.
Jameson, J. Franklin, ed. 1909. Narratives of New Netherland, 1609–1664. New York: Charles Scribner's Sons.
——. 1956. The American Revolution Considered as a Social Movement. Boston, Mass.: Beacon.
Jefferson, Thomas. Writings. See Ford, P. L.
——. 1799. "The Description of a Mould-board of the least resistance, and of the easiest and most certain construction. . . ." American Philosophical Society Transactions, 4, 313–322.
——. 1801. Notes on the State of Virginia. Philadelphia, Penn.: R. T. Rawle.
——. 1944. Thomas Jefferson's Garden Book, 1766–1824 with Relevant Extracts from His Other Writings, ed. Edwin Morris Betts. Philadelphia, Penn.: American Philosophical Society.
Jellicoe, Geoffrey A., and Susan Jellicoe. 1975. The Landscape of Man: Shaping the Environment from Prehistory to the Present Day. New York: Viking Press.
Johnson, Cuthbert William. 1848. British Husbandry; exhibiting the Farming Practice in Various Parts of the United Kingdom. London: Robert Baldwin.
Johnson, Paul Cornelius. 1976. Farm Inventions in the Making of America. Des Moines, Ia.: Wallace-Homestead Book Co.
Johnston, Alexander, and James Albert Woodburn. 1896. American Orations: Studies in American Political History. New York: G. P. Putnam's Sons.
Jones, E. L., and S. J. Woolf, eds. 1969a. Agrarian Change and Economic Development. London: Methuen.
——. 1969b. "The Historical Role of Agrarian Change in Economic Development." In their Agrarian Change and Economic Development. London: Methuen, 1–22.
Jones, Howard Mumford. 1964. O Strange New World. New York: Viking.
Jones, Hugh. 1956. The Present State of Virginia. Chapel Hill, N.C.: University of North Carolina Press. (Original 1724.)
Jones, Lewis R. 1977. "The Mechanization of Reaping and Mowing in American Agriculture, 1833–1870: Comment." Journal of Economic History, 27 (June), 451–455.
Jope, E. M. 1956a. "Agricultural Implements." In Charles Singer, E. H. Holmyard, and A. R. Hall, eds. A History of Technology. Oxford: Clarendon, Vol. 2, 81–102.
——. 1956b. "Vehicles and Harnesses." In Charles Singer, E. H. Holmyard, and A. R. Hall, eds. A History of Technology. Oxford: Clarendon, Vol. 2, 537–562.
Jordan, Weymouth T. 1969. "Some Problems of Colonial Tobacco Planters: A Critique." Agricultural History, 43 (January), 83–86.
Kalm, Peter. See Adolph Benson.
Keller, A. C. 1950. "Zilsel, the Artisans, and the Idea of Progress in the Renaissance." Journal of the History of Ideas, 11 (April), 235–240.

Kelly, Francis M., ed. 1936. The Seasons of the Year: Life in the Middle Ages as Depicted in Illuminated Miniatures. London: B. T. Batsford.
Kelsey, Darwin, ed. 1972. Farming in the New Nation: Interpreting American Agriculture, 1790–1840. Washington, D.C.: Agricultural History Society.
Kendall, Norman Festus. 1944. History and Geneology: Kendalls, Cunninghams, Snodgrasses. Grafton, West Virg.: Grafton Sentinel Publishing.
Kenyon, Cecilia M. 1968. "Republicanism and Radicalism in the American Revolution: An Old-Fashioned Interpretation." In Jack P. Greene, ed. The Reinterpretation of the American Revolution, 1763–1789. New York: Harper and Row, 291–320.
Kerr, Homer L. 1964. "Introduction of Forage Plants into Ante-Bellum United States." Agricultural History, 38 (April), 87–95.
Kerridge, Eric. 1967. The Agricultural Revolution. London: George Allen & Unwin.
——. 1969. "The Agricultural Revolution Reconsidered." Agricultural History, 43 (October), 463–475.
Kiger, Joseph C. 1963. American Learned Societies. Washington, D.C.: Public Affairs Press.
Kilgour, Frederick G. 1949. "The Rise of Scientific Activity in Colonial New England." Yale Journal of Biology and Medicine, 22 (December), 123–138.
Killebrew, J. B. 1883. "Request on the Culture and Curing of Tobacco in the United States." U.S. Bureau of the Census, Tenth Census, 1880, Report on the Productions of Agriculture.
King, F. H. 1907a. A Text Book of the Physics of Agriculture. Madison, Wisc.: F. H. King.
——. 1907b. "Tillage: Its Philosophy and Practice." In L. H. Bailey, ed. Cyclopedia of American Agriculture. New York: Macmillan, Vol.1, 378–387.
Kirk, Caleb. 1808. "Substitute for Trench Ploughing, and new Mode of putting in Winter Grain, and on live Fences." Philadelphia Society for Promoting Agriculture, Memoirs, 1, 85–92.
Klingman, David. 1971. "Food Surpluses and Deficits in American Colonies, 1768–1772." Journal of Economic History, 31 (September), 553–569.
Knight, Edward H. 1880. "Agricultural Implements." Reports of the United States Commissioners to the Paris Exposition, 1878, Vol. 5. Washington, D.C.: Government Printing Office.
Knight, Franklin, ed. 1847. Letters on Agriculture from his Excellency George Washington . . . to Arthur Young. Washington, D.C.: Franklin Knight.
Knowlton, Harry E., Robert B. Elwood, and Eugene G. McKibben. 1938. "Changes in Technology and Labor Requirements in Crop Production: Potatoes." Studies of Changing Techniques and Employment in Agriculture, Report No. A-4, WPA National Research Project, Philadelphia.
Kornwolf, James D. 1983. "The Picturesque in the American Garden and Landscape before 1800." Eighteenth-Century Life, 8 (January), 93–106.
Kramer, Fritz L. 1966. Breaking Ground: Notes on the Distribution of Simple Tilling Tools. Sacramento, Calif.: Sacramento Anthropological Society.
Kraus, Michael. 1928. Intercolonial Aspects of American Culture on the Eve of the Revolution. Ph.D. dissertation, Columbia University, New York.
——. 1942. "Scientific Relations Between Europe and America in the Eighteenth Century." Scientific Monthly, 55 (September), 259–272.
——. 1959. The United States to 1865. Ann Arbor, Mich.: University of Michigan Press.
Kren, Thomas, ed. 1983. Renaissance Painting in Manuscripts: Treasures from the British Library. New York: Hudson Hills.
Krussmaul, Ann. 1990. A General View of the Rural Economy of England, 1538–1840. Cambridge, England: Cambridge University Press.
Krythe, Maymie R. 1968. What So Proudly We Hail: All About Our American Flag, Monuments and Symbols. New York: Harper and Row.

Kuznets, Simon. 1966. Modern Economic Growth: Rate, Structure, and Spread. New Haven, Conn.: Yale University Press.

La Rochefoucauld Liancourt, [François Alexandre Frédéric], duc de. 1800. Travels Through the United States of North America, the Country of the Iroquois, and Upper Canada in the Years 1795, 1796, and 1797. London: T. Davidson.

Laing, Wesley N. 1955. "Cattle in Seventeenth-Century Virginia." Virginia Magazine of History and Biography, 67 (April), 143–163.

Land, Aubrey C. 1967. "Economic Behavior in a Planting Society: The Eighteenth-Century Chesapeake." Journal of Southern History, 33 (November), 469–485.

——. 1969. "The Tobacco Staple and the Planter's Problems: Technology, Labor, and Crops." Agricultural History, 43 (January), 69–89.

Lane, Wheaton J. 1939. From Indian Trail to Iron Horse: Travel and Transportation in New Jersey, 1620–1860. Princeton, N.J.: Princeton University Press.

——. 1950. "The Early Highway In America, to the Coming of the Railroad." In Jean Labatut and Wheaton J. Lane, eds. Highways In Our National Life. Princeton, N.J.: Princeton University Press, 66–76.

Lang, John. 1811. "On Harrowing Wheat in the Spring." Philadelphia Society for Promoting Agriculture, Memoirs, 1, 9–11.

Langdon, William Chauncey. 1941. Everyday Things in American Life, 1776–1876. London: Charles Scribner's Sons.

Larson, C. W. 1922. "The Dairy Industry." U.S. Department of Agriculture Yearbook, 281–394.

Lathem, Edward C. 1972. Chronological Tables of American Newspapers, 1690–1820; Being a Tabular Guide to Holdings of Newspapers Published in America Through the Year 1820. Worcester, Mass.: American Antiquarian Society.

Lathrop, Elise. 1926. Early American Inns and Taverns. New York: Robert M. McBride.

Latrobe, Benjamin Henry. 1905. The Journal of Latrobe. New York: D. Appleton.

Lawson, Douglas. 1982. Hand to Plough: Old Farm Tools and Machinery in Pictures. Sevenoaks: Ashgove.

Layton, Edwin. 1972. "Mirror-Image Twins: The Communities of Science and Technology." In George H. Daniels, Nineteenth-Century American Science: A Reappraisal. Evanston, Ill.: Northwestern University Press, 210–230.

Leaveritt, E. T. 1930. "George Washington and Power Farming: His Efforts Were to Reduce His Labor Expense Through Utilization of More Power." Farm and Ranch, 49 (February), 2, 14.

Leavitt, Charles T. 1933. "Attempts to Improve Cattle Breeds in the United States, 1790–1860." Agricultural History, 7 (April), 51–67.

Lee, Norman E. 1960. Harvests and Harvesting through the Ages. Cambridge, England: Cambridge University Press.

Lehner, Ernst, and Johanna Lehner. 1966. How They Saw The New World. New York: Tudor.

Lemmer, George F. 1947. "The Spread of Improved Cattle Through the Eastern United States to 1850." Agricultural History, 21 (April), 79–93.

——. 1957. "Early Agricultural Editors and Their Farm Philosophies." Agricultural History, 31 (October), 3–22.

Lemon, James T. 1966. "The Agricultural Practices of National Groups in Eighteenth-Century Southeastern Pennsylvania." Geographical Review, 56 (October), 467–496.

——. 1972. The Best Poor Man's Country: A Geographical Study of Early Southeastern Pennsylvania. Baltimore, Md.: Johns Hopkins.

Lender, Mark E. 1975. "The New Jersey Soldier." In Larry R. Gerlach, ed. New Jersey's Revolutionary Experience. Trenton: New Jersey Historical Commission, Number 5.

——. 1976. "The Mind of the Rank and File: Patriotism and Motivation in the Continental Line." In William C. Wright, ed. New Jersey in the American Revolution. Trenton, N.J.: New Jersey Historical Commission, 21–34.

Lerche, Grith, and Axel Steensberg. 1983. "Tools and Tillage in Iran: Observations Made in 1965 in the Province of Kermán." Tools and Tillage, 4, No. 4, 217–248.

Leser, Paul. 1931. Entstehung und Verbreitung des Pfluges. Munster: Aschendorffsche Verlagsbuchhandlung.

Lewis, Jan. 1983. The Pursuit of Pleasure: Family Life and Values in Jefferson's Virginia. Cambridge: Cambridge University Press.

Lewis, R. W. B. 1955. The American Adam: Innocence, Tragedy and Tradition in the Nineteenth Century. Chicago, Ill.: University of Chicago Press.

Lewton, F. L. 1943. "Notes on the Old Plows in the United States National Museum." Agricultural History, 17 (January), 62–64.

Liger, Louis. 1805. La Nouvelle Maison Rustique. Paris: Deterville.

Lindemans, Paul. 1952. Geschiedenis van de Landbouw in België. Antwerp: De Sikkel.

Lindgren, A. C., and O. B. Zimmerman. 1922. "Coordination of Theory and Practice in Plow Design and Operation." Agricultural Engineering, 3 (January), 3–10.

Lipscomb, Andrew A., and Alfred Ellery Bergh. 1903. The Writings of Thomas Jefferson. Washington, D.C.: The Thomas Jefferson Memorial Association.

Livingston, Robert. 1832. "American Agriculture." New Edinburgh Encyclopedia. Philadelphia, Penn.: Joseph and Edward Parker, Vol. 1, 332–341.

Lockridge, Kenneth A. 1974. Literacy in Colonial New England. New York: W. W. Norton.

Lodge, Henry Cabot, ed. 1904. The Works of Alexander Hamilton. New York: G. P. Putnam's Sons.

Loeb, Robert H. 1976. New England Village: Everyday Life in 1810. Garden City, N.Y.: Doubleday.

Loehr, Rodney C. 1937. "The Influence of English Agriculture on American Agriculture, 1775–1825." Agricultural History, 11 (January), 3–15.

——. 1969. "Arthur Young and American Agriculture." Agricultural History, 43 (January), 43–56.

Logan, George. 1797a. Agricultural Experiments on Gypsum and Experiments to Ascertain the Best Rotation of Crops. Philadelphia, Penn.: F. & R. Bailey.

——. 1797b. Fourteen Agricultural Experiments, to ascertain the best rotation of crops: addressed to the Philadelphia Agricultural Society. Philadelphia, Penn.: Francis and Robert Bailey.

——. 1818. "Account of Bennett's clover and turnip sowing machine." Philadelphia Society for Promoting Agriculture, Memoirs, 4, 44–45.

The London Encyclopaedia, or Universal Dictionary of Science, Art, Literature and Practical Mechanics. . . . London: [n.p.], 1829.

Longnon, Jean, et al. 1969. The Très Riches Heures of Jean, Duke of Berry, Musée Condé, Chantilly. New York: George Braziller.

Lorain, John. 1814. "On a simple Wheat Drill." Philadelphia Society for Promoting Agriculture, Memoirs, 3, 32–37.

——. 1825. Nature and Reason Harmonized in the Practice of Husbandry. Philadelphia, Penn.: Carey and Lea.

Lord, Russell. 1962. The Care of the Earth: A History of Husbandry. New York: Thomas Nelson and Sons.

Lossing, Benson J. 1851–52. The Pictorial Field-Book of the Revolution. New York: Harper and Brothers, Vol. 1.

Loudon, J. C. 1839. An Encyclopaedia of Agriculture. London: Longman, Orme, Brown, Green, and Longmans.

Lovejoy, Arthur O. 1957. The Great Chain of Being. Cambridge, Mass.: Harvard University Press.

Lucas, A. T. 1978. "The 'Gowl-Gob': An Extinct Spade Type from County Mayo, Ireland." Tools and Tillage, 3, No. 3, 191–198.

Lukes, Steven. 1973. Individualism. New York: Harper and Row.

Luttrell Psalter. See E. G. Millar.

Lyell, Charles. 1845. Travels in North America. London: John Murray.

M'Robert, Patrick. 1776. A Tour Through Part of the North Provinces of America . . . in the Years 1774 and 1775. . . . Edinburgh: The author.

Macdonald, Stuart. 1978. "The Early Threshing Machine in Northumberland." Tools and Tillage, 3, No. 3, 168–183.

MacGill, Caroline E. 1917. History of Transportation in the United States before 1860. Washington, D.C.: Carnegie Institution.

MacInnes, Charles Malcolm. 1926. The Early English Tobacco Trade. London: Kegan Paul, Trench, Trubner.

Mackay, Charles. 1859. Life and Liberty in America: Sketches of a Tour in the United States and Canada in 1857–8. New York: Harper & Brothers.

Macklay, William B. 1836. An Oration delivered before the Literacy Association on Monday, the Fourth of July, 1836. New York: J. P. Wright.

MacNeven, William James. 1825. Introductory discourse to a few lectures on the application of chemistry to agriculture, delivered before the New-York Athanaeum in the winter of 1825. New York: G. and C. Carvill.

Macy, Loring K., Lloyd E. Arnold, and Eugene G. McKibben. 1938. "Changes in Technology and Labor Requirements in Crop Production: Corn." Studies of Changing Techniques and Employment in Agriculture. Report No. A-5, WPA National Research Project, Philadelphia.

Mahn, Herbert A. 1966. Flags and their History. New York: Carleton.

Main, Jackson Turner. 1968. "Government by the People: The American Revolution and the Democratization of the Legislatures." In Jack P. Greene, ed. The Reinterpretation of the American Revolution, 1763–1789. New York: Harper and Row, 322–338.

Mairs, Thomas I. 1928. Some Pennsylvania Pioneers in Agricultural Science. University Park, Penn.: Pennsylvania State College Press.

Malin, Donald F. 1923. The Evolution of Breeds: An Agricultural Study of Breed Building as Illustrated in Shorthorn, Hereford, and Aberdeen Angus Cattle, Poland China and Duroc Jersey Swine. Des Moines, Ia.: Wallace.

Manning, W. H. 1970. "Mattocks, Hoes, Spades and Related Tools in Roman Britain." In Alan Gailey and Alexander Fenton, eds. The Spade in Northern and Atlantic Europe. Belfast: Ulster Folk Museum, 18–29.

Markham, Gervase. 1613. The English Husbandman. London: John Browne.

Markley, Benjamin A. 1811. An Oration delivered on the Fourth of July, 1811, in Commemoration of American Independence, at Charleston, S.C. Charleston, S.C.: J. Hoff.

Marshall, Geoffrey. 1978. "The 'Rotherham' Plough: A Study of a Novel 18th Century Implement in Agriculture." Tools and Tillage, 3, No. 3, 150–167.

——. 1982. "The 'Rotherham' Plough: New Evidence on its Original Manufacture and Method of Distribution." Tools and Tillage, 4, No. 3, 131–138.

Marshall, T. H. 1929. "Jethro Tull and the 'New Husbandry' of the Eighteenth Century." Economic History Review, 2 (January), 41–60.

——. 1967. "Early Agricultural Societies in New York: The Foundations of Improvement." New York History, 48 (October), 313–331.

Marti, Donald B. 1971. "The Purposes of Agricultural Education: Ideas and Projects in New York State, 1819–1865." Agricultural History, 45 (October), 271–283.

——. 1979. To Improve the Soil and the Mind: Agricultural Societies, Journals, and Schools in the Northeastern States, 1791–1865. Ann Arbor, Mich.: University Microfilms International.

——. 1980. "Agricultural Journalism and the Diffusion of Knowledge: The First Half-Century in America." Agricultural History, 54 (January), 28–37.

Martin, George A., ed. 1892. Farm Appliances: A Practical Manual. New York: Orange Judd.

Martin, J. Lynton. 1986. The Ross Farm Story. Halifax: Nova Scotia Museum.

Martineau, Harriet. 1837. Society in America. New York: Saunders and Otley.

Masefield, G. B. 1972. A Short History of Agriculture in the British Colonies. Oxford: Clarendon Press.

Mason, John, Jr. 1976. "A Commencement Oration delivered on the Fourth of July, 1818, at the College of William & Mary, Williamsburg, Virginia." Reprinted in Henry A. Hawken, ed. Trumpets of Glory: Fourth of July Orations, 1786–1861. Granby, Conn.: Salmon Brook Historical Society, 300–306.

Massachusetts Board of Agriculture. 1854. First Annual Report. 1853. Boston, Mass.: William White.

Massachusetts Society for Promoting Agriculture. 1942. An Outline of the History of the Massachusetts Society for Promoting Agriculture. Boston, Mass.: Meador.

Matteson, David M. 1931. Washington the Farmer. Washington, D.C.: George Washington Bicentennial Commission, Pamphlet No. 4.

Matthews, Albert. 1907. "The Snake Devices, 1754–1776, and the Constitutional Courant, 1765." Publications of the Colonial Society of Massachusetts, Transactions, 11 (December), 408–453.

Maxey, Edward A. 1601. A New Instruction of Ploughing and Settin of Corn. London: Felix Kyngston.

Maxwell, Robert, ed. 1843. Select Transactions of the Honourable Society of Improvers in the Knowledge of Agriculture in Scotland. Edinburgh: Sands, Brymer, Murray and Cochran.

May, Henry F. 1976. The Enlightenment in America. New York: Oxford University Press.

McCarr, Ken. 1969. "Early Horses in America: Colonial and Ohio Horses." Hoof Beats, 37 (October), 19.

McClelland, Peter D., and Richard J. Zeckhauser. 1982. Demographic Dimensions of the New Republic: American Interregional Migration, Vital Statistics, and Manumissions, 1800–1860. Cambridge, England: Cambridge University Press.

McCloskey, Donald N. 1981. "The Industrial Revolution: A Survey." In R. C. Floud and D. N. McCloskey, eds. The Economic History of Britain since 1700. Cambridge, England: Cambridge University Press, Vol. 1, 103–127.

McCorkle, Samuel E. 1976. "A Sermon for the Anniversary of American Independence, July 4th, 1786, delivered at Salisbury, Rowan County, North Carolina." Reprinted in Henry A. Hawken, ed. Trumpets of Glory: Fourth of July Orations, 1786–1861. Granby, Conn.: Salmon Brook Historical Society, 292–298.

McCullough, M. E. 1976. "Milestones in Dairy Cattle Feeding." Hoard's Dairyman, 121 (July), 793–812.

McDonald, Donald. 1908. Agricultural Writers, from Sir Walter of Henley to Arthur Young, 1200–1800. London: Horace Cox.

McMahon, Sarah F. 1985. "A Comfortable Subsistence: The Changing Composition of Diet in Rural New England, 1620–1840." William and Mary Quarterly, 42 (January), 26–65.

McNall, Neil A. 1952. An Agricultural History of the Genesee Valley, 1790–1860. Philadelphia, Penn.: University of Pennsylvania Press.

McRae, Robert. 1957. "The Unity of the Sciences: Bacon, Descartes, and Leibniz." Journal of the History of Ideas, 18 (January), 27–28.

Mead, Elwood. 1907. "Irrigation Engineering in Practice." In L. H. Bailey, ed. Cyclopedia of American Agriculture. New York: Macmillan, Vol. 1, 419–435.

Meade, R. K. 1818. "Mode of cultivating Indian Corn. Harrows." Philadelphia Society for Promoting Agriculture, Memoirs, 4, 184–188.

Mease, James. 1812. Archives of Useful Knowledge. Philadelphia, Penn.: David Hogan.

——. 1818. "On the Progress of Agriculture, with hints for its improvement in the United States." Philadelphia Society for Promoting Agriculture. Memoirs, 4, v–xxviii.

Meek, Alexander Beaufort. 1976. "Oration delivered in Tuscaloosa, July 4, 1833." Reprinted in Henry A. Hawken, ed. Trumpets of Glory: Fourth of July Orations, 1786–1861. Granby, Conn.: Salmon Brook Historical Society, 307–318.

Meinig, D. W. 1986. The Shaping of America, Vol 1: Atlantic America 1492–1800. New Haven, Conn.: Yale University Press.

Meriwether, Robert L., ed. 1959. The Papers of John C. Calhoun. Columbia, S.C.: University of South Carolina Press.

Meurers-Balke, Jutta, and Charlotte Loennecken. 1984. "Notes on the Use of Hand-and-Finger Protectors during Harvesting with Sickle." Tools and Tillage, 5, No. 1, 27–42.

Michaux, François André. 1805. Travels to the West of the Allegheny Mountains . . . and back to Charleston . . . Undertaken in the Year 1802. London: D. N. Shury.

Middlekauff, Robert. 1980. "Why Men Fought in the American Revolution." Huntington Library Quarterly, 43 (Spring), 135–148.

Millar, Eric George. 1932. The Luttrell Psalter. London: Bernard Quaritch.

Millar, John. 1862. A Description of the Province and City of New York . . . in the Year 1695. New York: W. Gowans.

Miller, August C., Jr. 1942. "Jefferson as an Agriculturalist." Agricultural History, 16 (April), 65–78.

Miller, Edward, and Frederic P. Wells. 1913. History of Ryegate, Vermont. St. Johnsbury, Vt.: Caledonian.

Miller, Frederick K. 1936. "The Farmers at Work in Colonial Pennsylvania." Pennsylvania History, 3 (April), 115–123.

Miller, Lewis. 1966. Sketches and Chronicles: The Reflections of a Nineteenth Century Pennsylvania German Folk Artist. York, Penn.: Historical Society of York County.

Miller, Merritt Finley. 1902. The Evolution of Reaping Machines. Washington, D.C.: U.S. Government Printing Office.

Miller, Perry. 1941. "Declension in a Bible Commonwealth." American Antiquarian Society Proceedings, 51 (April), 37–94.

——. 1953. The New England Mind: From Colony to Province. Cambridge, Mass.: Harvard University Press.

——. 1965. The Life of the Mind in America, from the Revolution to the Civil War. New York: Harcourt, Brace and World.

——. 1968. "The Moral and Psychological Roots of American Resistance." In Jack P. Greene, ed. The Reinterpretation of the American Revolution, 1763–1789. New York: Harper and Row, 251–274.

Miller, Samuel. 1803. A Brief Retrospect on the Eighteenth Century, Containing a Sketch of the Revolutions and Improvements in Science, Arts and Literature During that Period. New-York: T. and J. Swords.

Mingay, G. E. 1963. "A Reconsideration." Agricultural History, 37 (April), 123–133.

——. 1969. "Dr. Kerridge's 'Agricultural Revolution': A Comment." Agricultural History, 43 (October), 477–482.

Mitchell, Isabel S. 1933. Roads and Road-Making in Colonial Connecticut. Tercentenary Commission of the State of Connecticut. New Haven, Conn.: Yale University Press.

Mitchell, John. 1767. The Present State of Great Britain and North America with Regard to Agriculture, Population, Trade, and Manufacture. London: T. Becket.

Mokyr, Joel. 1993. The British Industrial Revolution: An Economic Perspective. Oxford: Westview Press.
Mokyr, Joel, ed. 1990. The Lever of Riches: Technological Creativity and Economic Progress. New York: Oxford University Press.
Mommsen, Theodor E. 1951. "St. Augustine and the Christian Idea of Progress." Journal of the History of Ideas, 12 (June), 346–374.
Moore, Frank, ed. 1862. The Patriot Preachers of the American Revolution, 1766–1783. New York: L. A. Osbourne.
Moore, Thomas. 1802. The Great Error of American Agriculture Exposed and Hints for Improvements Suggested. Baltimore, Md.: Bonsal and Niles.
Morgan, Edmond S. 1968. "The Puritan Ethic and the Coming of the American Revolution." In Jack P. Greene, ed. The Reinterpretation of the American Revolution, 1763–1789. New York: Harper and Row, 235–251.
Morison, Samuel Eliot. 1980. The Intellectual Life of Colonial New England. New York: New York University Press.
Mortimer, John. 1708. The Whole Art of Husbandry. London: H. Matlock.
Morton, John C., ed. 1856. A Cyclopedia of Agriculture. London: Blackie and Son.
Morton, Louis. 1946. "Robert Wormeley Carter of Sabine Hall: Some Notes on the Life of a Virginia Planter." Journal of Southern History, 12 (August), 345–365.
Mukhiddinov, Ikromiddin. 1979. "Traditional Tilling Implements Utilized by the Pamir Nationalities in the Nineteenth and Early Twentieth Centuries." Tools and Tillage, 3, No. 4, 215–226.
Mumford, Herbert W. 1907. "Capital Required for the Stock-Farm." In L. H. Bailey, ed. Cyclopedia of American Agriculture. New York: Macmillan, Vol. 1, 172–173.
Mumford, Lewis. 1934. Technics and Civilization. New York: Harcourt, Brace.
Murray, Chalmers S. 1949. This Our Land: The Story of the Agricultural Society of South Carolina. Charleston, S.C.: Carolina Art Association.
Narrett, David E. 1992. Inheritance and Family Life in Colonial New York City. Ithaca, N.Y.: Cornell University Press.
Neely, Wayne Caldwell. 1935. The Agricultural Fair. New York: Columbia University Press.
Nelson, Richard R. 1968. "Innovation." In David L. Sills, ed. International Encyclopedia of the Social Sciences. New York: Macmillan, Vol. 7, 339–345.
Nevins, Allan. 1962. The Gateway to History. Garden City, N.Y.: Doubleday [Anchor Books].
Newton, Hester W. 1934. "The Agricultural Activities of the Salzburgers in Colonial Georgia." Georgia Historical Quarterly, 18 (September), 248–263.
New York State Agricultural Society. 1868. Report on the Trial of Plows Held at Utica, Commencing September 8th, 1867. Albany, N.Y.: Van Benthuysen & Sons.
Nicholson, John. 1814. The Farmer's Assistant. Albany, N.Y.: H.C. Southwick.
Niemsewicz, Juljan U. 1965. Under the Vine and Fig Tree: Travels in America in 1797–1799, 1805, With Some Further Account of Life in New Jersey. Elizabeth, N.J.: New Jersey Historical Society.
Niermeyer, Jan Frederik. 1968. Bronnen voor de Economische Geschiedenis van het Beneden-Maasgebied. The Hagae: Martinus Nijhoff.
"Notes on Compton, A Township in Newport County, State of Rhode-Island, September, 1803." Massachusetts Historical Society, Collections, Vol. 9. Boston, Mass.: Hall & Hiller, 1804, 199–206.
Novak, Barbara. 1969. American Painting of the Nineteenth Century. New York: Praeger.
——. 1980. Nature and Culture: American Landscape and Painting, 1825–1875. New York: Oxford University Press.
Nye, Russel Blaine. 1960. The Cultural Life of the New Nation, 1776–1830. New York: Harper and Row.

O'Brien, Patrick K., and Roland Quinault, eds. 1993. The Industrial Revolution and British Society. Cambridge, England: Cambridge University Press.
Ohio Department of Agriculture. 1904. The Farmers' Centennial History of Ohio, 1803–1903. Springfield, Oh.: Springfield Publishing.
Oleson, Alexandra. 1976. "Introduction: To Build a New Intellectual Order." In Alexandra Oleson and Sanborn C. Brown, ed. The Pursuit of Knowledge in the Early American Republic. Baltimore, Md.: Johns Hopkins University Press, xv–xxv.
Oleson, Alexandra, and Sanborn C. Brown, eds. 1976. The Pursuit of Knowledge in the Early American Republic: American Scientific and Learned Societies from Colonial Times to the Civil War. Baltimore, Md.: Johns Hopkins University Press.
Oliver, John W. 1956. History of American Technology, 1607–1955. New York: Ronald.
Olmstead, Alan L. 1975. "The Mechanization of Reaping and Mowing in American Agriculture, 1833–1870." Journal of Economic History, 35 (June), 327–352.
——. 1979. "The Diffusion of the Reaper: One More Time!" Journal of Economic History, 39 (June), 475–476.
Olson, A. L. 1935. Agricultural Economy and the Population in Eighteenth-Century Connecticut. New Haven, Conn.: Yale University Press.
"On the Flemish Scythe: Report of a Committee of the Highland Society of Scotland, in regard to the Experiments made in Scotland, in Harvest 1825, with the Flemish Scythe." Transactions of the Highland Society of Scotland, 7 (1829), 244–249.
"On the Superiority of the Cradle-Framed Scythe to the Common One for Reaping Corn." Quarterly Journal of Agriculture, 5 (March, 1834–March, 1835), 106–111.
Ornduff, Donald R. 1960. The Herefords in America. Kansas City, Mo.: Privately Printed.
Palmer, R. R. 1968. "The American Revolution: The People As Constituent Power." In Jack P. Greene, ed. The Reinterpretation of the American Revolution, 1763–1789. New York: Harper and Row, 338–363.
Parker, Richard. 1789. "An Account of the Culture of Tobacco." American Museum, 5, 537–540.
Parkinson, Richard. 1805. A Tour in America in 1798, 1799, and 1800. London: J. Harding and J. Murray.
Parrington, Vernon Louis. 1927a. The Colonial Mind, 1620–1800. New York: Harcourt, Brace.
——. 1927b. The Romantic Revolution in America, 1800–1860. New York: Harcourt, Brace.
Partridge, Michael. 1973. Farm Tools through the Ages. Reading, England: Osprey.
Passmore, J. B. 1930. The English Plough. London: Oxford University Press.
Patterson, Richard S., and Richardson Dougall. 1976. The Eagle and the Shield: A History of the Great Seal of the United States. Washington, D.C.: U.S. Government Printing Office.
Paullin, Charles O. 1932. Atlas of the Historical Geography of the United States. Baltimore, Md.: A. Hoen.
Paulsen, Gary. 1977. Farm: A History and Celebration of the American Farmer. Englewood Cliffs, N.J.: Prentice-Hall.
Paxson, Frederic L. 1930. When the West Is Gone. New York: Henry Holt.
Payne, F. G. 1957. "The British Plough: Some Stages in Its Development." Agricultural History Review, 5, Part 2, 74–84.
Peale, Charles W. 1814. "Account of a Corn Shelling Machine." Philadelphia Society for Promoting Agriculture, Memoirs, 3, 248–251.
Perkins, Charles Elliott. 1937. The Pinto Horse. Santa Barbara, Calif.: Fisher and Skofield.
Perkins, J. A. 1977. "Harvest Technology and Labour Supply in Lincolnshire and the East Riding of Yorkshire, 1750–1850." Tools and Tillage, Part One, 3, No. 1, 47–58; Part Two, 3, No. 2, 125–135.

Perl, Lila. 1975. Slumps, Grunts, and Snickerdoodles: What Colonial America Ate and Why. New York: Seabury Press.
Persons, Stow. 1968. "Cyclical Theory of History." In Cushing Strout, ed. Intellectual History in America. New York: Harper and Row, Vol. I, 47–63.
Peters, Richard. 1797. Agricultural Enquiries on Plaister of Paris. . . . Philadelphia, Penn.: C. Cist.
——. 1811a. "Corn Grubs, or Cut Worms. Fall-Ploughing." Philadelphia Society for Promoting Agriculture, Memoirs, 2, Appendix, 89–94.
——. 1811b. "Observations on Colonel Taylor's Letter." Philadelphia Society for Promoting Agriculture, Memoirs, 2, 63–74.
——. 1811c. "Plan for Establishing a Manufactory of Agricultural Instruments; and a Warehouse and Repository for receiving and vending them." Philadelphia Society for Promoting Agriculture, Memoirs, 2, 113–119.
——. 1814. "Rotation and Changes of Crops defended." Philadelphia Society for Promoting Agriculture, Memoirs, 3, 252–256.
——. 1816. A discourse on agriculture: Its antiquity, and importance, to every member of the community . . . Delivered before the Philadelphia Society for Promoting Agriculture . . . 9th of January, 1816. Philadelphia, Penn.: Johnson and Warner.
——. 1818a. "Hill-side Plough. American Ploughs." Philadelphia Society for Promoting Agriculture, Memoirs, 4, 13–15.
——. 1818b. "Letter to Robert Vaux, Esq., August 3, 1817." Philadelphia Society for Promoting Agriculture, Memoirs, 4, 231–232.
——. 1818c. "On Hotchkiss's Straw Cutter." Philadelphia Society for Promoting Agriculture, Memoirs, 4, 104–106.
Peterson, Martin, and Marvin Grim. 1928. "How Are the Crops at Mt. Vernon? Washington Knew the Answer; His Diaries Show How Carefully He Farmed." Wallace's Farmer, 53 (February), 297.
Phelps, Charles Shepherd. 1917. Rural Life in Litchfield County. Norfolk, Conn.: Litchfield County University Club.
Philadelphia Society for Promoting Agriculture. 1826. "Memorial to the Senate and House of Representatives of the Commonwealth of Pennsylvania, in General Assembly met." Presented January 1819. Reprinted in Philadelphia Society for Agriculture, Memoirs, 5, 315–320.
——. 1854. Minutes, February, 1785 to March, 1810. Philadelphia, Penn.: John C. Clark and Son.
——. 1939. Sketch of the History of the Philadelphia Society for Promoting Agriculture. Philadelphia, Penn.: Philadelphia Society for Promoting Agriculture.
Phillips, Deane. 1922. Horse Raising In Colonial New England. Ithaca, N.Y.: Cornell University Agriculture Experiment Station.
Phillips, Ulrich Bonnell. 1908. A History of Transportation in the Eastern Cotton Belt to 1860. New York: Columbia University Press.
Pickering, Joseph. 1832. Inquiries of An Emigrant. . . . London: Effingham Wilson.
Piers Plowman. See Thomas Wright.
Pinkett, Harold T. 1950. "The American Farmer, A Pioneer Agricultural Journal, 1819–1834." Agricultural History, 24 (July), 146–151.
Pirtle, T. R. 1926. History of the Dairy Industry. Chicago, Ill.: Majonnier Bros. Co.
Pitkin, Timothy. 1817. Statistical View of the Commerce of the United States. New York: Eastburn.
"Ploughs and Ploughing." In C. E. Green and D. Young, eds. Encyclopedia of Agriculture. London: William Green, 1908, Vol. 3, 285–294.
Plumb, Charles S. 1906. Types and Breeds of Farm Animals. Boston, Mass.: Ginn & Company.

Poel, J. M. G. van der. 1960–61. "De Landbouw in Het Verte Verleden." Berichten van de Rijksdienst voor Het Oudheidkundig Bodemonderzoek, 10–11, 125–194.

——. 1964. "De Zeeuwse Voetploeg in de Late Middeleeuwen," Volkskunde, 65, N. R. 23, 181–188.

——. 1967. Oude Nederlandse Ploegen. Arnhem: Rijksmuseum voor Volkskunde "Het Nederlands Openluchtmuseum."

——. 1971. "Verkeerd Geïnterpreteerd: de Middeleeuwse Karploeg Volgens Lindemans." In P. J. Meertens and H. W. M. Plettenburg, eds. Vriendenboek voor A. J. Bernet Kempers. Arnhem: Vereniging Vrienden van het Nederlands Openluchtmuseum, 98–104.

——. 1983. Honderd Jaar Landbouwmechanisatie in Nederland. Wageningen, Netherlands: Vereniging voor Landbouwgeschiedenis.

Pollard, Sidney. 1968. The Idea of Progress. New York: Basic Books.

Pomfret, John E. 1964. The New Jersey Proprietors and Their Lands, 1664–1776. Princeton, N.J.: Van Nostrand.

Poore Ben Perley. 1867. "History of Agriculture of the United States." Report of the Commissioner of Agriculture for the Year 1866. Washington, D.C.: Government Printing Office, 498–527.

Pope, John. 1888. A Tour Through the Southern and Western Territories of the United States of North-America. . . . New York: Charles L. Woodward. (Original 1792.)

Potter, David M. 1969. People of Plenty: Economic Abundance and the American Character. Chicago, Ill.: University of Chicago Press.

Powell, John Hare. 1824. "On Cultivators—and the best Modes of American Husbandry; as fitted to our Climate and the Circumstances of the Country." Pennsylvania Agricultural Society, Memoirs, 229–234.

Power, Thomas. 1840. An Oration delivered by request of the City Authorities before the Citizens of Boston, on the Sixty-Fourth Anniversary of American Independence, July 4, 1840. Boston, Mass.: John H. Eastburn.

Prentice, Ezra P. 1942. American Dairy Cattle: Their Past and Future. New York: Harpers.

Price, Richard. 1785. Observations on the Importance of the American Revolution and the Means of making it a Benefit to the World. London: T. Cadell.

"The Progress of Society," Democratic Review, 8 (July 1840), 67–87.

Pursell, Carroll W., Jr. 1959. "E. I. du Pont, Don Pedro, and the Introduction of Merino Sheep into the United States, 1801: A Document." Agricultural History, 33 (April), 86–88.

——. 1962. "E. I. du Pont and Merino Mania in Delaware, 1805–1815." Agricultural History, 36 (April), 91–100.

Pusey, Ph. 1851. "Report to H. R. H. the President of the Commission for the Exhibition of the Works of Industry of all Nations: On Agricultural Implements, Class IX." Journal of the Royal Agricultural Society of England, 12, 587–648.

Pyne, W. H. 1806. Microcosm: or a Picturesque Delineation of the Art, Agriculture, Manufactures, Etc., of Great Britain. London: Gosnell.

Quaintance, H. W. 1904. "The Influence of Farm Machinery on Production and Labor." American Economic Association Quarterly, Third Series, 5 (November), 1–106.

Quennell, Marjorie, and C. H. B. Quennell. 1961. A History of Everyday Things in England. New York: G. P. Puntam's Sons.

Quick, Graeme R., and Wesley F. Buchele. 1978. The Grain Harvesters. St. Joseph, Mich.: American Society of Agricultural Engineers.

Ramsay, David. 1990. The History of the American Revolution. Indianapolis, Ind.: Liberty Classics.

Ramsey, Robert W. 1964. Carolina Cradle: Settlement of the Northwest Carolina Frontier, 1747–1762. Chapel Hill, N.C.: University of North Carolina Press.

Randolph, Wayne. 1991. "Preliminary Report on an Early American Plowshare." Tools and Tillage, 6, No 4, 221–231.

Range, Willard. 1947. "The Agricultural Revolution in Royal Georgia, 1752–1775." Agricultural History, 21 (October), 250–255.

Rasmussen, Wayne D. 1960. Readings in the History of America Agriculture. Urbana, Ill.: University of Illinois Press.

——. 1975a. "Experiment or Starve: The Early Settlers." U.S. Department of Agriculture Yearbook, 10–14.

——. 1975b. "Jefferson, Washington . . . and Other Farmers." U.S. Department of Agriculture Yearbook, 15–22.

——. 1982. "The Mechanization of Agriculture." Scientific American, 247 (September), 76–89.

Razzell, Peter. 1993. "The Growth of Population in Eighteenth-Century England: A Critical Reappraisal." Journal of Economic History, 53 (December), 743–771.

Reall, Joseph H. 1882. Dairying and Dairy Improvements. New York: J. H. Reall.

Rees, Abraham. 1810–1824. The Cyclopedia, or Universal Dictionary of Arts, Sciences, and Literature. Philadelphia, Penn.: Samuel F. Bradford and Muray Fairman.

Rees, Sian E. 1979. "Agricultural Implements in Prehistoric and Roman Britain." British Archeological Reports. British Series, Part I, 69 (i), Part II, 69 (ii).

Reynolds, John. 1879. My Own Times. Chicago, Ill.: B. H. Perryman and H. L. Davidson.

Rham, W. L. 1841. "On the Agriculture of the Netherlands." Journal of the Royal Agricultural Society of England, 2, 43–63.

Ridgely, Nicholas. 1818. "On Threshing out Wheat by a Roller." Philadelphia Society for Promoting Agriculture, Memoirs, 4, 29–44.

Riley, Franklin L. 1909. "Diary of a Mississippi Planter, January 1, 1840 to April, 1863." Mississippi Historical Society, Publications, 10, 305–481.

Robert, Joseph C. 1949. The Story of Tobacco in America. New York: Alfred A. Knopf.

——. 1965. The Tobacco Kingdom: Plantation, Market, and Factory in Virginia and North Carolina, 1800–1860. Gloucester, Mass.: Peter Smith.

Roberts, Isaac Phillips. 1897. The Fertility of the Land. New York: Macmillan.

Roberts, Kenneth D. 1976. Tools for the Trades and Crafts: An Eighteenth Century Pattern Book of R. Timmons and Sons, Birmingham. Hartford, Conn.: Bond Press.

Robin, Claude C. 1784. New Travels through North-America. Boston, Mass.: E. E. Pais and N. Willis.

Robinson, Solon. 1869. Facts for Farmers. New York: A. J. Johnson.

Rochefoucauld. See La Rochefoucauld.

Rogin, Leo. 1931. The Introduction of Farm Machinery in its Relation to the Productivity of Labor in the Agriculture of the United States During the Nineteenth Century. University of California Publications in Economics, Vol. 9. Berkeley, Calif.: University of California Press.

Romans, Bernard. 1961. A Concise Natural History of East and West Florida. New York: the author, 1775. Reprinted New Orleans: Pelican.

Rose, P. S. 1907. "Farm Motors." In L. H. Bailey, ed. Cyclopedia of American Agriculture. New York: Macmillan, Vol. 1, 217–230.

Ross, Earle D., and Robert L. Tontz. 1948. "The Term 'Agricultural Revolution' as Used by Economic Historians." Agricultural History, 22 (January), 32–38.

Rossiter, Margaret W. 1976. "The Organization of Agricultural Improvement in the United States, 1785–1865." In Alexandra Oleson and Sanborn C. Brown, ed. The Pursuit of Knowledge in the American Republic. Baltimore, Md.: Johns Hopkins University Press, 279–298.

Rostovtzeff, M. 1926. The Social and Economic History of the Roman Empire. Oxford: Clarendon.

Rothenberg, Marc. 1982. The History of Science and Technology in the United States: A Critical and Selective Bibliography. New York: Garland.
Rothenberg, Winifred B. 1984. "Farm Account Books: Problems and Possibilities." Agricultural History, 58 (April), 106–112.
——. 1992. "Structural Change in the Farm Labor Force: Contract Labor in Massachusetts Agriculture, 1750–1865." In Claudia Goldin and Hugh Rockoff, eds. Strategic Factors in Nineteenth Century American History. Chicago, Ill.: University of Chicago Press, 105–134.
Royall, Mrs. Anne. 1830. Mrs. Royall's Southern Tour, or Second Series of the Black Book. Washington, D.C.: Anne Royall.
Royster, Charles. 1979. A Revolutionary People at War: The Continental Army and American Character, 1775–1783. Chapel Hill, N.C.: University of North Carolina Press.
Ruffin, Edmund. 1832. An Essay on Calcareous Manures. Petersburg, Virg.: J. W. Campbell.
——. 1851. "Management of Wheat Harvest." In Report of the Commissioner of Patents for the Year 1850, Part 2: Agriculture. Washington, D.C.: Office of Printers to the House of Representatives, 102–113.
——. 1857. "Agricultural Features of Virginia and North Carolina." De Bow's Review, 23 (July), 1–20.
Runnals, Paul March. 1815. An Oration delivered at New-Durham, N. H., July 4th, 1815. Portsmouth, N.H.: Charles Turell.
Rupnow, John, and Carol W. Knox. 1975. The Growing of America: Two Hundred Years of U.S. Agriculture. Fort Atkinson, Wisc.: Johnson Hill Press.
Rush, Richard. 1976. "An oration delivered on the Fourth of July, 1812, in the Hall of the House of Representatives, D.C." Reprinted in Henry A. Hawken, ed. Trumpets of Glory: Fourth of July Orations, 1786–1861. Granby, Conn.: Salmon Brook Historical Society, 25–46.
Russell, Charles T. 1850. Agricultural Progress in Massachusetts for the Last Half Century. Boston: The author.
Russell, Howard S. 1976. A Long, Deep Furrow: Three Centuries of Farming in New England. Hanover, N.H.: University Press of New England.
Rutman, Darrett B. 1967. Husbandmen of Plymouth: Farms and Villages in the Old Colony, 1620–1692. Boston, Mass.: Beacon.
Šach, František. 1968. "Proposal for the Classification of Pre-Industrial Tilling Implements." Tools and Tillage, 1, No. 1, 3–27.
Sachs, Williams S. 1953. "Agricultural Conditions in the Northern Colonies before the Revolution." Journal of Economic History, 13 (Summer), 274–290.
Salley, Alexander S. 1911. Narratives of Early Carolina, 1650–1708. New York: Charles Scribner's Sons.
Saloutos, Theodore. 1946. "Efforts at Crop Control in Seventeenth Century America." Journal of Southern History, 12 (February), 45–64.
Sampson, R. V. 1956. Progress in the Age of Reason. Cambridge, Mass.: Harvard University Press.
Sanders, Alvin H. 1914. The Story of the Herefords. Chicago, Ill.: Breeder's Gazette.
——. 1918. Short-Horn Cattle: A Series of Historical Sketches. Chicago, Ill.: Sanders Publishing Co.
——. 1928. A History of Aberdeen-Angus Cattle. Chicago, Ill.: New Breeder's Gazette.
Savage, James. 1811. An Oration delivered July 4, 1811, at the request of the Selectmen of Boston in Commemoration of American Independence. Boston, Mass.: John Eliot, Jr.
Schafer, Joseph. 1922. A History of Agriculture in Wisconsin. Madison, Wisc.: State Historical Society of Wisconsin.

Scheer, George F., and Hugh F. Rankin. 1987. Rebels and Redcoats: The American Revolution Through the Eyes of Those Who Fought and Lived It. New York: Da Capo.
Schermerhorn, Frank Earle. 1948. American and French Flags of the Revolution, 1775–1783. Philadelphia, Penn.: Pennsylvania Society of the Sons of the Revolution.
Schjellerup, Inge. 1986. "Ploughing in Chuquibamba, Peru." Tools and Tillage, 5, No. 3, 180–189.
Schlebecker, John T. 1967. A History of American Dairying. Chicago, Ill.: Rand McNally.
——. 1972. Agricultural Implements and Machines in the Collection of the National Museum of History and Technology. Washington, D.C.: Smithsonian Institution.
——. 1973. The Use of the Land: Essays on the History of American Agriculture. Lawrence, Kan.: Coronado Press.
——. 1975. Whereby We Thrive. Ames, Ia.: Iowa State University Press.
——. 1981. The Changing American Farm, 1831–1981. Washington, D.C.: Smithsonian Institution.
Schlebecker, John T., and Andrew W. Hopkins. 1957. A History of Dairy Journalism in the United States, 1810–1950. Madison, Wisc.: University of Wisconsin Press.
Schlesinger, Arthur M. 1969. The Birth of the Nation: A Portrait of the American People on the Eve of Independence. New York: Alfred A. Knopf.
Schob, David E. 1975. Hired Hands and Plowboys: Farm Labor in the Midwest, 1815–60. Urbana, Ill.: University of Illinois Press.
Schoepf, Johann David. 1911. Travels in the Confederation, 1783–1784. Trans. Alfred J. Morrison. Philadelphia, Penn.: William J. Campbell.
Schorger, Arlie W. 1966. The Wild Turkey: Its History And Domestication. Norman, Okla.: University of Oklahoma Press.
Schweitzer, Mary McKinney. 1980. "Economic Regulation and the Colonial Economy: The Maryland Tobacco Inspection Act of 1747." Journal of Economic History, 40 (September), 551–570.
Scoville, Warren C. 1953. "Did Colonial Farmers 'Waste' Our Land?" Southern Economic Journal, 20 (October), 178–181.
Sears, Robert. 1816. A pictorial description of the United States; embracing the history, geographical position, agricultural and mineral resources. . . . Boston, Mass.: J. A. Lee.
Séguin, Robert Lionel. 1989. L'Équipement aratoire et horticole du Québec ancien (XVIIe, XVIIIe et XIXe siècles). Montreal: Guérin littérature.
Sellars, Charles. 1991. The Market Revolution: Jacksonian America, 1815–1846. New York: Oxford University Press.
Sellers, John R. 1976. "The Common Soldier in the American Revolution." In Stanley J. Underdal, ed. Military History of the American Revolution: Proceedings of the South Military History Symposium, USAF Academy, 1974. Washington, D.C.: Office of Air Force History, 151–161.
Seward, William Henry. 1976. "Address delivered at ceremonies commemorating the first centennial anniversary of Cherry Valley, New York, July 4, 1840." Reprinted in Henry A. Hawken, ed. Trumpets of Glory: Fourth of July Orations, 1786–1861. Granby, Conn.: Salmon Brook Historical Society, 194–199.
Seybert, Adam. 1818. Statistical Annals of the United States, 1789–1818. Philadelphia, Penn.: Thomas Dobon & Son.
Shaw, Ronald E. 1990. Canals for a Nation: The Canal Era in the United States, 1790–1860. Lexington, Ky.: University Press of Kentucky.
Sheffield, J. L. 1783. Observations of the Commerce of the United States. Philadelphia. Penn.: R. Bell.
Shepard, Silas M. 1886. The Hog In America. Indianapolis, Ind.: Swine Breeders' Journal.

Shirreff, Patrick. 1835. A Tour through North America together with a Comprehensive View of the Canadas and United States. Edinburgh: Oliver and Boyd.

Shryock, Richard H. 1939. "British Versus German Traditions in Colonial Agriculture." Mississippi Valley Historical Review, 26 (June), 39–54.

Shy, John. 1990. A People Numerous and Armed. Ann Arbor, Mich.: University of Michigan Press.

Sigaut, François. 1989. "Storage and Threshing in Pre-industrial Europe: Additional Notes." Tools and Tillage, 6, No. 2, 119–124.

Singer, Charles. 1956. "East and West in Retrospect." In Charles Singer, E. J. Holmyard, and A. R. Hall, eds. A History of Technology. Oxford: Clarendon, Vol. 2, 753–776.

Singleton, Arthur [pseud.] (Henry Cogswell Knight). 1824. Letters from the South and West. Boston, Mass.: Richardson and Lord.

Sinkler, Wharton. 1958. "Washington as Agriculturalist." Bulletin of the Garden Club of America, 46 (January), 7–11.

Slicher van Bath, B. H. 1960. "The Rise of Intensive Husbandry in the Low Countries." In J. Bromley and E. H. Kossmann, eds. Britain and the Netherlands. London: Chatto & Windus, 130–153.

——. 1969. "Eighteenth-Century Agriculture on the Continent of Europe: Evolution or Revolution?" Agricultural History, 43 (January), 169–179.

Sloane, Eric. 1955. Our Vanishing Landscape. New York: Wilfred Funk.

——. 1958. The Seasons of America Past. New York: Wilfred Funk.

——. 1964. A Museum of Early American Tools. New York: Ballantine.

Small, James. 1784. Treatise on Ploughs and Wheel Carriages. Edinburgh: W. Creech and C. Elliot.

Smit, Homme Jakob. 1924–1939. De Rekeningen der Graven en Gravinnen vit het Henegouwsche Huis. Amsterdam: J. Muller.

Smith, Adam. 1976. An Inquiry into the Nature and Causes of the Wealth of Nations. Oxford: Clarendon Press. (Original 1776.)

Smith, Harrison Pearson. 1955. Farm Machinery and Equipment. New York: McGraw-Hill.

Smith, J. Gray. 1974. A Brief Historical, Statistical and Descriptive Review of East Tennessee . . . London: J. Leath, 1842. Reprinted Spartanburg, S.C.: The Reprint Company.

Smith, John. 1966. The Generall Historie of Virginia, New-England and the Summer Isles. Ann Arbor, Mich.: University Microfilms. (Original 1624.)

Smith, Maryanna S. 1979. Chronological Landmarks in American Agriculture. Washington, D.C.: U.S. Department of Agriculture, Economics, Statistics, and Cooperative Service.

Smith, R. E. F. 1983. "Some Tillage Implement Parts in the Zausailov Collection, National Museum of Finland." Tools and Tillage, 4, No. 4, 205–215.

Smith, Richard. 1906. A Tour of Four Great Rivers: The Hudson, Mohawk, Susquehanna, and Delaware in 1769, ed. Francis W. Hasey. New York: Charles Scribner's Sons.

Smyth, Albert Henry, ed. 1906. The Writings of Benjamin Franklin. New York: Macmillan.

Smyth, J. F. D. 1784. A Tour in the United States of America, containing . . . Improvements in Husbandry that may be adopted with great Advantage in Europe. London: G. Robinson, J. Robson and J. Sewell.

Sommerville, Robert. 1795. Report on the Subject of Manures, with appendix. London: Board of Agriculture.

Sordinas, Augustus. 1978. "The Ropas Plow from the Island of Corfu, Greece." Tools and Tillage, 3, No. 3, 139–149.

Southard, Samuel L. 1830. Address delivered before the Newark Mechanics' Association, July 5, 1830. Newark, N.J.: W. Tuttle.

Spadafora, David. 1990. The Idea of Progress in Eighteenth-Century Britain. New Haven, Conn.: Yale University Press.

Spafford, Hoartio Gates. 1813. A Gazetteer of the State of New-York. Albany, N.Y.: H. C. Southwick.

Sparks, Jared. 1837. The Writings of George Washington. Boston, Mass.: John B. Russell.

Spencer, A. J., and J. B. Passmore. 1930. Handbook of the Collection Illustrating Agricultural Implements and Machinery. London: His Majesty's Stationery Office.

Spinden, Herbert J. 1950. Tobacco Is American: The Story of Tobacco before the Coming of the White Man. New York: New York Public Library.

Sprague, Charles. 1825. An Oration delivered on Monday, Fourth of July, 1825, in Commemoration of American Independence, before the Supreme Executive of the Commonwealth, and the City Council and Inhabitants of the City of Boston. Boston, Mass.: True and Greene.

Spurrier, John. 1793. The Practical Farmer. . . . Wilmington, Del.: Brynberg and Andrews.

Stabler, Edward. 1879. Overlooked Pages of Reaper History: A Brief Narrative of the Invention of Reaping Machines. Chicago, Ill.: W. B. Conkey.

Steensberg, Axel. 1943. Ancient Harvesting Implements: A Study in Archaeology and Human Geography. Copenhagen: Bianco Lunos Bogtrykkeri.

——. 1977. "SULA: An Ancient Term for the Wheel Plough in Northern Europe?" Tools and Tillage, 3, No. 2, 91–98.

Stephens, Henry. 1855. The Book of the Farm. London: William Blackwood.

Stetson, Sarah P. 1949. "The Traffic in Seeds and Plants from England's Colonies in North America." Agricultural History, 23 (January), 45–56.

Steward, John F. 1931. The Reaper. New York: Greenberg.

Stokely, B. 1850. "Remarks and General Observations on Mercer County, Pennsylvania." Memoirs of the Historical Society of Pennsylvania, 4, Part 2, 65–82.

Stone, Archie A. 1945. Farm Machinery. New York: John Wiley and Sons.

Stover, Stephen L. 1962. "Early Sheep Husbandry in Ohio." Agricultural History, 36 (April), 101–107.

Stowe, E. J. 1948. Crafts of the Countryside. London: Longmans, Green.

Stowell, Marion B. 1977. Early American Almanacs: The Colonial Weekday Bible. New York: B. Franklin.

Strickland, William. 1801. Observations on the Agriculture of the United States of America. London: W. Bulmer.

——. 1971. Journal of a Tour in the United States of America, 1794–1795, ed. J. E. Strickland. New York: New-York Historical Society.

Stromberg, R. N. 1951. "History in the Eighteenth Century." Journal of the History of Ideas, 12 (April), 295–304.

Struik, Dirk J. 1948. Yankee Science in the Making. Boston, Mass.: Little, Brown.

Sutcliffe, Alice Clary. 1909. Robert Fulton and the "Clermont." New York: Century.

Sweeney, Kevin M. 1988. "Gentlemen Farmers and Inland Merchants: The Williams Family and Commercial Agriculture in Pre-Revolutionary Western Massachusetts." In Peter Benes, ed. The Farm. Boston, Mass.: Boston University, 60–73.

Sweet, Orville K. 1975. Birth of a Breed: The History of Polled Herefords, America's First Beef Breed. Kansas City, Mo.: Lowell Press.

Swem, Earl G. 1957. Brothers of the Spade: Correspondence of Peter Collinson of London, and of John Custis, of Williamsburg, Virginia, 1734–1746. Barre, Mass.: Barre Gazette.

Symon, J. A. 1959. Scottish Farming, Past and Present. Edinburgh: Oliver and Boyd.

Sypher, G. Wylie. 1965. "Similarities Between the Scientific and the Historical Revolutions at the End of the Renaissance." Journal of the History of Ideas, 26 (July–September), 353–368.

Tatham, William. 1800. An Historical and Practical Essay on the Culture and Commerce of Tobacco. London: Vernor and Hood.

Taylor, Carl Cleveland. 1971. The Farmer's Movement, 1620–1920. Westport, Conn.: Greenwood Press.

Taylor, John. 1811. "On Gypsum." Philadelphia Society for Promoting Agriculture, Memoirs, 2, 51–62.

——. 1813. Arator. Georgetown, D.C.: J. M. & J. B. Carter.

Taylor, Paul S. 1954. "Plantation Agriculture in the United States: Seventeenth to Twentieth Centuries." Land Economics, 30 (May), 141–152.

Teggart, Frederick J. 1925. Theory of History. New Haven, Conn.: Yale University Press.

Teggart, Frederick J., ed. 1949. The Idea of Progress: A Collection of Readings. Berkeley, Calif.: University of California Press.

Thirsk, Joan. 1985. "Agricultural Innovations and Their Diffusion." In Joan Thirsk, ed. The Agrarian History of England and Wales, Vol. 5: 1640–1750, 533–589. Cambridge, England: Cambridge University Press.

——. 1987. Agricultural Regions and Agrarian History in England, 1500–1750. London: Macmillan.

Thoen, Erik. 1988. Landbouwekonomie en bevolking in Vlaanderen gedurende de Late Middleleeuwen en het begin van de Moderne Tijden. Ghent: Belgisch Centrum voor Landelijke Geschiedenis.

Thomas, Arline H. 1952. "The American Eagle: Living Symbol of the Nation." American Heritage, new series, 3 (Winter), 3, 82.

Thomas, Isaiah. 1810. The History of Printing in America. Worcester, Mass.: Isaiah Thomas, Jr.

Thomas, John J. 1855. Farm Implements and the Principles of Their Construction and Use. New York: Harper & Brothers. (2d ed. 1869.)

Thompson, James W. 1942. A History of Livestock Raising in the United States, 1607–1860. Agricultural History Series No. 5. U.S. Department of Agriculture, Washington, D.C.

Thornton, John Wingate. 1860. The Pulpit of the American Revolution. Boston, Mass.: Gould and Lincoln.

Tocqueville, Alexis De. 1841. Democracy in America. New York: J. and H. G. Langley.

Todd, S. Edwards. 1867a. "Improved Farm Implements." In U.S. Department of Agriculture, Report of the Commissioner of Agriculture for the Year 1866. Washington, D.C.: Government Printing Office, 225–288.

——. 1867b. The Young Farmer's Manual. New York: George E. Woodward.

Towne, Charles W., and Edward N. Wentworth. 1950. Pigs: From Cave to Corn Belt. Norman, Okla.: University of Oklahoma Press.

Toynbee, Arnold. 1964. Lectures on the Industrial Revolution in England. Boston, Mass.: Beacon Press.

Trevor-Roper, Hugh. 1987. The Golden Age of Europe, from Elizabeth I to the Sun King. London: Thames and Hudson.

Trevor-Roper, Hugh, ed. 1968. The Age of Expansion: Europe and the World, 1559–1660. London: Thames and Hudson.

True, Alfred C. 1928. A History of Agricultural Extension Work In the United States, 1785–1923. Washington, D.C.: U. S. Department of Agriculture Miscellaneous Publication 15.

——. 1929. A History of Agricultural Education in the United States, 1785–1925. U.S. Department of Agriculture, Misc. Pub. No. 36. Washington, D.C.: Government Printing Office.

——. 1937. A History of Agricultural Experimentation and Research In The United States, 1607–1925: Including A History Of The United States Department of Agriculture. Washington, D.C.: U.S. Department of Agriculture Miscellaneous Publication 251.

True, Rodney H. 1920a. "Beginnings of Agricultural Literature in America." American Library Association Bulletin, 14 (July), 186–194.

——. 1920b. "The Early Development of Agricultural Societies in the United States." American Historical Association, Annual Report, 295–306.

——. 1928. "Jared Eliot, Minister, Physician, Farmer." Agricultural History, 2 (October), 185–212.

——. 1935. Sketch of the History of the Philadelphia Society for Promoting Agriculture. Philadelphia, Penn.: The Society.

——. 1938. "Some Pre-Revolutionary Agricultural Correspondence of Jared Eliot." Agricultural History, 12 (April), 107–117.

Tryon, Rolla Milton. 1917. Household Manufactures in the United States, 1640–1840. Chicago, Ill.: University of Chicago Press.

Tucher, Andrea J. 1984. Agriculture in America, 1622–1860: Printed Works in the Collections of the American Philosophical Society, The Historical Society of Pennsylvania and the Library Company of Philadelphia. New York: Garland.

Tucker, Gilbert M. 1909. American Agricultural Periodicals: An Historical Sketch. Albany, N.Y.: The author.

Tull, Jethro. 1733. The Horse-Hoeing Husbandry. Dublin: A. Rhames.

Tunis, Edwin. 1957. Colonial Living. New York: World.

——. 1961. Frontier Living. New York: Thomas Y. Crowell.

——. 1966. Shaw's Fortune: The Picture Story of a Colonial Plantation. New York: World.

——. 1969. The Young United States. New York: Thomas Y. Crowell.

Tureson, Ernest Lee. 1968. Redeemer Nation: The Idea of America's Millennial Role. Chicago, Ill.: University of Chicago Press.

Turner, Charles W. 1964. "Virginia State Agricultural Societies, 1811–1860." Agricultural History, 39 (July), 167–177.

Turner, Frederick Jackson. 1906. Rise of the New West, 1819–1829. New York: Harper and Brothers.

Turner, J. A. 1857. The Cotton Planter's Manual. New York: Orange Judd.

Tusser, Thomas. 1878. Five Hundred Pointes of Good Husbandrie. London: Lackington, Allen. (Original 1577.)

U.S. Bureau of the Census. 1862. Preliminary Report on the Eighth Census. Washington, D.C.: U.S. Government Printing Office.

——. 1864. Eight Census of the United States. 1860. Agriculture. Washington, D.C.: Government Printing Office.

——. 1975. Historical Statistics of the United States, Colonial Times to 1970. Washington, D.C.: Government Printing Office.

U.S. Commissioner of Agriculture. 1865. Report for the Year 1864. Washington, D.C.: U.S. Government Printing Office.

U.S. Commissioner of Patents. 1860. Report for the Year 1859. Agriculture. Washington, D.C.: George W. Bowman.

U.S. Congress. 1833. House. Documents Relative to Manufacturing in the United States. 22nd Cong., 1st Sess., Doc. No. 308.

U.S. Department of Agriculture. 1912. Abridged Agricultural Records, Vol. 7: Miscellaneous Farm Topics. Washington, D.C.: Agricultural Service Company.

U.S. Department of Labor. 1898. Thirteenth Annual Report of the Commissioner of Labor, 1898. Hand and Machine Labor. Washington, D.C.: U.S. Government Printing Office.

U.S. Department of State. 1909. The History of the Seal of the United States. Washington, D.C.: U.S. Government Printing Office.

——. 1960. The Great Seal of the United States. Washington, D.C.: U.S. Government Printing Office.

U.S. Senate. 1878. Committee on Patents. Arguments before the Committee. August. 45th Congress, 2nd Session, Misc. Doc. No. 50.

United Nations. 1973. The Determinants and Consequences of Population Trends. New York: United Nations.

——. 1992. Long-range World Population Projections. New York: United Nations.

Updyke, Frank A. 1915. The Diplomacy of the War of 1812. Baltimore, Md.: Johns Hopkins Press.

Van der Donck, Adriaen. 1968. A Description of the New Netherlands, ed. Thomas F. O'Donnell. Syracuse, N.Y.: Syracuse University Press. (Original 1655.)

van der Poel, J. M. G. See Poel, J. M. G. van der.

van Wagenen, Jared, Jr. 1953. The Golden Age of Homespun. Ithaca, N.Y.: Cornell University Press. (Expanded book version of 1927 article.)

Venable, William Henry. 1911. A Buckeye Boyhood. Cincinnati, Oh.: Robert Clarke.

Verhulst, Adriaan. 1990. Précis d'histoire rurale de la Belgique. Brussels: Editions de l'Université de Bruxelles.

Vince, John. 1983. Old Farms: An Illustrated Guide. New York: Schocken.

Voorhees, E. B. 1907. "The Use of Green-Manures in Soil Improvement." In L. H. Bailey, ed. Cyclopedia of American Agriculture. New York: Macmillan, Vol. 1, 503–509.

Voskuil, J. J. 1972. "De Sikkel, De Zeis Of De Zicht Voor Het Oogsten Van Het Graan." Mededelingen van het Instituut voor Dialectologie, Volkskunde en Naamkunde, 24, 12–22.

Voss, John. 1976. "Foreword: The Learned Society in American Intellectual Life." In Alexandra Oleson and Sanborn C. Brown, eds. The Pursuit of Knowledge in the Early American Republic. Baltimore, Md.: Johns Hopkins University Press, vii–x.

Wagar, W. Warren. 1967. "Modern Views of the Origins of the Idea of Progress." Journal of the History of Ideas, 28 (January–March), 55–70.

Wailes, B. L. C. 1854. Report on the Agriculture and Geology of Mississippi. Philadelphia, Penn.: Lippincott, Grambo and Co.

Walcott, Robert. 1936. "Husbandry in Colonial New England." New England Quarterly, 9 (June), 218–252.

Walker, Cornelius Irvine. 1919. History of the Agricultural Society of South Carolina. Charleston, S.C.: Agricultural Society of South Carolina.

Wallace, John H. 1897. The Horse in America in His Derivation, History, and Development. New York: The author.

Ward, James E. 1945. "Monticello: An Experimental Farm." Agricultural History, 19 (July), 183–185.

Waring, George E., Jr. 1868. The Elements of Agriculture. New York: Tribune Association.

Washington, George. 1803. Letters . . . to Arthur Young and Sir John Sinclair, containing an Account of His Husbandry, with his opinions on various Questions in Agriculture. . . . Alexandria, Virg.: Cottom and Stewart.

——. 1919. The Agricultural Papers of George Washington, ed. Walter Edwin Brooke. Boston, Mass.: R. G. Badger.

Washington, H. A., ed. 1869. The Writings of Thomas Jefferson. Philadelphia, Penn.: J. B. Lippincott.

Watkins, James L. 1908. King Cotton: A Historical and Statistical Review, 1790–1908. New York: J. L. Watkins & Sons.

Watson, Elkanah. 1811. Address, delivered to the members of the Berkshire Agri-

cultural Society, at the town-house in Pittsfield, September 24, 1811. Pittsfield, Mass.: P. Allen.
——. 1814. Address of Elkanah Watson, esq. delivered before the Berkshire Agricultural Society . . . 7th October, 1814. Pittsfield, Mass.: P. Allen.
——. 1820. History of the rise, progress, and existing condition of the western canals in the state of New-York, from September 1788, to . . . 1819. Together with the rise, progress, and existing state of modern agricultural societies, on the Berkshire system. Albany, N.Y.: D. Steele.
——. 1857. Men and Times of the Revolution. New York: Dana.
Weber, Max. 1958. The Protestant Ethic and the Spirit of Capitalism. New York: Charles Scribner's Sons.
Webster, Daniel. 1825. An Address delivered at the laying of the Corner Stone of the Bunker Hill Monument, 1825. Boston, Mass.: Cummings, Hilliard.
——. 1882. Fourth of July Oration delivered at Fryeburg, Maine, in the year 1802. Boston, Mass.: A. Williams.
——. 1976. "Oration delivered on the Fourth of July, 1800, at Hanover, New Hampshire." Reprinted in Henry A. Hawken, ed. Trumpets of Glory: Fourth of July Orations, 1786–1861. Granby, Conn.: Salmon Brook Historical Society, 1–15.
Webster, James Carson. 1938. The Labors of the Months in Antique and Mediaeval Art to the End of the Twelfth Century. Evanston, Ill.: Northwestern University Press.
Webster, Noah. 1814. An Oration Pronounced before the Knox and Warren Branches of the Washington Benevolent Society, at Amherst, on the celebration of the anniversary of the Declaration of Independence, July 4, 1814. Northampton, Mass.: William Butler.
Weeden, William B. 1963. Economic and Social History of New England, 1620–1789. New York: Hillary House.
Weiss, Roger W. 1954. "Mr. Scoville on Colonial Land Wastage." Southern Economic Journal, 21 (July), 87–90.
Weiss, Thomas. 1992. "U.S. Labor Force Estimates and Economic Growth, 1800–1860." In Robert E. Gallman and John Joseph Wallis, eds. American Economic Growth and Standards of Living before the Civil War. Chicago, Ill.: University of Chicago Press, 19–75.
Weld, Isaac, Jr. 1807. Travels through the States of North America, and the Provinces of Upper and Lower Canada, during the Years 1795, 1796, and 1797. 4th ed. London: J. Stockdale.
Welter, Rush. 1955. "The Idea of Progress in America." Journal of the History of Ideas, 16 (June), 401–415.
Wentworth, Edward N. 1951. "A Livestock Specialist Looks at Agricultural History." Agricultural History, 25 (April), 49–53.
Wheeler, John T. 1948. Two Hundred Years of Agricultural Education In Georgia. Danville, Ill.: Interstate Publishing Co.
Whitaker, James W. 1964. "A Venture in Jack Stock." Agricultural History, 38 (October), 217–225.
White, E. A. 1918. "A Study of the Plow Bottom and Its Action upon the Furrow Slice." Journal of Agricultural Research, 12 (January), 149–181.
White, John B. 1815. An Oration delivered before the inhabitants of Charleston, S.C., on the 4th March, 1815, in commemoration of the adoption of the Federal Constitution. Charleston, S.C.: Southern Patriot.
White, K. D. 1963. "Wheat-Farming in Roman Times." Antiquity, 37 (September), 207–212.
——. 1967. Agricultural Implements of the Roman World. Cambridge, England: Cambridge University Press.
——. 1984. Greek and Roman Technology. Ithaca, N.Y.: Cornell University Press.

White, Lynn, Jr. 1962. Medieval Technology and Social Change. Oxford: Clarendon.
Whitehill, Walter Muir. 1976. "Early Learned Societies in Boston and Vicinity." In Alexandra Oleson and Sanborn C. Brown, eds. The Pursuit of Knowledge in the Early American Republic. Baltimore, Md.: Johns Hopkins University Press, 151–173.
Whitney, Andrew Griswold. 1976. "An Oration delivered in Detroit, Michigan, July 4, 1818." Reprinted in Henry A. Hawken, ed. Trumpets of Glory: Fourth of July Orations, 1786–1861. Granby, Conn.: Salmon Brook Historical Society, 329–351.
Whitten, David O. 1982. "American Rice Cultivation, 1860–1980: A Tercentenary Critique." Southern Studies, 21 (Spring), 5–26.
Wiener, Philip P. 1957. "Leibniz's Project of a Public Exhibition of Scientific Inventions." In Philip P. Weiner and Aaron Noland, eds. Roots of Scientific Thought. New York: Basic Books, 460–468.
Wiener, Philip P., and Aaron Noland, eds. 1957. Roots of Scientific Thought: A Cultural Perspective. New York: Basic Books.
Wilkes, Peter. 1978. An Illustrated History of Farming. Bourne End, England: Spurbooks.
Willcox, Walter F. 1931. "Increase in the Population of the Earth and of the Continents since 1650." In Walter F. Wilcox, ed. International Migrations, Vol. 2: Interpretations. New York: National Bureau of Economic Research, 33–82.
Willcox, William B., ed. 1972. The Papers of Benjamin Franklin. New Haven, Conn.: Yale University Press.
Willich, A. F. M. 1803. The Domestic Encyclopaedia. Philadelphia, Penn.: William Young Birch and Abraham Small.
Wilson, Harold F. 1935. "The Rise and Decline of the Sheep Industry in Northern New England." Agricultural History, 9 (January), 12–40.
Wilson, Leonard. 1928. The Coat of Arms, Crest and Great Seal of the United States of America. San Diego, Calif.: N. Francis Maw.
Wilson, M. L. 1942. "Survey of Scientific Agriculture." Proceedings of the American Philosophical Society, 86 (September), 52–62.
Wiltze, Charles Maurice. 1935. The Jeffersonian Tradition in American Democracy. Chapel Hill, N.C.: University of North Carolina Press.
Winslow, Hubbard. 1838. An Oration delivered by request by The Municipal Authorities of the City of Boston, July 4, 1838, in the Old South Church, in celebration of American Independence. Boston, Mass.: John H. Eastburn.
Wiser, Vivian, ed. 1976. "Two Centuries of American Agriculture: Bicentennial Symposium." Agricultural History, 50 (January), 1–309.
Wojtilla, Gyula. 1989. "The Ard-Plough in Ancient and Early Medieval India." Tools and Tillage, 6, No. 2, 94–106.
Wolfinger, J. F. 1867. "Indian Corn Culture." Report of the Commissioner of Agriculture for the Year 1866. Washington, D.C.: Government Printing Office, 215–224.
Wood, Frederic J. 1919. The Turnpikes of New England. Boston, Mass.: Marshall Jones.
Wood, Gordon S. 1969. The Creation of the American Republic, 1776–1787. Chapel Hill, N.C.: University of North Carolina Press.
——. 1992. The Radicalism of the American Revolution. New York: Alfred A. Knopf.
Wood, Henry Trueman. 1913. A History of The Royal Society of Arts. London: John Murray.
Woodward, Carl R. 1927. The Development of Agriculture in New Jersey, 1640–1880. New Brunswick, N.J.: New Jersey Agricultural Experimental Station.
——. 1929. "Agricultural Legislation in Colonial New Jersey." Agricultural History, 3 (January), 15–28.
——. 1939. "Memoir 6 of the Philadelphia Society for Promoting Agriculture." Agricultural History, 13 (July), 157–160.

——. 1941. Ploughs and Politicks: Charles Read of New Jersey and His Notes on Agriculture, 1715–1774. New Brunswick, N.J.: Rutgers University Press.
——. 1969. "A Discussion of Arthur Young and American Agriculture." Agricultural History, 43 (January), 57–67.
Worlidge, John. 1697. Systema Agriculturae; The Mystery of Husbandry Discovered. London: Nathaniel Rolls.
Woude, Ad van der. 1987. "Boserup's Thesis and the Historian." In Antoinette Fauve-Chamoux, ed. Évolution agraire et croissance démographique. Liege: Derovaux Ordina, 381–384.
Wright, Chester Whitney. 1910. Wool-Growing and the Tariff. New York: Houghton Mifflin.
Wright, Louis B. 1957. The Cultural Life of the American Colonies, 1607–1763. New York: Harper and Row.
——. 1965. The Dream of Prosperity in Colonial America. New York: New York University Press.
Wright, Philip A. 1961. Old Farm Implements. London: Adam & Charles Black.
Wright, Thomas, ed. 1856. The Vision and Creed of Piers Ploughman. London: John Russell Smith.
Wroth, Lawrence C. 1958. Abel Buell of Connecticut: Silversmith, Type Founder and Engraver. Middletown, Conn.: Wesleyan University Press.
Wyckoff, Vertrees J. 1936. Tobacco Regulation in Colonial Maryland. Baltimore, Md.: Johns Hopkins University Press.
Wylie, William N. T. 1990. The Blacksmith in Upper Canada, 1784–1850. Gananoque, Ontario: Langdale.
Xing-quang, Wang. 1989. "On the Chinese Plough." Tools and Tillage, 6, No. 2, 63–93.
Young, Arthur. 1771. A Course of Experimental Agriculture. Dublin: J. Exshaw.
——. 1787. Annals of agriculture and other useful arts. London: Bury St. Edmunds.
——. 1804. General View of the Agriculture of Hertfordshire. London: Richard Phillips.
Zeichner, Oscar. 1949. Connecticut's Years of Controversy: 1750–1776. Williamsburg, Virg.: University of North Carolina Press.
Zeuner, F. E. 1954a. "Cultivation of Plants." In Charles Singer, E. J. Holmyard, and A. R. Hall, eds. A History of Technology. Oxford: Clarendon Press, Vol. 1, 353–375..
——. 1954b. "Domestication of Animals." In Charles Singer, E. J. Holmyard, and A. R. Hall, eds. A History of Technology. Oxford: Clarendon Press, Vol. 1, 327–352.
Zilsel, Edgar. 1945. "The Genesis of the Concept of Scientific Progress." Journal of the History of Ideas, 6 (June), 325–349.
Zintheo, C. J. 1907. "Machinery in Relation to Farming." In L. H. Bailey, ed. Cyclopedia of American Agriculture. New York: Macmillan, Vol. 1, 208–216.
Zirkle, Conway. 1952. "Early Ideas on Inbreeding and Cross Breeding." In J. W. Gowen, ed. Heterosis. Ames, Ia.: Iowa State College Press.
——. 1959. "John Clayton and Our Colonial Botany." Virginia Magazine of History and Biography, 17 (July), 284–294.
——. 1969. "To Plow or Not to Plow: Comment on the Planters' Problems." Agricultural History, 43 (January), 87–89.

Index

Peter D. McClelland is Professor of Economics at Cornell University.